UNIT OPERATIONS - I

[FLUID FLOW AND MECHANICAL OPERATIONS]

For

SECOND YEAR DIPLOMA IN CHEMICAL ENGINEERING, CHEMICAL TECHNOLOGY AND DEGREE COURSE IN CHEMICAL ENGINEERING

K. A. GAVHANE

Former Vice-Principal & Head of Chemical Engg. Dept.

S.E. Society's Satara Polytechnic,

Satara.

Price ₹ 400.00

AF529653

NIRALI PRAKASHAN

ADVANCEMENT OF KNOWLEDGE

N0767

UNIT OPERATIONS - I **ISBN 978-81-96396-11-4**

Thirty First Edition : October 2022

Published By :
NIRALI PRAKASHAN
Abhyudaya Pragati, 1312, Shivaji Nagar,
Off J.M. Road, Pune – 411005
Tel - (020) 25512336/37/39
Email : niralipune@pragationline.com

Polyplate

Printed By :
YOGIRAJ PRINTERS AND BINDERS
Survey No. 10/1A, Ghule Industrial Estate
Nanded Gaon Road
Nanded, Pune - 411041

DISTRIBUTION CENTRES

PUNE

Nirali Prakashan
(For orders outside Pune)
S. No. 28/27, Dhayari Narhe Road, Near Asian College
Pune 411041, Maharashtra
Tel : (020) 24690204; Mobile : 9657703143
Email : bookorder@pragationline.com

Nirali Prakashan
(For orders within Pune)
119, Budhwar Peth, Jogeshwari Mandir Lane
Pune 411002, Maharashtra
Tel : (020) 2445 2044; Mobile : 9657703145
Email : niralilocal@pragationline.com

MUMBAI
Nirali Prakashan
Rasdhara Co-op. Hsg. Society Ltd., 'D' Wing Ground Floor, 385 S.V.P. Road
Girgaum, Mumbai 400004, Maharashtra
Mobile : 7045821020, Tel : (022) 2385 6339 / 2386 9976
Email : niralimumbai@pragationline.com

DISTRIBUTION BRANCHES

DELHI
Nirali Prakashan
Room No. 2 Ground Floor
4575/15 Omkar Tower, Agarwal Road
Darya Ganj, New Delhi 110002
Mobile : 9555778814/9818561840
Email : delhi@niralibooks.com

BENGALURU
Nirali Prakashan
Maitri Ground Floor, Jaya
Apartments, No. 99, 6th Cross, 6th
Main, Malleswaram, Bengaluru
560003 Karnataka; Mob : 9686821074
Email : bengaluru@niralibooks.com

NAGPUR
Nirali Prakashan
Above Maratha Mandir, Shop No. 3,
First Floor, Rani Jhanshi Square,
Sitabuldi Nagpur 440012 (MAH)
Tel : (0712) 254 7129
Email : nagpur@niralibooks.com

KOLHAPUR
Nirali Prakashan
New Mahadvar Road, Kedar Plaza,
1st Floor Opp. IDBI Bank
Kolhapur 416 012 Maharashtra
Mob : 9850046155
Email : kolhapur@niralibooks.com

JALGAON
Nirali Prakashan
34, V. V. Golani Market, Navi Peth,
Jalgaon 425001, Maharashtra
Tel : (0257) 222 0395
Mob : 94234 91860
Email : jalgaon@niralibooks.com

SOLAPUR
Nirali Prakashan
R-158/2, Avanti Nagar, Near Golden
Gate, Pune Naka Chowk
Solapur 413001, Maharashtra
Mobile 9890918687
Email : solapur@niralibooks.com

marketing@pragationline.com | www.pragationline.com

Also find us on www.facebook.com/niralibooks

“ श्री स्वामी समर्थ चरणी”

समर्पित

PREFACE TO THE THIRTY FIRST EDITION

I am very happy to present this thirty first edition of the book : **"Unit Operations–I"** to all students of diploma course in chemical engineering (entirely in SI units).

The salient features of this revised and enlarged edition include simplified figures, SI units, addition of required new matter wherever necessary to make the book more complete, a simple and lucid language, proper approach to the presentation of the subject matter and a fairly large number of solved examples.

I am very thankful to all staff members and students of Chemical Engineering for giving a warm welcome to my book right from the first edition.

I hope and believe that students and staff members will find this edition very useful and helpful.

K. A. GAVHANE

Mob. No. 9850242440

kagavhane @ yahoo.in

CONTENTS

❑❑❑

CHAPTER ONE

INTRODUCTION

- Chemical Engineering is the branch of engineering which is concerned with the design and operation of industrial chemical plants. A chemical or process plant is required to carry out transformation of raw materials into desired products efficiently, economically and safely.
- Chemical Engineering is that branch of engineering which deals with the production of bulk materials from basic raw materials in a most economical way by chemical means.
- A chemical engineer is the one who develops, designs, constructs, operates and controls any physical and/or chemical or biochemical changing process.

A chemical engineer plays a vital role in the production and processing of food, in the manufacture of fertilizers, insecticides, herbicides, pesticides, plastics, synthetic fibres, elastomers, drugs, pharmaceuticals, pulp, paper, etc. He is also engaged in the processing of petroleum crude, the production of synthetic fuels, the utilisation of bio-mass and wind energy. He has to devise methods and equipments to control the release of harmful substances to protect the environment.

Chemical engineers work in four main segments of the chemical process industries - research and development, design, production and sales.

1.1 UNIT OPERATIONS

The operations carried out in the chemical process industry involving physical changes in the materials handled/in the system under consideration are regarded as unit operations.

Features of unit operations :

(i) They are physical in nature. (A physical change results in a change in property of matter and it requires some sort of driving force.)

(ii) They are common to all types of diverse industries (with no change in concept, merely change in conditions).

(iii) Individual operations have common techniques and are based on the same scientific principles irrespective of/regardless of the materials being processed.

(iv) Practical methods of carrying them out/conducting them may be more or less different in different industries.

(v) They are independent of industries in which they are carried out.

The operations involving physical changes are termed as unit operations to indicate that each single operation, such as distillation, is used in a wide range of industries and normally under varying operating conditions (temperature, pressure).

The unit operation distillation is employed to separate or purify alcohol in the beverage industry and hydrocarbons in the petroleum industry. Therefore, distillation of hydrocarbons in the petroleum industry is similar to distillation in the beverage industry. Evaporation of salt solutions in the chemical industry is similar to evaporation of sugar solutions in the food industry.

Broadly, unit operations are : (i) Mechanical operations e.g., size reduction, conveying, filtration, etc. (ii) Fluid flow operations in which the pressure difference acts as a driving force, (iii) Heat transfer (operations) in which the temperature difference acts as a driving force e.g., evaporation and (iv) Mass transfer operations in which the concentration difference acts as a driving force e.g., distillation.

They are basically used to conduct the physical steps of preparing the reactants, separating and purifying the products, recycling the unconverted reactants and controlling the energy transfer in and out of the chemical reactor.

A physical change resulting in a change in any property of matter occurs by the action of some sort of driving force. The normal driving force in heat transfer is a temperature difference. Hence, any unit operation results in a change in some property of the system under consideration. For example, in the screening operation it is the size.

Usually a large number of unit operations of chemical engineering are directed towards separating a substance into its component parts. For heterogeneous mixtures, such separations may be entirely mechanical, e.g., the separation of solid particles according to their size or the filtration of a solid from a suspension in a liquid.

Many unit operations involve particulate solids as well as fluids. In many cases, the solids are an integral part of the material being processed/treated. For example, feeding pulverised coal (in air) to a burner.

Mechanical operations involving particulate solids are :

(1) Size reduction-crushing and grinding.
(2) Mixing-solid-solid and liquid-liquid, etc.
(3) Classification-screening, froth flotation, magnetic separation, jigging, tabling and electrostatic separation (classification - it comprises of techniques of classifying a solid mixture into fractions, which differ from each other in some property) and wet classification.
(4) Solid-fluid separations - filtration, sedimentation and centrifugal separation.
(5) Gas-solid separations - dust collection, bag filtration, electrostatic precipitation.
(6) Solid handling - storage, feeding and conveying.
(7) Size enlargement - pelletization, agglomeration, granulation and extrusion.

1.1.1 Mechanical Separations

These separation techniques are applicable to heterogeneous mixtures and grouped into five headings as : (i) separation of solids from solids, (ii) separation of solids from solids in liquids, (iii) separation of solids from liquids, (iv) separation of solids or liquid drops from gases and (v) separation of liquids from liquids.

Solids are separated in the dry state by screening, electrostatic separation and magnetic separation.

Solids are separated from gases by gravity settling, centrifugal separation/settling filtration and electrostatic precipitation.

Solids are separated from liquids by filtration and sedimentation.

Various methods that are used for the separation of solids from solids in liquids include jigging, flotation, classification and tabling.

Immiscible liquids are separated from one another by using either a gravity decanter or a centrifugal decanter.

1.1.2 Fluid Transportation

A fluid is a substance which is capable of flowing if allowed to do so.

A fluid is a substance which undergoes continuous deformation when subjected to shear force.

Fluid flow : It is concerned with the principles that determine the flow or transportation of any fluid from one location to another.

Gases, vapours and liquids possessing the above characteristics are referred to as fluids.

Handling fluids is much simpler, cheaper and less difficult than handling of solids. Therefore, whenever possible, it is desirable to handle everything in the form of liquids, solutions or suspensions.

In every chemical process industry, pumps, fans, blowers and compressors, pipelines, ducts, valves and fittings are the essential components of a system used for transportation of fluids from one location to another.

A pump is a machine used for handling liquids, solutions, and slurries, while fan, blower, and compressor are used for handling gases. In these machines, mechanical work is transformed into fluid energy and energy input to a fluid by means of these machines causes the fluid to be transported through piping systems from one location to another. The machines commonly used in the chemical process industries include centrifugal pumps, rotary pumps (e.g. gear pumps), and reciprocating pumps (e.g. piston pump etc.) for liquids and corresponding fans, blowers, and compressors for gases.

Centrifugal pumps are used for handling thin liquids and suspension of solids in liquids, gear pumps are used for handling high viscosity liquids, diaphragm pumps are used for handling corrosive liquids, and screw pumps are used for handling slurries containing higher proportions of solids. Fans are used for moving gases when pressure heads of less than 30 kPa are involved. Fans are commonly employed for ventilation works, supplying air to dryers, removal of fumes, etc. Blowers are used for conveying a gas stream upto pressure of about 250 kPa and compressors are used for pressures as high as 240 MPa.

1.2 UNITS AND DIMENSIONS

The student of chemical engineering will find that data which he uses are expressed in a great variety of different units. Thus, a thorough knowledge of all the four systems of units and conversion factors is must to convert these quantities into a single system of units before proceeding with his calculations.

A unit is an arbitrarily selected standard of measure for a physical quantity. Therefore, volume may be measured in litres, cubic metres, etc.

Any physical quantity that can be measured or counted is always expressed as a product of its numerical value and units thereof. For example, when the distance between two stations is expressed as 100 m, 100 is a numerical value and the metre is the unit of distance (length).

Physical quantities are divided into two types. Certain quantities are regarded as basic quantities, such as length, mass, time, etc. and others are regarded as derived quantities, such as area, velocity, acceleration, etc. The derived quantities are obtained from the basic quantities. The derived units are formed from the base units. For example, velocity is defined as distance per unit time. Therefore, the unit of velocity is m/s. Acceleration is velocity per unit time, so the unit of acceleration is m/s^2 (in the SI system).

The dimension of a physical quantity is defined *as the power to which the fundamental quantities are raised in order to represent that quantity.*

The fundamental quantities such as mass, length, time, etc. are enclosed in square brackets [ ] to represent their dimensions.

For example, the velocity of a body is the distance covered by it in a specified time and is given as :

$$\text{Velocity} = \frac{\text{Distance}}{\text{Time}}$$

$$\text{Dimension of distance} = [L]$$

$$\text{Dimension of time} = [T]$$

$$\text{Dimension of velocity} = \frac{[L]}{[T]}$$

$$= [LT^{-1}]$$

Various systems of units and the basic/fundamental quantities associated with them are given below :

Basic Quantity	System of Units				Dimension
	CGS	MKS	FPS	SI	
Length	Centimetre (cm)	Metre (m)	Foot (ft)	Metre (m)	L
Mass	Gram (gm)	Kilogram (kg)	Pound (*l*b)	Kilogram (kg)	M
Temperature	Celsius (°C)	Celsius (°C)	Fahrenheit (°F)	Kelvin (K)	T
Time	Second (s)	Second (s)	Second (s)	Second (s)	θ

Symbolic abbreviations of units are given in brackets.

°C ⇒ degrees Celsius

°F ⇒ degrees Fahrenheit

K ⇒ degrees Kelvin

SI ⇒ International system of units

MKS ⇒ Metric system

Basic SI Units :

Mass : kilogram (kg)
Length : meter (m)
Time : second (s)
Temperature : kelvin (K)
Mole : kilogram mole (kmol)
Force : newton (N)
Pressure : newtons per square meter [N/m^2] ... Pa (pascals)
Energy : newtons meter [(N.m) = 1 N.m = 1 J] or joules
Power : newton meter per second [(N.m)/s = J/s = W]

Considering the increasing adoption of the International System of Units, abbreviated as SI units in India, particularly in the field of engineering and technology, this system of units will be adopted in this text book.

Derived Quantities :

Derived Quantity	Units in SI system	Abbreviated units	Dimension
Area	square meters	m^2	L^2
Volume	cubic meters	m^3	L^3
Density	kilograms per cubic meter	kg/m^3	ML^{-3}
Linear velocity	meters per second	m/s	$L\theta^{-1}$
Linear acceleration	meters per second per second	m/s^2	$L\theta^{-2}$
Capacity	cubic meters	m^3	L^3
Force	newtons	N	$ML\theta^{-2}$
Pressure	newtons per square meter (pascal)	N/m^2 (Pa)	$ML^{-1}\theta^{-2}$
Mass flow rate	kilograms per hour	kg/h	$M\theta^{-1}$
Volumetric flow rate	cubic meters per hour	m^3/h	$L^3\theta^{-1}$
Work/energy	joules	J	$ML^2\theta^{-2}$
Heat/Enthalpy	joules	J	$ML^2\theta^{-2}$
Power	kilowatts	kW	$ML^2\theta^{-3}$
Viscosity	newtons second per square meter [kilograms per meter per second]	(N·s)/m^2 or Pa·s [kg/(m·s)]	$ML^{-1}\theta^{-1}$

1.2.1 Force

The SI unit of force is newton (abbreviated as N). This unit is derived from Newton's second law of motion. *The law states that force (F) is proportional to the product of mass (m) and acceleration (a).*

$$F \propto m \times a$$

$$F = kma$$

where k is a proportionality constant.

In the SI system, the constant k is fixed at or taken as unity and we have

$$F = m \times a \text{ or } ma$$

Velocity, v, is defined as dL/dt, where L is the length (distance) and t is the time. The unit of velocity is m/s. The acceleration (a) is defined as a = dv/dt, so the unit of acceleration is m/s^2.

When a body of mass 1 kg is accelerated by 1 m/s^2, the force on the body is 1 (kg.m)/s^2, which is denoted as 1 newton (N).

Newton is the force which when applied to a body of mass 1 kg gives it an acceleration of 1 m/s^2. The unit of force, newton (N) has been named after Newton.

$$1 \text{ N} = 1 \text{ (kg.m)/s}^2$$

Dyne is the force which when applied to a body of mass 1 kg gives it an acceleration of 1 cm/s^2. The unit of force in the CGS system is dyne (dyn).

$$1 \text{ N} = 10^5 \text{ dyn}$$

1.2.2 Pressure

Pressure is defined as the force per unit area.

Therefore, the pressure of a fluid (gas or liquid) on a surface is the force (normal) exerted by the fluid on the surface.

The unit of pressure in the SI system is newton per square meter, N/m^2 (when force is measured in N and area in m^2).

The MKS unit of pressure is kilogram-force per square centimeter (kgf/cm^2).

The FPS unit of pressure is pound-force per square inch (lbf/in^2) (it is commonly known as psi-pound-force or poundal per square inch.)

In the SI system, the unit N/m^2 is called the pascal, symbol Pa (in honour of the scientist Pascal). As the unit of pressure Pa is small in magnitude, the pressure is usually expressed in SI in kilopascal (kPa). A multiple of pascal is called the bar and it is also used as a unit of pressure.

$$1 \text{ bar} = 10^5 \text{ Pa} = 10^5 \text{ N/m}^2$$

The air of the atmosphere exerts pressure on all bodies that are exposed to it. At the mean sea level, the pressure exerted by the air is 101325 N/m^2 (101325 Pa = 101.325 kPa). This pressure is known as the standard or normal atmospheric pressure. One standard atmosphere (1 atm) balances a column of 760 mm of mercury at 0°C.

$$1 \text{ atm} = 760 \text{ mm Hg} = 101325 \text{ Pa or N/m}^2 = 101.325 \text{ kPa or kN/m}^2$$
$$= 1.103 \text{ bar} = 760 \text{ torr} = 10.33 \text{ m } H_2O = 1.033 \text{ kgf/cm}^2 = 14.7 \text{ psi}$$

Sub-atmospheric pressure (pressure below the actual/normal atmospheric pressure) is expressed in torr (in honour of the scientist Torricelli). This unit is used for systems under vacuum.

Normally, pressure is measured with the help of a pressure gauge. The pressure gauge registers the difference between the pressure prevailing in the vessel (actual pressure) and the local atmospheric pressure. The pressure registered by the pressure gauge is called the gauge pressure and therefore the letter 'g' follows the unit of pressure [1.2 (kgf/cm^2) g, 2 atmg, 300 kPag, 30 psig].

The gauge pressure does not indicate the true total pressure (absolute pressure). For obtaining the prevailing true pressure or pressure above reference zero, we have to add the local atmospheric or barometric pressure in consistent units to the gauge pressure. If no letter follows the pressure units, it is taken as absolute pressure.

The relationship between absolute and gauge pressure is

$$\text{Absolute pressure} = \text{gauge pressure} + \text{atmospheric pressure}$$

Vacuum refers to sub-atmospheric pressure (pressure below the atmospheric pressure).

The relationship between absolute pressure and vacuum is

$$\text{Absolute pressure} = \text{atmospheric pressure} - \text{vacuum}$$

Vacuum is generally expressed in torr or mmHg (1 torr = 1 mmHg).

1.2.3 Work / Energy

Work (energy) is defined as *the product of the force acting on a body and the distance travelled by the body in the direction of the applied force.*

If force F acting on a body moves it through a distance L, then the work done on the body is given by

$$W = F \times L \Rightarrow \text{newton} \times \text{metre} \Rightarrow \text{N.m} \Rightarrow \text{joule (J)}$$

The unit of work (energy) in the SI system is newton-metre (N.m) or joule (abbreviated as J).

The MKS unit of work is meters kilogram force (m.kgf).

The CGS unit of work is erg.

When the point of application of one newton force moves a distance of one meter in the direction of the applied force, then the work done is one joule.

When the point of application of one dyne force moves a distance of one centimeter in the direction of the applied force, then the work done is one erg.

$$1\ J = 10^7\ erg$$

Energy is defined as *the capacity of a body for doing work.*

Mechanical work can produce a property of the matter called energy, e.g., when a body is forced upward it results in an increase in its potential energy.

Energy is present in a system in different forms, e.g., mechanical-potential, kinetic, chemical, thermal and electrical.

Any form of energy can be converted into work. For example, electrical energy (electricity) is utilised to run mechanical machinery, potential energy can be utilised to run hydraulic turbine, etc. So the units of work and energy are the same.

The energy possessed by a body by virtue of its motion is called kinetic energy. If a body of mass m is moving at a velocity v, then the kinetic energy of the body is

$$\text{Kinetic energy} = \frac{1}{2} mv^2$$

In the SI system with mass in kg and velocity in m/s, kinetic energy has the units of $(kg.m^2)/s^2$.

As the newton is the composite unit $(kg.m)/s^2$, kinetic energy is measured in newton-meters or joules.

$$\text{Units of K.E.} \Rightarrow kg\ (m^2/s^2) \Rightarrow (kg.m/s^2) \cdot m$$

$$\Rightarrow N.m \Rightarrow J$$

The unit of energy in the SI system is joule (J) ... it may be mechanical/thermal/chemical energy.

1.2.4 Power

Power is defined as *the work done per unit time or the rate at which work is done.*

$$P = \frac{W}{t} \Rightarrow \frac{\text{joule}}{\text{second}} \Rightarrow \text{watt}$$

The SI unit of power is watt, abbreviated as W.

Watt is the power that gives rise to the production of energy at the rate of one joule per second.

$$1\ W = 1\ J/s$$

$$1\ \text{Metric horse power (hp)} = 75\ (m.kgf)/s = 735.5\ W$$

$$1\ \text{British horse power} = 745.7\ W = 550\ (ft.lbf)/s.$$

1.2.5 Heat

It is defined as a *form of energy which is in transit between a hot source to a cold receiver.* It flows from higher temperature to lower temperature, so the temperature difference acts as a driving force for the transfer of energy as heat. It cannot be stored as such within the system. The exchange of energy occurs either as heat or as work. The unit of heat in the SI system is joule (J).

The unit of heat in the CGS system is calorie (cal).

The unit of heat in the MKS system is kilocalorie (kcal).

$$1\ cal = 4.1868\ J$$

For each kind of quantity there is one and only one unit in the SI system. For example, the joule $[J \equiv N.m \equiv (kg.m^2)/s^2]$ is the derived unit of energy whether it is kinetic, potential, chemical or in transition (i.e., energy in transit) as work and heat.

1.2.6 Volume

The SI unit of volume is cubic meter (m^3).

The MKS unit of volume is litre (*l*).

Volume V is a quantity that represents the product of three lengths and it depends on the amount or quantity of the material.

A litre is the volume occupied by pure air free water of mass 1 kg at the temperature of its maximum density and under normal atmospheric pressure.

$$1 \text{ m}^3 = 1000 \, l$$

Specific volume is defined as the volume per unit mass. It is the reciprocal of density. Molar volume is defined as volume per mole. The SI unit of specific volume is m^3/kg and that of molar volume is m^3/kmol.

1.2.7 Temperature

It is a measure of the *degree of hotness or coldness of a body (the degree of heat present in a body or substance)*. The SI unit for temperature is kelvin (abbreviated as K, the unit of temperature has been named as kelvin in the honour of the scientist Kelvin).

1.3 CONVERSION FACTORS

It is often necessary to convert the units of a particular quantity from one system to some other system. For doing this we have to make use of the appropriate conversion factors. The conversion factors are simply the ratio of the magnitude of the unit in one system to the magnitude of the same unit in the other system. The following table gives the conversion factors to the SI units.

Conversion factors to SI units :

To convert from	To	Multiply by
1. Mass, M :		
*l*b	kg	0.4536
t	kg	1000
g	kg	10^{-3}
2. Length, L :		
ft	m	0.3048
cm	m	0.01
in	mm	25.4
in	m	6.0254
3. Area, L^2 :		
ft^2	m^2	0.0929
cm^2	m^2	10^{-4}
in^2	m^2	6.0452×10^{-4}
4. Volume, L^3		
ft^3	m^3	0.02832
cm^3	m^3	10^{-6}
l	m^3	10^{-3}
in^3	m^3	16387.1
5. Density, M/L^3 :		
*l*b/ft^3	kg/m^3	16.019
g/cm^3	kg/m^3 = g/*l*	1000
kg/*l*	kg/m^3	1000

... Contd.

6. Viscosity, M/Lθ :		
P [poise] ≡ g/(cm.s)	kg/(m.s) = (N.s)/m^2	0.10
cP	kg/(m.s) = (N.s)/m^2 = Pa.s	0.001
*l*b/(ft.s)	kg/(m.s)	1.488
*l*b/(ft.h)	(mN.s)/m^2	0.414
1 stoke	m^2/s	10^4
7. Force, F or ML/θ^2 :		
*l*bf	N	4.448
kgf	N	9.807 (9.81)
dyn	N	10^{-5}
8. Velocity, L/θ :		
ft/s	m/s	0.3048
ft/h	m/s	8.467×10^{-5}
9. Volumetric flow rate, L^3/θ :		
ft^3/s	m^3/s	0.02832
ft^3/h	m^3/s	7.867×10^{-6}
l/h	m^3/s	2×10^{-7}
l/s	m^3/s	10^{-3}
10. Pressure F/L^2 or $ML^{-1} \theta^{-2}$:		
*l*bf/ft^2	N/m^2 = Pa	47.88
std.atm	N/m^2 = Pa	1.01325×10^5
std.atm	kPa	101.325
in Hg	N/m^2 = Pa	3.386×10^3
torr (mm Hg)	N/m^2 = Pa	133.3
kgf/cm^2	N/m^2 = Pa	9.808×10^4
dyn/cm^2	N/m^2 = Pa	10^{-1}
bar	N/m^2 = Pa	10^5
*l*bf/in^2	N/m^2 = Pa	6894.76
m H_2O	N/m^2 = Pa	9806.65
11. Mass flow rate, M/θ :		
*l*b/s	kg/s	0.4536
*l*b/h	kg/s	1.26×10^{-4}
12. Mass velocity, M/L^2θ :		
*l*b/(ft^2.h)	kg/(m^2.s)	1.356×10^{-3}
g/(cm^2.s)	kg/(m^2.s)	10
13. Power, FL/θ or $ML^2 \theta^{-3}$:		
(ft.*l*bf)/s	(N.m)/s = W	1.356
hp (British)	(N.m)/s = W	745.7
kcal/h	(N.m)/s = W	1.163
hp (metric)	(N.m)/s = W	735.5

... Contd.

14. Energy, Work, Heat, FL or $ML^2\ \theta^{-2}$:		
cal	N.m = J	4.187
kcal	N.m = J	4187
erg	N.m = J	10^{-7}
kW.h	N.m = J	3.6×10^6
Btu	N.m = J	1055

Some of the Prefixes for SI units :

Prefix	Symbol	Factor (Multiply by)
tera	T	10^{12}
giga	G	10^9
mega	M	10^6
kilo	k	10^3
deci	d	10^{-1}
centi	c	10^{-2}
milli	m	10^{-3}
micro	μ	10^{-6}
nano	n	10^{-9}
pico	p	10^{-12}

While writing the units of a fundamental or derived quantity, please remember the following :

1. Correct : 20 N.m Incorrect : 20 Nm

The unit symbol should be written in lower case letters but when the unit is named after a scientist the first letter of the unit is written in the upper case (capital letter). For example, meter ⇒ m, newton ⇒ N, pascal ⇒ Pa.

2. Correct : 100 kg Incorrect : 200 kgs
3. Correct : 10 cm Incorrect : 10 cm.
4. Correct : 10000 $W/(m^2.K)$ Incorrect : 10000 $W/m^2.K$ or $W/m^2/K$
5. Correct : 10 kW Incorrect : 10 k W
6. Correct : 1 mg Incorrect : 1 μkg

For 1 : There should be a space or dot between N and m. The unit symbols should be separated by a gap or a dot (·).

For 2 : The unit symbols should not be written in the plural form. kgs – incorrect - s should not be written.

For 3 : The unit symbols should not be followed by a full stop (period) unless it is at the end of a sentence.

For 4 : Oblique stroke must be repeated on the same line unless ambiguity is avoided by parantheses.

For 5 : There should not be any space between the prefix symbol and the unit symbol.

1.4 MATERIAL BALANCES

- A material balance is an accounting of all the material that enters, leaves, accumulates, or is depleted in a process or in a process unit in a given time.

The basis for material balance calculations of any unit process or unit operation is the law of conservation of mass.

- The law of conservation of mass states that *matter can neither be created nor destroyed.* Therefore, the material entering in any process must either leave or accumulate within it.
- The law of conservation of mass states that *the total mass of various components taking part in any unit operation or unit process is constant.* Thus, for any unit operation,

$$\text{Input} = \text{Output} + \text{Accumulation}$$

For steady-state operations where accumulation is constant or nil, the law of conservation of mass takes the simple form as $\text{Input} = \text{Ouput}$

i.e., for any operation carried out under steady-state conditions (conditions that do not vary with time), the mass of the material entering is equal to the mass of the material leaving. The material balances must hold over the entire process, or process equipment or over any part of it. Material balances must apply to all the material that enters and leaves the process or to any one material.

1.5 ENERGY BALANCES

The energy balance of a particular process can be achieved from the first law of thermodynamics.

It states that *energy can neither be created nor destroyed during a process, although it may change from one form to another.*

The energy balance must include all types of energy.

The energy balance equation for a steady-state flow process is

$$\text{Input energy} = \text{Output energy.}$$

1.6 MOLECULAR UNITS (MOLAR UNITS)

In problems of material balances involving chemical reactions, molecular/molar units are often used instead of weight units for simplification of calculations.

A mole is defined as the *amount of substance that is numerically equal to its molecular weight.*

A gram mole of a substance is defined as *the mass in grams of the substance that is equal numerically to its molecular weight.*

Thus, one gram mole of methanol is equivalent to 32 grams of methanol. The number of moles of any substance are calculated by knowing its mass or weight using the relation :

$$N_A = \frac{W_A}{M_A} \qquad \text{... (1.1)}$$

where N_A = number of moles (either mole or kmol) of A

W_A = weight of A in g or kg

M_A = molecular weight of A either in g/mol or kg/kmol

In this book, gram mole and kilogram mole are specified as mol and kmol respectively.

1.7 WEIGHT FRACTION

In a mixture of substances, the composition may be expressed in terms of weight fraction or weight percent.

The weight fraction of a component in a mixture is defined as *the ratio of the weight of the individual component to the total weight of the mixture.*

$$\therefore \quad \text{Weight fraction of A} = \frac{\text{Weight of A}}{\text{Weight of the mixture}} \qquad \text{... (1.2)}$$

$$= \frac{W_A}{W_A + W_B + W_C + \ldots} \qquad \text{... (1.3)}$$

where W_A, W_B, W_C are the weights of A, B, C ...... respectively.

The weight percent of a component in a mixture is defined as *the weight of the component expressed as a percentage of the total weight of the mixture.*

$$\therefore \quad \text{Weight \% A} = \left(\frac{W_A}{W_A + W_B + W_C + ...}\right) \times 100 \quad ... (1.4)$$

From Equations (1.3) and (1.4), we get,

$$\text{Weight \% of A} = (\text{Weight fraction of A}) \times 100 \quad ... (1.5)$$

Note that : *The sum of the weight fractions of all the components present in the mixture is equal to unity.*

1.8 MOLE FRACTION

The composition of a mixture of substances can also be expressed in terms of mole fraction or mole percent.

The mole fraction of a component in a mixture is defined as *the ratio of the moles of the individual component to the total moles of the mixture.*

$$\therefore \quad \text{Mole fraction of A} = \frac{N_A}{N_A + N_B + N_C + ...} \quad ... (1.6)$$

where N_A, N_B, N_C ...... are the moles of substances A, B, C ...... respectively.

The mole percent of a component in a mixture is defined *as the moles of the component expressed as a percentage of the total moles of the mixture.*

$$\text{Mole \% of A} = \left(\frac{N_A}{N_A + N_B + N_C + ...}\right) \times 100 \quad ... (1.7)$$

From Equations (1.6) and (1.7), we get,

$$\text{Mole \% of A} = \text{Mole fraction of A} \times 100 \quad ... (1.8)$$

Note that : *The sum of the mole fractions of all the components present in the mixture is equal to unity.*

1.9 GAS LAWS

(a) Ideal gas law : An ideal gas law is a relationship of considerable utility in the engineering practice since it is sufficiently accurate for a great majority of gases and vapours at ordinary temperatures and pressures. Mathematically, the ideal gas law for 'n' moles of gas is given by

$$PV = nRT \quad ... (1.9)$$

where P = absolute pressure in kPa in SI

V = volume (m^3)

n = number of moles of gas (kmol)

T = absolute temperature in Kelvin (K)

R = universal gas constant

R = 8.31451 m^3.kPa/(kmol.K) = 8.31451 J/(mol.K)

= 0.08206 *l*.atm/(mol.K) or m^3.atm/(kmol.K)

This law states three facts : (i) the volume of a gas is directly proportional to the number of moles of the gas, (ii) the volume of a gas directly proportional to the absolute temperature, and (iii) the volume of a gas inversely proportional to the pressure.

When the mass of a gas is not known and if we know the volume occupied by the gas at a specified temperature and pressure, and conditions are changed and if we know two of three variables in the final state, then the third one can be calculated by means of the proportionality indicated by the ideal gas law.

Let V_1, T_1 and P_1 be the volume, temperature and pressure respectively of 'n' moles of gas at state 1.

Let V_2, T_2 and P_2 be the volume, temperature and pressure respectively of 'n' moles of gas at state 2.

Then,
$$P_1V_1 = n\,RT_1 \quad ... (1.10)$$

and
$$P_2V_2 = n\,RT_2 \quad ... (1.11)$$

Combining the above two equations, we get,

$$\frac{P_1V_1}{T_1} = \frac{P_2V_2}{T_2} \quad \text{... (1.12)}$$

Molar volume : It is clear from the equation (1.9) that a mole of an ideal gas under definite conditions of temperature and pressure always occupies a definite volume regardless of the nature of the gas. This volume occupied by one mole of gas is called molar volume. At 273 K (0 °C) and 101.325 kPa pressure, one kmol (kilogram mole) of an ideal gas occupies a volume of 22.41 cubic metres. The concept of the molar volume can be applied to the mixture of gases just as to a pure gas.

(b) Dalton's Law : It states that the *total pressure exerted by a gas mixture is equal to the sum of the partial pressures of the component gases present in the mixture.*

Mathematically,

$$P = p_A + p_B + p_C \quad \text{... (1.13)}$$

where P is the total pressure and p_A, p_B, p_C ... are the partial pressures of the component gases A, B, C ...... respectively.

This law expresses the additive nature of partial pressures.

Partial Pressure : The partial pressure of a component gas that is present in a mixture of gases is *the pressure that would be exerted by that component in the same volume and at the same temperature.*

(c) Amagat's Law : It states that *the total volume occupied by a gas mixture is equal to the sum of the pure component volumes of the component gases present in the mixture.*

Mathematically,

$$V = V_A + V_B + V_C + ... \quad \text{... (1.14)}$$

where V is the total volume and V_A, V_B, V_C are pure component volumes of the component gases A, B, C ...... respectively.

For gases behaving ideally (i.e., gases following the ideal gas equation), we have

$$\text{Volume \%} = \text{Mole \%} = \text{Pressure \% (for any component)} \quad \text{... (1.15)}$$

1.10 MECHANICAL LAWS

The basic mechanical equation is Newton's second law of motion. This law may be given as

$$F = k\,m.a \quad \text{... (1.16)}$$

It states that force is proportional to the product of mass and acceleration.

where F = resultant of all forces acting on a body.

m = mass of body

a = acceleration of body in the direction of resultant force

k = proportionality constant

By using a = du/dt in equation (1.16), an equivalent statement of the law is

$$F = k \cdot \frac{d}{dt}(mu) \quad \text{... (1.17)}$$

where 'u' is the velocity of the body and 't' is the time. Equation (1.17) states that the resultant of all forces acting on a body of mass 'm' is proportional to the time rate of momentum change of the body in the direction of the resultant force.

1.11 DIMENSIONAL FORMULAE

The dimensional formula of a physical quantity measured in derived units expresses the way in which the basic units enter into the operation by which the quantity is measured. Thus, acceleration is defined as the velocity per unit time, and velocity is defined as the distance per unit time. Therefore, the dimensional formula of acceleration is

$$[a] = \frac{L}{\theta^2} = L\theta^{-2} \quad \text{... (1.18)}$$

The dimensional formula of the quantity 'a' is denoted by the symbol [a].

The dimensional formula of any secondary quantity has the form $M^{\alpha} L^{\beta} \theta^{\gamma}$ ...... where M, L, θ ...... are dimensions of mass, length, time, etc. The exponents α, β, γ ...... of the dimensions may be positive or negative integers (whole numbers), fractions, or zero.

1.12 DIMENSIONLESS EQUATIONS

Equations derived directly from the basic laws of the physical sciences consist of terms that have the same dimensions because of the fact that the basic laws themselves are used to define secondary/derived quantities. An equation in which all terms have the same dimensions is called a dimensionally homogeneous equation. If a dimensionally homogeneous equation is divided by any one of its terms, then the dimensions of each term cancel and only numerical magnitudes remain. Such an equation is called a dimensionless equation.

A dimensionally homogeneous equation can be used with any set of basic units, provided that the same basic units of mass, length, time, force and temperature are used throughout. Units meeting this requirement are referred to as consistent units.

For example, consider an equation of motion of a freely falling body

$$Z = u_o t + \frac{1}{2} g t^2 \quad \text{... (1.19)}$$

If the dimensions are substituted for the variables in terms of dimensions of mass, length etc., it will be seen that the dimension of each term is length.

$$(L) = (L\,\theta^{-1})\,(\theta) + \frac{1}{2}\,(L\theta^{-2})\,(\theta)^2 \quad \text{... (1.20)}$$

or dividing Equation (1.19) by Z, we get

$$1 = \frac{u_o t}{Z} + \frac{\left(\frac{1}{2}\right)(gt^2)}{Z} \quad \text{... (1.21)}$$

If we check the dimensions of each term of Equation (1.21) we see that the dimensions of each term cancel, and thus, each term is dimensionless. A combination of variables which is dimensionless is called a dimensionless group. The terms on the right hand side of Equation (1.21) are dimensionless groups.

1.13 DIMENSIONAL EQUATIONS

An equation containing terms of varying dimensions is called a dimensional equation or dimensionally non-homogeneous equation. In case of such equations there is no point in using consistent units, and two or more time units, such as hours and minutes or two or more length units, such as metres and centimetres may appear in the same equation.

For example, an empirical relation to find the rate of heat loss by conduction and convection from a horizontal pipe to the surrounding atmosphere is

$$\frac{Q}{A} = 0.055 \frac{(\Delta T)^{1.25}}{(D_o)^{0.25}} \quad \text{... (1.22)}$$

where
- Q is the rate of heat loss, Btu/h
- A is the area of heat transfer, ft^2
- ΔT is the temperature difference between pipe wall and atmosphere, ^oF
- D_o is the outside diameter of pipe, in

In the above equation, the dimensions of Q/A are different from those of the RHS of Equation (1.22) and hence, Equation (1.22) is dimensional.

1.14 DIMENSIONAL ANALYSIS

Many physical problems that are important in chemical engineering in the area of fluid flow, heat transfer, and mass transfer, which cannot be completely solved by theoretical or mathematical methods, can be handled by dimensional analysis.

Dimensional analysis is a method of correlating the number of variables into a single equation stating clearly an effect. When the value of a given physical quantity is influenced/affected by a number of variables, then it is not possible to determine their individual effects by experimental methods. In such cases, the problem can be made more easily handled by making use of the method of dimensional analysis wherein the variables are arranged in dimensionless groups which are considerably less than the number of variables.

This method is based on the fact that, if a theoretical equation does exist among the variables affecting a physical quantity, then the equation must be dimensionally homogeneous. Due to this a large number of variables can be grouped into a smaller number of dimensionless groups of variables. It does not provide a numerical equation and experiment is required to complete the solution of the problem.

1.15 DIMENSIONLESS GROUP

Several important dimensionless groups have been found by using a method of dimensional analysis. For example, $N_{Re} = \frac{Du\rho}{\mu}$ = Reynolds number, $N_{Pr} = \frac{C_p\mu}{k}$ = Prandtl number, etc.

Each of the above mentioned groups of variables is having no dimensions, i.e., is dimensionless. The numerical value of a dimensionless group is the same for any consistent set of units used for the variables within the groups. The units chosen for one group need not be consistent with those for another groups.

1.16 THE RAYLEIGH METHOD

If $Q_1, Q_2, ..., Q_n$ are 'n' dimensional variables, out of which Q_1 is the dependent variable and $Q_2, Q_3, ..., Q_n$ are the independent variables, then according to this method of dimensional analysis :

Q_1 varies as $Q_2^a\ Q_3^b$ ...... The dimensionless groups are obtained by evaluating the powers/exponents so that the relationship among the variables is dimensionally homogeneous. Similar/like powers or exponents of the quantities are grouped together in order to get dimensionless groups.

The following problem will give a clear idea regarding the Rayleigh method.

1.17 APPLICATION OF DIMENSIONAL ANALYSIS TO FLUID FLOW

Statement of a problem : The pressure drop/loss due to friction for the flow of a fluid through a pipe depends on the following variables.

(i) Diameter of the pipe (D),

(ii) Length of the pipe (L),

(iii) Velocity of the fluid (u),

(iv) Density of the fluid (ρ),

(v) Viscosity of the fluid (μ).

Using Rayleigh method of dimensional analysis, obtain a relation between the pressure drop (ΔP) and the given variables.

Solution : The relationship between the frictional pressure drop and variables on which it depends may be written as

$$\Delta P = f(D\ u\ \rho\ \mu\ L) \quad ... (1.23)$$

$$\Delta P = k\ D^a\ u^b\ \rho^c\ \mu^d\ L^e \quad ... (1.24)$$

where k is a dimensionless constant.

The dimensions of the variables in terms of M, L and θ are :

Parameter	Units	Dimensions
Pressure drop, ΔP	N/m^2, $kg/(m.s^2)$	$ML^{-1}\,\theta^{-2}$
Diameter, D	m	L
Length, L	m	L
Density, ρ	kg/m^3	ML^{-3}
Viscosity, μ	kg/(m.s)	$ML^{-1}\,\theta^{-1}$
Velocity, u	m/s	$L\theta^{-1}$

According to Rayleigh's method, for the relationship to be dimensionally homogeneous, $D^a\, u^b\, \rho^c\, \mu^d\, L^e$ must have the same dimensions as ΔP.

Substituting the dimensions of each term in Equation (1.24) gives

$$[ML^{-1}\,\theta^{-2}] = k\,[L]^a\,[L\theta^{-1}]^b\,[ML^{-3}]^c\,[ML^{-1}\,\theta^{-1}]^d\,[L]^e \qquad \text{... (1.25)}$$

As Equation (1.23) is assumed to be dimensionally homogeneous, the exponents/power/indices of the individual primary/basis units on the left hand side of Equation (1.25) must be equal to those on the right hand side.

∴ Equating the powers/indices of M, L, and θ, we get following set of equations :

Exponents/powers of M : $1 = c + d$ ... (1.26)

Exponents of L : $-1 = a + b - 3c - d + e$ ... (1.27)

Exponents of θ : $-2 = -b - d$ ... (1.28)

Here, there are five unknowns, but we have only three equations. Three of these unknowns may be found in terms of the remaining two. Therefore, let us express the exponents a, b and c in terms of d and e.

From Equation (1.26), $c = 1 - d$ ... (1.29)

From Equation (1.28), $b = 2 - d$

From Equation (1.27), $-1 = a + b - 3c + e$

$-1 = a + 2 - d - 3 + 3d - d + e$

∴ $a = -d - e = -(d + e)$ ... (1.31)

Substituting for a, b and c from Equations (1.29), (1.30) and (1.31) in equation (1.24), we get

$$\Delta P = k\,[D]^{-d-e}\,[u]^{2-d}\,[\rho]^{1-d}\,[\mu]^d\,[L]^e$$

Collecting the terms into groups having the same exponents, we get

$$\frac{\Delta P}{\rho u^2} = k\left[\frac{\mu}{Du\rho}\right]^d\left[\frac{L}{D}\right]^e \qquad \text{... (1.32)}$$

The dimensions of each of the bracketed groups in Equation (1.32) are zero and thus all the groups are dimensionless.

The group Duρ/μ is known as the Reynold's number. It will occur frequently in the study of fluid flow and it gives idea regarding the type of flow in a given geometry. Equation (1.32) involves the reciprocal of the Reynolds number and thus Equation (1.32) can be rewritten as :

$$\frac{\Delta P}{\rho u^2} = k\left[\frac{Du\rho}{\mu}\right]^{-d}\left[\frac{L}{D}\right]^e \qquad \text{... (1.33)}$$

and more generally as, $$\frac{\Delta P}{\rho u^2} = \phi\left[\frac{Du\rho}{\mu}, \frac{L}{D}\right] \qquad \text{... (1.34)}$$

$\frac{\Delta P}{\rho u^2}$ is the Euler number, N_{Eu} and Duρ/μ is the Reynolds number, $N_{Re.}$

(II) Statement of a problem : The power required P for an agitator depends upon the propeller diameter D, the rotational speed N of the agitator, the liquid density ρ, the viscosity μ and the gravitational acceleration g. Find, by a dimensional analysis, the correct representation for the power requirement in terms of dimensionless groups.

Solution : The relationship between the power, P and the variables on which it depends is given as :

$$P = f(D, N, P, \mu, g) \qquad \text{... (1.35)}$$

$$P = k\, D^a\, N^b\, \rho^c\, \mu^d\, g^e \qquad \text{... (1.36)}$$

The dimensions of the parameters/variables in terms of M, L, and θ are :

Parameter	Units	Dimensions
Power, P	(N.m)/s = (kg.m²)/s²	$ML^2\theta^{-3}$
Diameter, D	m	L
Speed, N	1/s	θ^{-1}
Viscosity, μ	kg/(m.s)	$ML^{-1}\theta^{-1}$
Density, ρ	kg/m³	ML^{-3}
Gravitational acceleration, g	m/s²	$L\theta^{-2}$

In terms of dimensions, Equation (1.36) can be written as

$$[ML^2\theta^{-3}] = k\,[L]^a\,[\theta^{-1}]^b\,[ML^{-3}]^c\,[ML^{-1}\theta^{-1}]^d\,[L\theta^{-2}]^e \qquad \text{... (1.37)}$$

Equating the exponents / indices of M, L and θ, we get

$$\text{Exponents of M : } 1 = c + d \qquad \text{... (1.38)}$$

$$\text{Exponents of L : } 2 = a - 3c - d + e \qquad \text{... (1.39)}$$

$$\text{Exponents of } \theta : -3 = -b - d - 2e \qquad \text{... (1.40)}$$

There are three equations and five unknowns. Three of the unknowns may be found in terms of the remaining two. Therefore, express the exponents a, b and c in terms of d and e.

$$\text{From Equation (1.38) : } c = 1 - d \qquad \text{... (1.41)}$$

$$\text{From Equation (1.40) : } b = 3 - d - 2e \qquad \text{... (1.42)}$$

From Equation (1.39),

$$2 = a - 3c - d + e$$

$$2 = a - 3(1 - d) - d + e$$

$$2 = a - 3 + 3d - d + e$$

$$\therefore \quad a = 5 - 2d - e \qquad \text{... (1.43)}$$

Substituting the values of a, b and c in terms of d and e in Equation (1.35) gives

$$P = k\,(D)^{5-2d-e}\,(N)^{3-d-2e}\,(\rho)^{1-d}\,(\mu)^d\,(g)^e \qquad \text{... (1.44)}$$

Collecting the terms, into groups, having the same exponents, we get

$$\frac{P}{D^5 N^3 \rho} = k\left(\frac{D^2 N \rho}{\mu}\right)^{-d}\left(\frac{g}{N^2 D}\right)^e \qquad \text{... (1.45)}$$

$\frac{P}{D^5 N^3 \rho}$ is the Power number, N_{Np}; $\frac{D^2 N \rho}{\mu}$ is the Reynolds number for agitator, N_{Re}

$\frac{DN^2}{g}$ is the Froude number, N_{Fr}

Equation (1.45) is the required relationship.

1.18 BUCKINGHAM'S π THEOREM

Buckingham's π theorem states that the *number of dimensionless groups is equal to the number of variables minus the number of fundamental dimensions.*

If there are 'n' variables, Q_1, Q_2 ...... Q_n affecting a given physical process, the functional relationship between them can be written as f (Q_1, Q_2, ..., Q_n) = 0 ... (1.46)

If there are 'm' fundamental dimensions (such as M, L, θ, etc.) required to define the dimensions of Q_1, Q_2, ... , Q_n, then there will be n – m dimensionless groups (π_1, π_2, ... , π_{n-m}). The functional relationship between them can be written as

$$f(\pi_1, \pi_2, \ldots\ldots, \pi_{n-m}) = 0 \quad \ldots (1.47)$$

The groups π_1, π_2, etc. must be independent of each other.

With the help of Buckingham's π theorem, it is possible to obtain the dimensionless groups more easily than by solving the simultaneous equations for the exponents/indices.

Procedure :

(a) Select m of the original variables in order to form what is called a recurring set.

(b) The dependent variable, that is, the variable that depends upon (or gets affected by) the other variables, should not be included in the recurring set.

(c) Any set m of the variables may be selected with the following provisions.

  (i) Each of the fundamental/basic dimensions must appear in at least one of the m variables.

  (ii) Some or all of the variables within the recurring set must not form a dimensionless group.

(d) Take each of the remaining n – m variables on its own and form it into a dimensionless group by combining it with one or more members of the recurring set. In this manner, the n – m dimensionless groups are formed.

The following problem will give a clear idea regarding the procedure of obtaining dimensionless groups with the help of the Buckingham's π theorem.

1.18.1 Use of Buckingham's π Theorem for Dimensional Analysis

The relationship between the variables affecting the pressure drop for flow of a fluid in a pipe may be written as :

$$f(\Delta P, d, l, u, \rho, \mu) = 0 \quad \ldots (1.48)$$

In Equation (1.48), there are six variables and three fundamental dimensions (mass, length and time).

Therefore, number of dimensionless groups = 6 – 3 = 3.

The recurring set must contain three variables that cannot themselves form a dimensionless group.

(i) l and d cannot be selected as they can be themselves formed into the dimensionless group l/d.

(ii) ΔP, ρ and u cannot be selected as they can be formed into the dimensionless group $\Delta P/\rho u^2$.

(iii) ΔP is the affected variable. Therefore, it cannot be included in the recurring set.

If we select the variables d, u, ρ as the recurring set, then this set fulfils all the above conditions.

Dimensionally :
$$d = L$$
$$u = L\theta^{-1}$$
$$\rho = ML^{-3}$$

Each of the dimensions M, L and θ can then be obtained in terms of the variables d, u and ρ. Therefore,

$$L = D$$
$$M = \rho D^3$$
$$\theta = Du^{-1}$$

Thus, the three dimensionless groups are formed by taking each of the remaining variables ΔP, *l* and μ in turn.

(i) Taking ΔP to form one dimensionless group :

ΔP has dimensions $ML^{-1}\theta^{-2}$.

$\Delta P\, M^{-1} L\, \theta^2$ is therefore dimensionless.

Therefore, group π_1 is $\pi_1 = \Delta P\,(\rho D^3)^{-1}\,(D)\,(Du^{-1})^2 = \Delta P/\rho u^2$ ... first group

(ii) Taking *l* :

l has dimension L.

$l . L^{-1}$ is therefore dimensionless.

Therefore, group π_2 is $\pi_2 = l\,(D)^{-1} = l/D$ ... second group

(iii) Taking μ :

μ has dimensions $ML^{-1}\theta^{-1}$.

$\mu\, M^{-1} L\, T$ is therefore dimensionless.

Therefore, group π_3 is $\pi_3 = \mu\,(\rho D^3)^{-1}\,(D)\,(Du^{-1}) = \mu / Du\rho$ ... third group

Hence,
$$f\left(\frac{\Delta P}{\rho u^2}, \frac{l}{D}, \frac{\mu}{Du\rho}\right) = 0 \quad \text{... (1.49)}$$

or
$$\frac{\Delta P}{\rho u^2} = f\left(\frac{l}{D}, \frac{Du\rho}{\mu}\right) \quad \text{... (1.50)}$$

$\mu/Du\rho$ is an arbitrarily inverted group since the Reynolds number is usually expressed in the form $Du\rho/\mu$.

1.19 USEFUL MATHEMATICAL METHODS

The mathematical procedure of the calculations needed for the theory of unit operations involves the use of the most elementary parts of the calculus. Two simple mathematical techniques that are useful in the treatment of unit operations are :

1. Graphic methods of integration.
2. Graphical treatment of exponential functions.

1.19.1 Graphical Integration

The value of a definite integral $\int_{x_1}^{x_2} f(x)\,.\,dx$, where f(x) is a function of x, from the first principles of integral calculus is the area bounded by the curve of f (x) v/s x, the ordinates $x = x_1$ and $x = x_2$ and the x-axis.

Any definite integral can therefore be evaluated numerically by plotting f(x) v/s x, drawing the verticals/vertical lines through the given limits (i.e., through the values of x) and measuring the area under the curve, between the vertical lines through the given limits and the x-axis. Thus, in Fig. 1.1, the curve ABCD represents the plot of f(x) v/s x, and vertical lines BE and CF correspond to the values of x_1 and x_2, respectively and the area bounded by the points B, E, F and C is the desired integral.

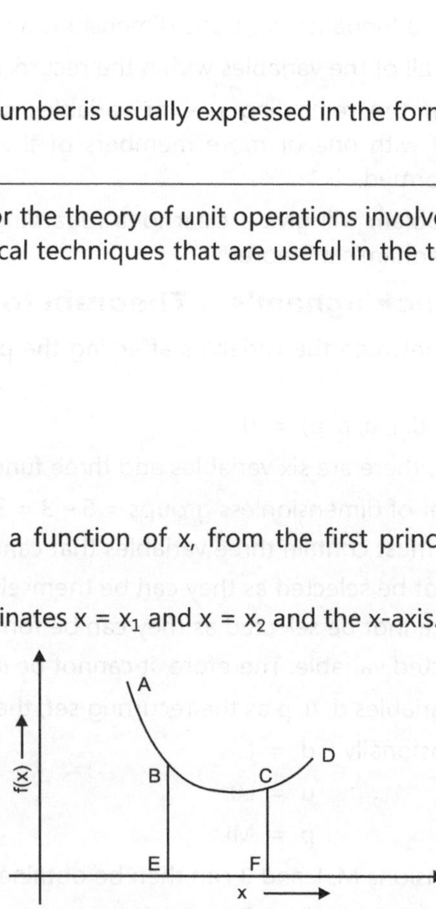

Fig. 1.1 : Principles of graphical integration

If the area under the curve is Z cm² (measured), the scale of x-axis is 1 cm = 2 units and that of y-axis is 1 cm = 0.1 units, then the value of the integral is obtained as

$$\int_{x_1}^{x_2} f(x)\, dx = \text{Area under the curve} \times (\text{Scale of x-axis}) \times (\text{Scale of y-axis}) \quad \text{... (1.51)}$$

$$= Z\ \text{cm}^2 \times \left(\frac{2}{1\ \text{cm}}\right)\left(\frac{0.1}{1\ \text{cm}}\right) \quad \text{... (1.52)}$$

1.19.2 Exponential Equations and Log-Log Plots

In many cases, experimental data involving the variables x and y fit an expression of the type

$$y = m x^n \quad \text{... (1.53)}$$

where m and n are constants.

Taking the logarithms of both sides of Equation (1.51), we get

$$\log y = \log m + n \log x \quad \text{... (1.54)}$$

It is clear from the equation (1.54) that a plot of log y against log x yields a straight line with a slope equal to 'n' and an intercept equal to log m.

In order to plot a set of data by this method we need the logarithms of the numerical values of the two variables but instead of following this procedure it is more convenient to use a log-log paper for the same.

Log-log paper is a coordinate paper on which the scales are logarithmic instead of being uniform. In other words, the intervals marked 1, 2, 3, etc. are not equidistant but are in proportion to the logarithms of the numbers 1, 2, 3, etc. One such a plot is drawn in Chapter 7.

Disadvantage of the log-log plot is that the scales normally used cannot be read too closely or linear interpolation cannot be used to read the scales.

['log' is used to represent logarithms to be base 10 and 'ln' is used to represent logarithms to the base e.]

Fig. 1.2 shows a log-log paper.

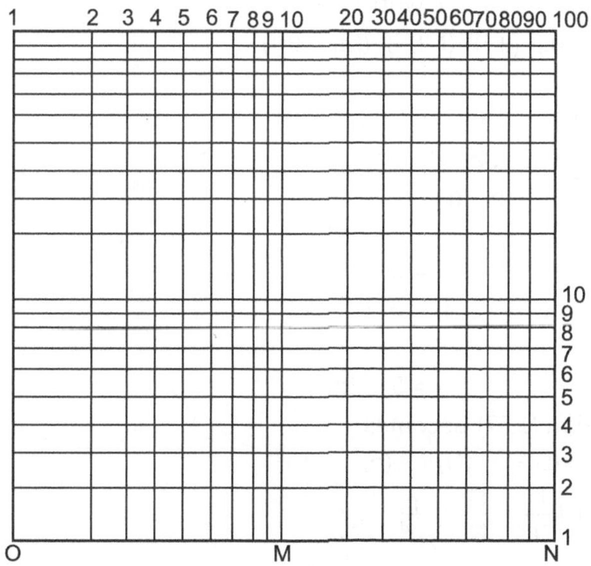

Fig. 1.2 : Log-Log Paper

From Fig. 1.2, it can be seen that the scales on x-axis and y-axis are not linear. Let us take point O as the origin. It must be kept in mind that the origin on the log-log paper is never zero on x-axis or y-axis as log (zero) is equal to (−∞). Suppose that the origin is chosen such that its x co-ordinate is 1, then point M on the x-axis is 10 and point N is 100. The distance OM is one cycle. Thus, the x-axis has two cycles, if we choose origin as x = 10. Then point M on the x-axis is 100 and point N is 1000. Thus, the span on the x-axis is 10^2 as there are two cycles. Similarly, there are two cycles on y-axis. The log-log paper is indicated by the number of cycles on the x-axis and y-axis. If one of the axes is logarithmic and the other is having an ordinary scale, then the graph paper is called semilogarithmic. The semilogarithmic papers are useful if experimental data fit to an equation of the type

$$y = m\,(a)^{nx} \text{ i.e. } \log y = \log m + nx \log a \quad \dots (1.55)$$

$$\text{i.e., } \log y = n'x + c \text{ where } c = \log m \text{ and } n' = n \log a \quad \dots (1.56)$$

where m, a, n, c, n' are constants.

SOLVED EXAMPLES

Example 1.1 : *The power (P) required by an agitator in a tank is a function of the following variables :*

(i) Diameter of the impeller (D).

(ii) Number of rotations of impeller per unit time (N).

(iii) Viscosity of the liquid (μ).

(iv) Density of the liquid (ρ).

From dimensional analysis, obtain a relation between the power and the given variables.

Solution : We have $P = f(D\, N \rho\, \mu)$

$$\therefore \quad P = k\, D^a N^b \rho^c \mu^d \quad \dots (1.57)$$

where k is a dimensionless constant.

The dimensions of each parameter/variable in terms of M, L and θ are :

Parameter	Units	Dimensions
Power, P	(N.m)/s = (kg.m²)/s³	$ML^2\theta^{-3}$
Diameter, D	m	L
Speed of rotation	l/s	θ^{-1}
Viscosity, μ	kg/(m.s)	$ML^{-1}\theta^{-1}$
Density, ρ	kg/m³	ML^{-3}

In terms of dimensions, Equation (1.57) can be written as

$$ML^2\theta^{-3} = k\,(L)^a (\theta^{-1})^b (ML^{-3})^c (ML^{-1}\theta^{-1})^d \quad \dots (1.58)$$

Equating the indices, we get

Exponents of M : $1 = c + d$

Exponents of L : $2 = a - 3c - d$

Exponents of θ : $-3 = -b - d$

Solving in terms of d :

$1 = c + d \quad \therefore \quad c = 1 - d$

$-3 = -b - d \quad \therefore \quad b = 3 - d$

$2 = a - 3c - d$

$2 = a - 3(1 - d) - d \quad \therefore \quad a = 5 - 2d$

Substituting for c, b and d, Equation (1.57) becomes

$$P = k(D)^{5-2d} (N)^{3-d} (\rho)^{1-d} (\mu)^d$$

$$P = k\left(\frac{D^5}{D^{2d}} \cdot \frac{N^3}{N^d} \cdot \frac{\rho}{\rho^d} \cdot \mu^d\right)$$

Collecting the terms into groups, we get

$$P / D^5 N^3 \rho = k\,(D^2 N \rho / \mu)^{-d}$$

or $N_p = P / D^5 N^3 \rho$ = power number

$N_{Re} = D^2 N \rho / \mu$ = Impeller Reynolds number.

Therefore, the power number is a function of the Impeller Reynolds number to the power m. In fact Np is also a function of the Froude number, DN^2/g.

(a) The power consumption is found experimentally to be proportional to the square of the speed of rotation of the impeller. By what factor the power would increase if the diameter of the impeller were doubled ?

The above equation may be written as :

$$P / D^5 N^3 \rho = k (D^2 N \rho / \mu)^B$$

From the above equation : $P \propto N^B N^3$

Experimentally, $P \propto N^2$

$\therefore \quad B + 3 = 2 \quad$ and $B = -1$

Therefore, for the same fluid, that is for the same density and viscosity,

$$P_1 / D_1^5 N_1^3 \rho = k (D_1^2 N_1 \rho/\mu)^{-1}$$

$$P_2/D_2^5 N_2^3 \rho = k (D_2^2 N_2 \rho/\mu)^{-1}$$

$$(P_2/P_1) (D_1^5 N_1^3 / D_2^5 N_2^3) = (D_2^2 N_2/D_1^2 N_1)^{-1}$$

$$(P_2/P_1) = N_2^2 D_2^3 / N_1^2 D_1^3$$

In this case, $N_1 = N_2$ and $D_2 = 2D_1$

$$\therefore \quad (P_2/P_1) = 8 D_1^3/D_1^3 = 8$$

$$P_2 = 8 P_1$$

$\therefore$ The power would increase by the factor of eight (i.e., the power consumption would increase by eight times by doubling the impeller diameter).

Example 1.2 : *Convert a viscosity of 1 poise to SI units.*

Solution : Viscosity = 1 P

We have : 1 P = 0.1 (N.s)/m²

$$\therefore \quad \text{Viscosity} = 1\text{ P} \times \frac{0.1\text{ (N.s)/m}^2}{1\text{P}} = \mathbf{0.10\ (N.s)/m^2} \text{ or } \mathbf{0.10\ Pa.s}$$ ... **Ans.**

Example 1.3 : *Convert a volumetric flow rate of 2000 l/s to m³/s.*

Solution : We have : $1000\ l = 1\ m^3$

$$\text{Volumetric flow rate} = 2000\ l/s = \frac{2000\ l}{1s} = \frac{2000\ l \times \left[\frac{1}{1000}\right]\left(\frac{m^3}{l}\right)}{1s} = \mathbf{2\ m^3/s}$$ ... **Ans.**

Example 1.4 : *A pressure gauge on a tank reads 50 psig on a day when the barometer reads a pressure of 28 in Hg (28 inches of mercury). Find the absolute pressure in the tank in psi.*

Solution : Given : Pressure recorded by the pressure gauge = 50 psig.

(as the letter g follows the unit, it is a gauge pressure, i.e., pressure recorded by the pressure gauge)

$$P = 50 \text{ psig}$$

We know that :

Atmospheric pressure = Barometric pressure ⇒ Pressure recorded by the barometer

= 28 in Hg

Let us convert the atmospheric pressure from Hg to psi.

We have for pressure :

$$29.92 \text{ in Hg} = 1 \text{ atm}$$

$$1 \text{ atm} = 14.7 \text{ psi}$$

$$\therefore \quad \text{Atmospheric pressure} = 28 \text{ in Hg} \times \frac{1 \text{ atm}}{29.92 \text{ in Hg}} \times \frac{14.7 \text{ psi}}{1 \text{ atm}} = 13.75668 \simeq 13.76 \text{ psi}$$

The relationship between absolute pressure and gauge pressure is

Absolute pressure = Gauge pressure + Atmospheric pressure = 50 + 13.76

Absolute pressure in the tank = **63.76 psi** ... **Ans.**

Example 1.5 : *A pressure gauge on a tower indicates a vacuum of 3.53 in Hg. The barometric pressure is 29.31 in Hg. Find the absolute pressure in the tower in mmHg.*

Solution : Vacuum = 3.53 in Hg

We know that :

Atmospheric pressure = Barometric pressure = 29.31 in Hg

The relationship between absolute pressure and vacuum is

Absolute pressure = Atmospheric pressure – Vacuum = 29.31 – 3.53 = 25.78 in Hg

We have for pressure :

29.92 in Hg = 760 mmHg

$$\therefore \quad \begin{bmatrix}\text{Absolute pressure} \\ \text{in the tower}\end{bmatrix} = 25.78 \text{ in Hg} \times \frac{760 \text{ mmHg}}{29.92 \text{ in Hg}} = 654.84 \approx \mathbf{655\ mmHg}$$ ... **Ans.**

Practice Questions

1. Define (i) Weight fraction, (ii) Mole fraction, (iii) Weight percent.
2. State Ideal gas law, Amagat's law and Dalton's law.
3. Explain briefly the Buckingham's π theorem.
4. Convert a pressure of 1.5 atm into kPa.

□□□

CHAPTER TWO

SIZE REDUCTION OF SOLIDS

2.1 CONCEPT OF SIZE REDUCTION

- **Size reduction** refers to *an operation wherein particles of solids are cut or broken into smaller pieces.*
- *Size reduction is a mechanical process of breakdown of solids into smaller size particles without altering the state of aggregation of solids. It is also called comminution.*
- Solids are reduced in size by compression, impact, attrition and cutting.

2.2 IMPORTANCE OF SIZE REDUCTION

In the process industries, this operation is usually carried out in order –

(i) to increase the surface in order to increase the rate of a physical or chemical process. In most reactions and unit operations (e.g., leaching) involving solid particles, the rate increases by increasing the area of contact between solid and second phase since the rate is proportional to the area of contact between the phases involved.

In combustion process, the rate of combustion is proportional to the area presented to the gas. Thus, the rate of combustion of solid particles is high if the particles are of small size.

In leaching, the rate of extraction increases because of the increased area of contact between the solid and the solvent.

(ii) to effect the separation of two constituents in cases where one is dispersed in small isolated pockets.

(iii) to meet stringent specifications regarding the size of commercial products.

(iv) to accomplish intimate mixing of solids in a solid-solid operation since the mixing is more complete if the particle size is small.

(v) to improve dissolution rate, solubility, binding strength and dispersion properties.

(vi) Many solid materials exist/present in sizes that are too large to be used directly. Thus, such materials must be reduced in size before use.

Size reduction machines more commonly reduce the size of solids by (a) compression, (b) impact, (c) attrition, or rubbing, and (d) cutting. In general, compression is used for the coarse reduction of hard solids (to yield relatively few fines), impact gives coarse, medium, or fine products, attrition gives very fine products from soft, non-abrasive materials and cutting produces a product of a definite particle size and sometimes a definite shape, with few or no fines.

Applications of Size Reduction (Examples) :

- Size reduction operation is carried out in coal washeries, ore processing industries, chemical industry, paint industry, cement industry and food processing industry.

2.3 ENERGY AND POWER REQUIREMENT FOR SIZE REDUCTION EQUIPMENTS

- The cost of power is a major expense in the crushing and grinding operations. Thus, an accurate estimation of the energy required is important in the design and selection of a size reduction equipment.
- During size reduction, the solid particles are first distorted and strained, work required to strain them is stored temporarily in the solid particles as mechanical energy of stress. By applying additional force, the stressed particles are distorted beyond their ultimate strength and suddenly break into smaller particles. Thus, new surface is generated.

- As a unit area of solid has a definite amount of surface energy, the generation of new surface requires work, which is provided by the release of energy of stress when the particles break. The energy of stress in excess of the new surface energy created appears as heat.
- It is not possible to estimate accurately the power requirement of crushing and grinding equipments to achieve the size reduction of a given material. Thus, for the same a number of empirical laws have been put forward, such as Rittinger's law, Kick's law and Bond's law.

2.3.1 Rittinger's Law

- It states that *the work required for the crushing operation is directly proportional to the new surface created.* Mathematically, the law can be written as

$$\frac{P}{\dot{m}} = K_r\left[\frac{1}{\bar{D}_{sb}} - \frac{1}{\bar{D}_{sa}}\right] \quad ...(2.1)$$

where P = power required by machine

$\dot{m}$ = feed rate to machine

K_r is constant (known as Rittinger's constant)

$\bar{D}_{sa}, \bar{D}_{sb}$ = volume-surface mean diameter of the feed and product respectively.

2.3.2 Kick's Law

- It states that *the work required for crushing a given material is proportional to the logarithm of the ratio between the initial and final diameters.*

$$\frac{P}{\dot{m}} = K_K \ln (D/d) \quad ... (2.2)$$

where K_K is a constant (known as Kick's constant) and D and d are the initial and final sizes respectively.

- Since the energy required is directly related to the reduction ratio (D/d), the energy required to crush a given quantity of material from a 100 mm size to a 50 mm size is the same as that required to reduce the particle size from 12 mm to 6 mm.
- Kick's law is more accurate than Rittinger's law for coarse crushing where the amount of surface produced is considerably less.

2.3.3 Bond's Law and Work Index

- Bond has proposed a law intermediate between Rittinger's and Kick's law for estimating the power required for crushing and grinding operations.
- It states that *the work required to form particles of size D_p from very large feed is proportional to the square root of the surface-to-volume ratio of the product (S_p/V_p), $S_p/V_p = 6/\phi_s D_p$.*

Thus,

$$\frac{P}{\dot{m}} = \frac{K_b}{\sqrt{D_p}} \quad ... (2.3)$$

where D_p is the particle size and K_b is a constant that depends on the type of machine and the material being crushed.

- The Bond's law is somewhat more realistic in estimating the power requirements of commercial size reduction machines.
- For using Equation (2.3), a work index W_i is defined as the *amount of energy in kilowatt-hours per ton of feed material, required to reduce a very large feed to such a size that 80 percent of the product passes through a 100 μm screen.*
- If D_p is in mm, P in kW, and $\dot{m}$ in tons per hour, then the relationship between K_b and W_i based on the definition of work index is

$$K_b = \sqrt{100 \times 10^{-3}}\ W_i = 0.3162\ W_i \quad ... (2.4)$$

- If 80 percent of the feed passes through a mesh of size D_{pa} mm and 80 percent of the product passes through a mesh of size D_{pb}, then from Equations (2.3) and (2.4) we get

$$\frac{P}{\dot{m}} = 0.3162\ W_i \left(\frac{1}{\sqrt{D_{pb}}} - \frac{1}{\sqrt{D_{pa}}}\right) \qquad \text{... (2.5)}$$

where D_{pa} and D_{pb} are the particle size of the feed and product in mm respectively.

- Since the work index include the friction in the crusher, the power given by Equation (2.5) is the gross power. Typical work indices for some common materials are given in Table 2.1.

Table 2.1 : Work indices for dry crushing* or wet grinding

Sr. No.	Material	Working Index (W_i)
1.	Clay	6.30
2.	Gypsum rock	6.37
3.	Bauxite	8.78
4.	Phosphate rock	9.92
5.	Cement raw material	10.51
6.	Limestone	12.74
7.	Coal	13
8.	Quartz	13.57
9.	Coke	15.13
10.	Gravel	16.06

* For dry grinding, multiply by 4/3.

2.3.4 Crushing Efficiency

- It is defined as the *ratio of the surface energy created by crushing to the energy absorbed by the solid.*

$$\eta_c = \frac{e_s\,(A_b - A_a)}{W_n} \qquad \text{... (2.6)}$$

where
η_c = crushing efficiency
W_n = energy absorbed by material, J/kg
e_s = surface energy per unit area, J/m^2
A_b = area of product, m^2
A_a = area of feed, m^2.

- The surface energy created by fracture is very small as compared to the mechanical energy stored in the material at the time of its rupture. Most of the mechanical energy stored in the material is converted into heat and thus crushing efficiencies are low.
- The energy absorbed by the solid (W_n) is less than the energy supplied to the machine (W). Part of the total energy input to the machine is utilised to overcome the friction in the bearings and other moving parts, and the remaining part is available for crushing. The mechanical efficiency is the *ratio of the energy absorbed to the energy input.*

$$\eta_m = \frac{W_n}{W} \qquad \text{... (2.7)}$$

where
η_m = mechanical efficiency
W = energy input to the machine
W_n = energy absorbed by the solid

From Equations (2.6) and (2.7), we get

$$W = \frac{W_n}{\eta_m} = \frac{e_s (A_b - A_a)}{\eta_m \cdot \eta_c} \quad \text{... (2.8)}$$

If $\dot{m}$ is the feed rate of solids to a machine, then the power required by the machine is given by

$$P = W. \dot{m} = \frac{\dot{m}\, e_s (A_b - A_a)}{\eta_m \cdot \eta_c} \quad \text{... (2.9)}$$

where A_a, A_b are the specific surface area per unit mass of feed and product respectively.

The specific surface is given by

$$A = 6/\phi\, \bar{D}_s\, \rho_p$$

where ϕ = sphericity, ρ_p = density, $\bar{D}_s$ = volume-surface mean diameter. Volume-surface mean diameter is used to specify particle size for a mixture of particles.

- A number of empirical laws that are mentioned earlier have been put forward to estimate the energy required for size reduction.

2.4 TYPES OF SIZE-REDUCTION EQUIPMENTS

- Size-reduction equipments are divided into four principal types as given in Table 2.2.
- Crushers are employed for breaking large pieces of solid materials into small lumps.
- A primary crusher is the one which crushes very large lumps to yield a product 150 to 250 mm in size. A secondary crusher is the one which takes the product from a primary crusher and reduces it to particles of about 6 mm size.
- Grinders are the machines which reduce crushed feed to powder. An intermediate grinder yield a product that might pass a 40 mesh screen. A fine grinder gives a product most of which would pass a 200 mesh screen. Ultrafine grinders are the machines which accept feed particles having a size less than 6 mm and yield a product of size 1 to 50 μm.
- Cutters are size-reduction machines which give particles of definite size and shape, usually 2 to 10 mm in length.

Table 2.2 : Principal types of size-reduction machines

1. Crushers (coarse and fine)
   (a) Jaw crusher, (b) Gyratory crusher, (c) Crushing rolls
2. Grinders (Intermediate and fine)
   (a) Hammer mills
   (b) Rolling-compression mills
      (i) Bowl mills, (ii) Rolling mills
   (c) Attrition mills
   (d) Revolving mills
      (i) Rod mills, (ii)Ball mills; pebble mills, (iii) Tube mills
3. Ultrafine grinders
   (a) Hammer mills with internal classification
   (b) Fluid-energy mills
   (c) Agitated mills
4. Cutting machines
   (a) Knife cutters, dicers, slitters.

- The size reduction machines perform their work in distinctly different ways. Crushers employ compression while grinders employ impact and attrition. Ultrafine grinders employ attrition. A cutting action is the characteristic of knife cutters, dicers and slitters.
- The factors to be considered while selecting the equipment for size reduction are :
  (i) Properties of the feed to be handled such as hardness, crushing strength, etc.
  (ii) Nature of the product required.
  (iii) Quantity of the material to be handled.
  (iv) Size of the material to be handled.
  (v) Speed of the size reduction equipment.

2.4.1 Crushers

- Crushers are slow-speed machines employed for the coarse reduction of large quantities of solids. Jaw crushers, gyratory crushers and smooth-roll crushers are different types of crushers. They operate by compression and can break large lumps of hard materials. They find application in rockary and mining industries.

2.4.1.1 Jaw Crushers

- Jaw crushers compress the feed between a stationary jaw and a movable jaw.

Types of jaw crushers :

1. Blake jaw crusher. 2. Dodge jaw crusher.

- In the Blake jaw crusher, the movable jaw is pivoted at the top, thus giving greatest movement at the bottom.
- In the Dodge jaw crusher, the movable jaw is pivoted at the bottom, thus giving greatest/maximum movement at the top.
- The Dodge crusher is less widely used because of its tendency to choke due to the minimum movement of the jaw at the bottom.

Difference between Blake Jaw Crusher and Dodge Jaw Crusher :

(i) Blake Jaw Crusher : (i) movable jaw is pivoted at the top, (ii) maximum movement is at the bottom, (iii) no tendency to choke/clog (freedom from choking), (iv) suitable for high production rates, (v) large reduction ratio is not possible, (vi) low maintenance, (vii) comparatively made in large sizes, (viii) does not give uniform product, (ix) commonly/widely used.

(ii) Dodge Jaw Crusher : (i) movable jaw is pivoted at the bottom, (ii) maximum movement is at the top, (iii) tendency to choke (no freedom from choking), (iv) suitable for low production rates, (v) large reduction ratio is possible, (vi) high maintenance, (vii) comparatively made in smaller sizes, (viii) gives uniform product, (ix) seldom/less widely used.

1. The Blake Jaw Crusher

Principle :

- It works on the principle of compression. It reduces size by compressive force. That is, it utilises compressive force for size reduction.

Construction :

- A schematic diagram of the Blake jaw crusher is shown in Fig. 2.1. It has a fixed jaw and a movable jaw. The movable jaw is pivoted at the top. The jaws are set to form a V open at the top. The swinging jaw (movable jaw) which reciprocates in a horizontal plane usually makes an angle of 20 to 30° with the fixed jaw (which is nearly vertical).

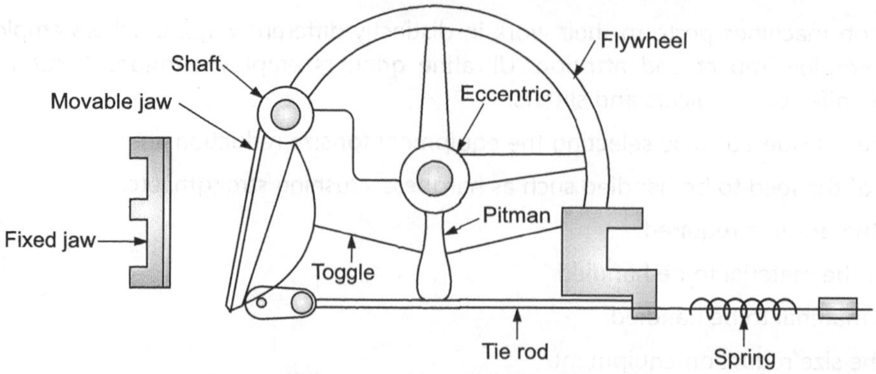

Fig. 2.1 : Blake Jaw Crusher

- The jaws are usually made of manganese steel or some other material that will withstand abrasion. The faces of the crushing jaws are usually corrugated for concentrating the pressure on relatively small areas.
- In addition to the jaws, crusher consists of a pitman, toggles, flywheel, eccentric shaft, drawback rod and springs and frame. In this machine, an eccentric causes the pitman to oscillate in a vertical direction, and this vertical movement is communicated horizontally (reciprocating motion) to the movable jaw by the toggles.
- The speed of operation should not be high or otherwise a large quantity of fines is produced as the material cannot escape quickly and gets repeatedly crushed. Since the crushing action is intermittent, the loading on the machine is uneven and due to this the crusher incorporates a heavy flywheel.
- Since the maximum movement of the jaw is at the bottom (discharge), there will be little tendency for the crusher to choke.

Protection of Machine :

- The machine is usually protected so that it is not damaged if accidental pieces of iron such as hammer heads, stray bolts, etc. enter into the crusher, by making one of the toggles in the driving mechanism relatively weak. That is, one particular toggle is made into two pieces which are held together with bolts that are purposely made the weakest part in the crusher so that, if stresses are set up, these bolts shear first.
- Thus, the failure is made at a point that can be easily and quickly repaired, instead of breaking some vital part of the machine.

Working :

- The material to be crushed is admitted between two jaws from the top. The material caught between the upper part of the jaws is crushed to a smaller size during forward motion by compression. The crushed material then drops/falls into the narrower space below during the backward motion and is recrushed as the jaws close next time. After sufficient reduction, the crushed material drops out the bottom of the machine.
- The jaws usually open and close 250 to 400 times per minute.

2.4.1.2 Gyratory Crusher

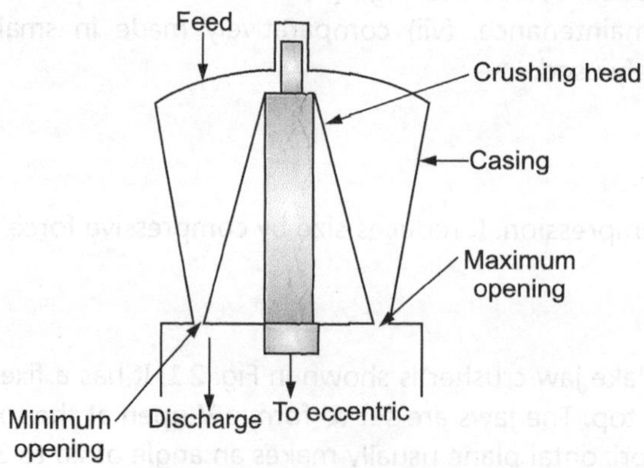

Fig. 2.2 : Gyratory Crusher

Principle :

- It works on the principle of compression.

Construction :

- It consists of a funnel shaped casing, open at the top. A conical crushing head, in the form of a truncated cone, gyrates inside a casing. The crushing head is mounted on a heavy shaft pivoted at the top of the machine.
- The upper end of the shaft is held in a flexible bearing and the lower end of the shaft is driven by an eccentric so as to trace a circle.
- Thus, at any point on the periphery of the casing the bottom of the crushing head moves towards, and then away from the stationary wall. The crushing action takes place around the whole of the cone.

Working :

- The material to be crushed is charged from the top. The conical head gyrates inside the casing. At any point on the periphery of the casing, the bottom of the crushing head moves towards and then away from the stationary wall.
- The solids caught in the V-shaped space between the head and the casing are broken and rebroken until they drop out from the bottom of the machine. The speed of the crushing head usually lies between 125 to 425 gyrations per minute.
- Since some part of the crushing head is working at all times, the discharge from this crusher is continuous instead of intermittent as in a Blake crusher.

Features :

- 1. continuous in action 2. fluctuations in stresses are smaller 3. load on the motor is nearly uniform 4. power consumption per ton of material crushed is smaller and 5. requires less maintenance than a jaw crusher.
- Since the capital cost of this crusher is high, it is suitable only where large quantities of materials are to be handled.

Jaw Crusher	Gyratory Crusher
(i) It is a reciprocating machine.	(i) It is a gyratory machine.
(ii) Intermittent in action, i.e., discharge is discontinuous.	(ii) Continuous in action, i.e., discharge is continuous.
(iii) It is a primary crusher. It takes a feed of larger size.	(iii) It is a secondary crusher. It takes a feed of smaller size.
(iv) The load on the motor is not uniform.	(iv) The load on the motor is nearly uniform.
(v) More maintenance is required.	(v) Less maintenance is required.
(vi) Power consumption per ton of material crushed is more.	(vi) Power consumption per ton of material crushed is lower.
(vii) Capital cost is relatively low.	(vii) Capital cost is high.
(viii) It has smaller capacity when used to produce/effect a small size reduction.	(viii) It has large capacity when used to produce/effect a small size reduction.

2.4.1.3 Crushing Rolls / Roll Crushers

Smooth Roll Crusher :

Principle :

Size reduction is achieved by compression (i.e., it employs compressive force for size reduction).

Construction :

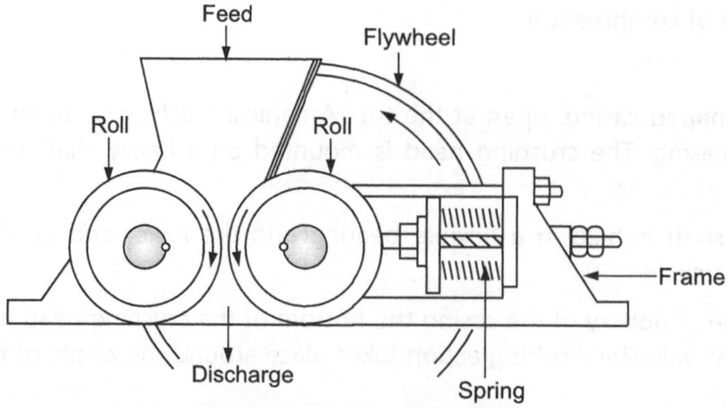

Fig. 2.3 : Smooth roll crusher

- Smooth-roll crusher consists of two heavy metal rolls (Fig. 2.3) of the same diameter placed side by side each other in the horizontal position. The rolls, mounted on shafts, are rotated towards each other at the same speed. One of the shafts moves in the fixed bearings while other moves in the movable bearings.
- The clearance between the rolls can be adjusted according to the size of feed and the size of product required. One of the rolls is driven directly and the other by friction with the solids being crushed. The rolls have relatively narrow faces and are large in diameter, therefore they can nip moderately large lumps. The material fed to the machine is reduced in size by compression and discharged from the bottom.
- The machine is protected by spring loading (i.e., by mounting the bearings of one of the roll shafts against coiled springs) against damage due to tramp and very hard material.
- The speed of rolls varies from 50 to 300 rev/min. Crushing rolls are secondary crushers accepting feed 12 to 75 mm in size and yielding products-12 mm to about 20 mesh.

Working :

- The material to be crushed is fed from the top. As the rolls rotate, the material gets caught between them and gets reduced in size by compression and discharged from the bottom.

Selection of Crushing rolls (Derivation of the angle of nip)

- In selecting the rolls for a certain duty, it is necessary to know the size of the feed and the size of the product.
- Consider a system as shown in Fig. 2.4 wherein the spherical particle B of a material is just being caught between the rolls. A and A' are the centres of two crushing rolls of radius r.

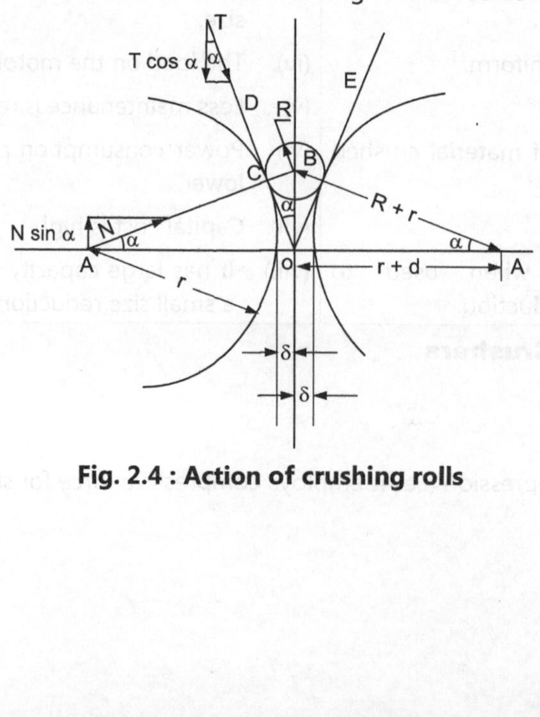

Fig. 2.4 : Action of crushing rolls

- The line AB passes through the centre of the left-hand roll, through the centre of particle, and through point C where the particle is in contact with the roll. Let the angle between line AB and the horizontal i.e. line joining two centres of rolls be 'α'. The line OD is tangent to the roll at point 'C'. As the line OD is perpendicular to line AB, it makes the same angle α with the vertical. Neglecting the gravity force, the two forces acting at point C are : the tangential frictional force T having a vertical component T cos α, and the radial force 'N', having a vertical component N sin α. The force T is related to the force N through the coefficient of friction μ, so T = μN.
- The vertical components of the forces T and N are opposed. Force N sin α (a resolved component of the force N) tends to expel the particle from the rolls, while force T cos α tends to draw the particle between the rolls. If the particle is to be drawn between the rolls and crushed,

$$T \cos \alpha \geq N \sin \alpha \quad \text{... (2.10)}$$

T and N are related through

$$T = \mu N \quad \text{... (2.11)}$$

$$\therefore \quad \mu N \cos \alpha \geq N \sin \alpha \quad \text{... (2.12)}$$

$$\mu \geq \tan \alpha \quad \text{... (2.13)}$$

- That is, the tangent of angle 'α' must be less than the coefficient of friction. The value of μ varies from material to material, but for all practical purposes, the value of the angle α is usually taken about 16°. The angle DOE, which is twice the angle α, is called the *angle of nip*.
- **Angle of nip** is the *angle formed by the tangents to the roll faces at a point of contact with a particle to be crushed.*
- Let R be the radius of the feed particle, r be the radius of the roll and 2d be the distance/gap between the rolls (the diameter of the largest particle in the product).
Then, in the triangle ABG (Fig. 2.5), the angle BAG is 'α' (half the angle of nip), AG is r + d, and AB is r + R. Then, from the simple geometry of the figure, the angle of nip is given by

$$\cos \alpha = \frac{AG}{AB} = \frac{r + d}{r + R} \quad \text{... (2.14)}$$

- Equation (2.14) gives the relationship between the size of the feed, radius of rolls, gap between the rolls, and the angle of nip. With this equation, the roll diameter can be determined by knowing values of the size of feed, size of product and angle of nip.

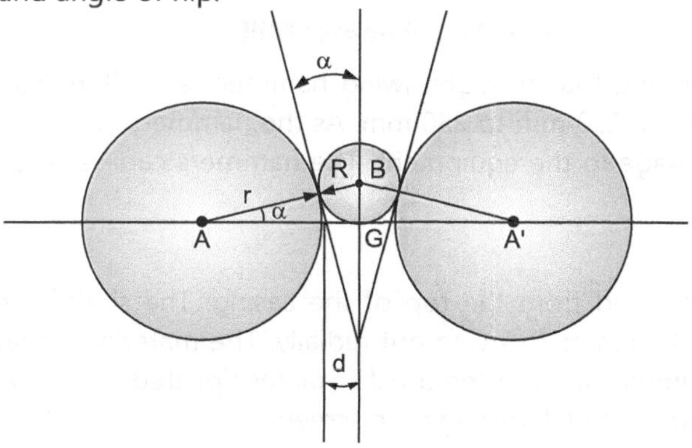

Fig. 2.5 : Capacity of crushing rolls

- If α = 16°, then cos α = 0.961 and we have

$$0.961 = (r + d)/(r + R) \quad \text{... (2.15)}$$

- Crushing rolls are widely used for crushing of oil seeds and in the gun powder industry.

2.5 GRINDERS

- Grinding *means sub-dividing the solids to a finer product than crushing.* The size reduction machines employed for an intermediate duty are referred to as grinders. A grinder is often charged with the product from a crusher which it reduces to powder. The commercial grinders described in this section are hammer mills and revolving mills.

2.5.1 Hammer Mill

Principle :

Size reduction is achieved by impact and attrition.

Construction :

- The hammer mill consists of a high speed rotor turning inside a cylindrical casing. The rotor is mounted on a shaft which is usually horizontal. The swing hammers are pinned to a rotor disk. The hammers are rectangular bars of metal with plain or enlarged ends. In this mill, the particles are broken by the sets of swing hammers. The product falls through a grate or screen which forms the lower portion of the casing.

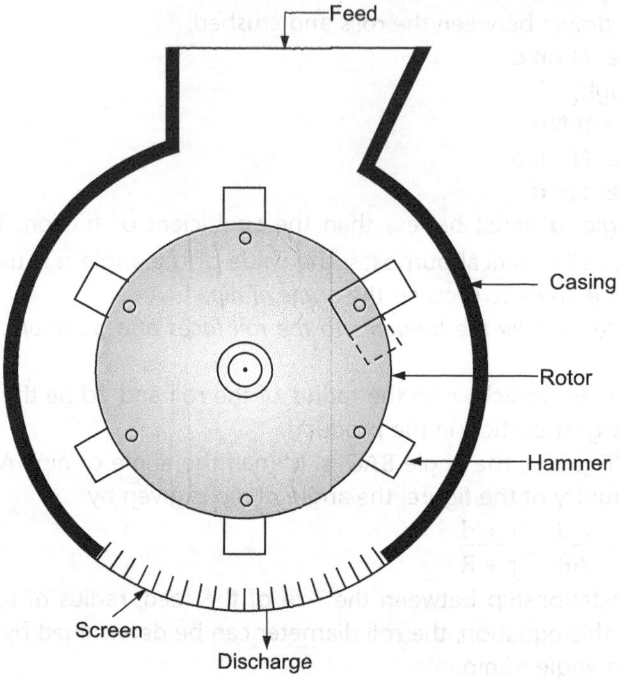

Fig. 2.6 : Hammer Mill

- Several rotor disks each carrying four to eight swing hammers are often mounted on a single shaft. The rotor disk diameter ranges from 150 mm to 250 mm. As the hammers are hinged, the presence of any hard material does not cause damage to the equipment. The hammers can be readily replaced when they worn out.

Working :

- The material to be crushed is fed from the top of the casing. The shaft is rotated at a high speed and centrifugal force causes the hammers to swing out radially. The material is beaten by the hammers around inside of the casing and by impact against the breaker plates (located on inside of the casing) or the screen is crushed until it is small enough to fall through the screen.
- Hammer mills are employed to grind tough fibrous solids like bark or leather, steel turnings, hard rock, sticky clay, etc.

2.5.2 Revolving Mills / Tumbling Mills

- *A revolving/tumbling mill is a cylindrical shell slowly rotating on a horizontal axis and charged with a grinding medium to about half of its volume.* The shell is usually made of steel and lined with abrasion resistant materials such as manganese steel, ceramic or rubber. The grinding medium is usually made of flint, ceramic or metal. Ball, pebble, tube and rod mills are the various types of revolving mills. The ball mill differs from the tube mill in that it is short in length; the length is approximately equal to its diameter.

- The grinding medium more commonly used in the ball mill is steel balls.
- The tube mill is usually long in comparison with its diameter; the length being twice the diameter or more. It employs smaller balls, and produces a finer product. The pebble mill is a tube mill employing flint or ceramic pebble as a grinding medium.
- The rod mill employs metal rods (steel rods) as a grinding medium and delivers more uniform product than any other revolving mills. In a ball mill, or pebble mill, much of the reduction is effected by impact, while in a rod mill, much of the reduction is effected by rolling, compression and attrition. The ball mill and pebble mill are very easy to operate and versatile in use.
- Revolving mills may be operated batchwise or continuously. In a batch machine, a known quantity of the material to be ground is charged into the mill through an opening in the shell. The opening is then closed and the mill is rotated for a certain time, and finally the product is discharged. In continuous mill, the material flows steadily through the revolving shell, entering and leaving through hollow trunions at opposite ends of the mill.

2.5.2.1 Ball Mill

- It is a mechanical device used to grind (crushed) materials (to powder form). The grinding medium is the balls (which are made of stainless steel, steel, ceramic or rubber).

Principle :

- It works on the principle of impact, i.e., size reduction is done by impact as the balls drop from near the top of the shell.

Construction :

- A ball mill consists of a hollow cylindrical shell rotating about its axis. The axis of the shell may be either horizontal or at a small angle to the horizontal. It is partially filled with balls. The grinding medium is the balls which may be made of steel, stainless steel or rubber.
- The inner surface of the cylindrical shell is usually lined with an abrasion-resistant material such as manganese steel or rubber. The length of the mill is approximately equal to its diameter.
- The balls occupy about 30 to 50 percent of the volume of the mill. The diameter of ball used is/lies in between 12 mm and 125 mm. The optimum diameter is approximately proportional to the square root of the size of the feed. The shell is rotated at low speed through a drive gear (60-100 rpm) and in a large ball mill, the shell might be 3 m in diameter and 4.25 m in length.

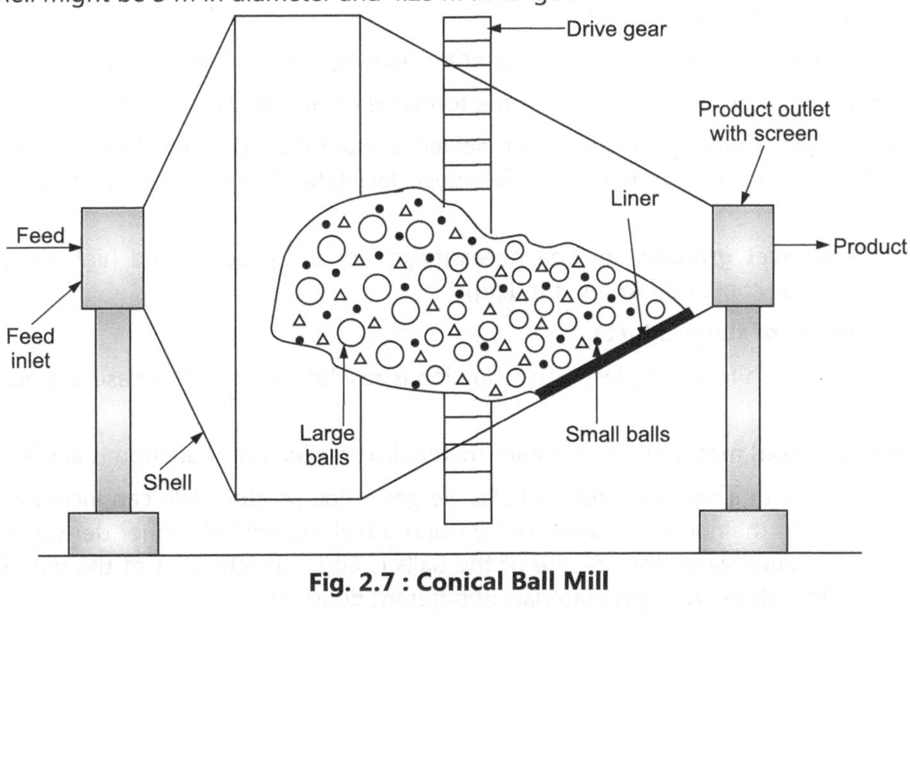

Fig. 2.7 : Conical Ball Mill

- The ball mill may be operated in a batch or continuous fashion, wet or dry. In a continuously operated mill as shown in Fig. 2.7, the outlet is normally covered with a coarse screen to prevent the escape of the balls.

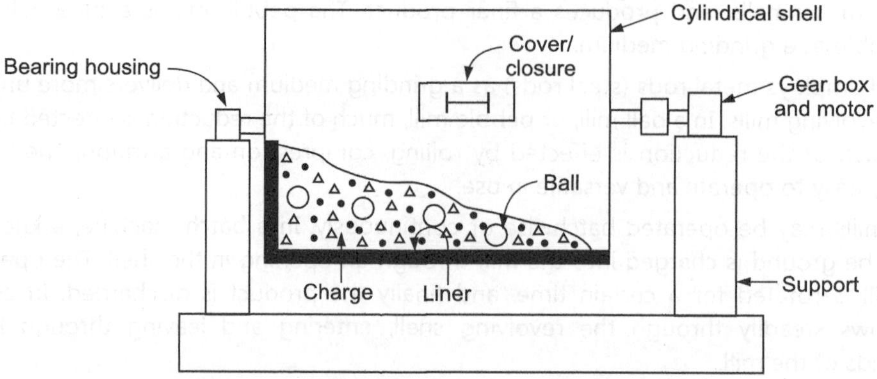

Fig. 2.8 : Batch operated Ball mill (Lab. scale)

Working :

- In case of continuously operated ball mill, the material to be ground is fed from the left through a 60° cone and the product is discharged through a 30° cone to the right. As the shell rotates, the balls are lifted up on the rising side of the shell and then they cascade down (or drop down on to the feed), from near the top of the shell. In doing so, the solid particles in between the balls are ground and reduced in size by impact.
- The mill contains balls of various ages and sizes since the balls continually wear by attrition and are replaced by new ones. As the shell rotates, the large balls segregate near the feed end and small balls segregate near the product end/discharge. The initial breaking of the feed particles is done by the largest balls dropping from the largest distance and small particles are ground by small balls dropping from a much smaller distance. If the rate of feed is increased, a coarser product will be obtained and if the speed of rotation is increased (less than critical speed), the fineness for a given capacity increases.
- During grinding, balls themselves wear and are constantly replaced by new ones so that mill contains balls of various ages and thus of various sizes.
- In case of batch operated mill, a known quantity of material to be ground is charged into the mill through the opening in the shell. The opening is then closed and the mill is rotated for a predecided time. It is then stopped and the product is discharged.

Applications : The ball mill is used for grinding materials such as coal, pigments, and felspar for pottery.

Grinding can be carried out either wet or dry but the former is carried at low speeds.

The advantages of wet grinding include lower power consumption (20-30% less than it for dry grinding), increased capacity, reduction in the formation of fines/dust, facilitates the removal of the product and no dust formation.

The disadvantages of wet grinding include necessity to dry the product and high wear on the grinding medium (about 20% higher as compared to dry grinding).

Factors influencing the size of the product :

(a) Feed rate : With a high feed rate, less size reduction is resulted since in this case the material is in the mill for a shorter time.

(b) Properties of the feed material : With a hard material, a smaller size reduction is achieved.

(c) Weight of balls : With a heavy charge of balls, we get a fine product. We can increase the weight of the charge by increasing the number of balls or by using a ball material of higher density. Optimum grinding conditions are obtained when the volume of the balls is equal to 50% that of the mill. So the variation in the weight of balls is done by using materials of different densities.

(d) Speed of rotation of the mill : At low speeds, the balls simply roll over one-another and little grinding is obtained, while at very high speeds, the balls are simply carried along the walls of the shell and little or no grinding takes place. So for an effective grinding, the ball mill should be operated at a speed (optimum speed) equal to 50 to 75 percent of the critical speed.

(e) Level of the material in the mill : A low level of material in the mill results into a reduction in the power consumption. If the level of material is increased, the cushioning action increases and power is wasted by the production of undersize material in an excessive quantity.

Advantages of the Ball Mill :

(i) The cost of installation is low.

(ii) The cost of power required is low.

(iii) It is suitable for materials of all degrees of hardness.

(iv) It is suitable for batch as well as continuous operation.

(v) It can be used for grinding of certain explosive materials since it can be used with an inert atmosphere.

(vi) It is suitable for open as well as closed circuit grinding.

(vii) The grinding medium is cheap.

Action in Revolving / Tumbling mills

- When the revolving mill is in operation, the balls are picked up by the mill wall and are carried near the top of the mill. The balls then break contact with the wall and drop down to the bottom. During the upward movement of the balls, centrifugal force keeps the balls in contact with the wall and with each other. The balls when in contact with the wall surface, perform some grinding by slipping and rolling over each other, but most of the grinding takes place when free falling balls strike the bottom of the mill (by impact).
- The balls are projected across the mill depending upon the speed of rotation. At low speeds of operation, the balls simply roll over each other resulting into little crushing action. If the mill is operated at slightly higher speeds, the balls will be carried up further inside the mill and greater will be the power consumption. But at the same time, as the balls fall down from higher distances, greater will be the impact at the bottom, and larger will be the capacity of the mill.
- If the mill is operated at very high speeds, the balls are carried right round in contact with the sides of the mill and the mill is said to be **centrifuging**.
- *The minimum speed at which centrifuging occurs* is called the **critical speed** of the mill, and under these conditions, centrifugal force will be exactly balanced by the weight of the ball. Little or no grinding takes place when the mill is centrifuging.
- If the mill is to operate practically, the **operating speed** must be less than the **critical speed.**
- The critical speed of a grinding mill (e.g., a ball mill) is the rotation speed of the mill where a centrifugal force is exactly balanced by a gravitational force at the inside shell surface of the mill. It is the speed at which balls will not fall away from the inside surface of the shell of the mill.

Derivation of critical speed of a ball mill

- The speed at which the outermost balls break contact with the wall depends on the balance between centrifugal force and gravitational force. This can be shown with the help of Fig. 2.9. Consider the ball at point B on the periphery of the ball mill. Let R be the radius of the mill and r be the radius of the ball. R–r represents the distance between the centre of the ball and the axis of the mill. Let 'α' be the angle between OB and vertical through the point O. The forces acting on the ball are :

  1. The force of gravity, mg where 'm' is the mass of the ball and
  2. The centrifugal force, $mv^2/(R - r)$, where 'v' is the peripheral speed.

- The component of gravity opposing the centrifugal force (centripetal component) is (mg) cos α. As long as the centrifugal force exceeds the centripetal component of the force of gravity, the particle will not lose contact with the wall. As the angle α decreases, the centripetal force increases. Unless the speed crosses the critical value, a stage is reached where the above opposing forces are equal and the ball is ready to fall away from the wall. The angle at which the said phenomenon occurs is found out by equating the two opposing forces. Thus,

$$mg \cos \alpha = \frac{mv^2}{(R - r)} \quad \text{... (2.16)}$$

$$\cos \alpha = \frac{v^2}{(R - r)\, g} \quad \text{... (2.17)}$$

The relationship between the peripheral speed and the speed of rotation is given by

$$v = 2\pi N (R - r) \quad \text{... (2.18)}$$

Substituting the value of v from equation (2.18) into equation (2.16), we get

$$\cos \alpha = \frac{4\pi^2 N^2 (R - r)}{g} \quad \text{... (2.19)}$$

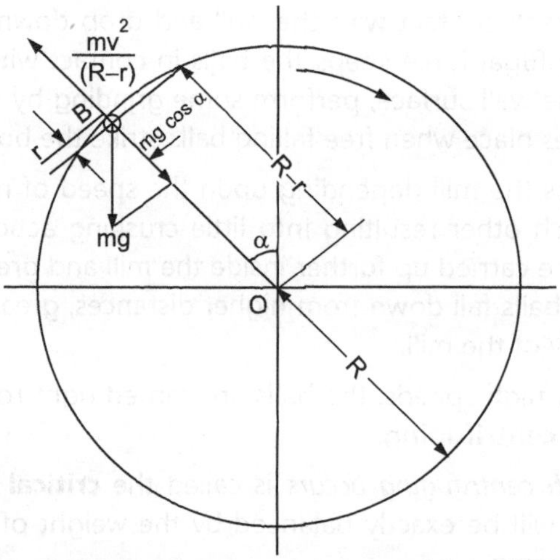

Fig. 2.9 : Forces on ball in Ball mill

At the critical speed : α = 0, and thus cos α = 1 and N becomes the critical speed N_c.

$$\therefore \quad \cos \alpha = 1 = \frac{4\pi^2 N_c^2 (R - r)}{g} \quad \text{... (2.20)}$$

$$N_c^2 = \frac{g}{4\pi^2 (R - r)} \quad \text{... (2.21)}$$

$$N_c = \frac{1}{2\pi} \sqrt{\frac{g}{R - r}} \quad \text{... (2.22)}$$

The operating speed/optimum speed of the ball mill is between 50 and 75 percent of the critical speed.

2.6 COMPARISON OF CRUSHING AND GRINDING OPERATION

Crushing as well as grinding are aimed at size reduction. But there are certain differences in these operations which are cited below.

Crushing	Grinding
1. The solid particles are reduced in size by compression.	1. The solid particles are reduced in size by impact and attrition.
2. It is aimed at breaking large pieces of solid material into small lumps.	2. It is aimed at reducing crushed feed to powders.
3. Mostly crushing equipments are operated in open-circuit.	3. Grinding equipments are always operated in closed-circuit.
4. When crushers are operated in closed-circuit, dry screens are used as a size separation unit.	4. In grinders operated in closed-circuit, some sort of classifier is used as a size separation unit.
5. This operation is performed on dry feed material.	5. This operation can be performed on dry as well as wet feed.
6. In crushing operation, the reduction ratio seldom exceeds 6 to 8.	6. In grinding operation, the reduction ratio as high as 100 is possible.
7. Crushing is usually one shift operation as residence time in the crusher is less and throughput is large.	7. Grinding is carried out in all shifts as in the grinding machines the residence time is larger and throughput is smaller.
8. Crushers are of two types, e.g., primary crushers and secondary crushers.	8. Grinders are of two types, e.g., fine grinders and ultrafine grinders.
9. Crushers are heavy duty, low speed machines.	9. Grinders are relatively light duty, high speed machines.
10. In coarse crushing, the feed size is 1500 to 40 mm and product size is 50 to 5 mm.	10. In fine grinding, the size of feed is 5 to 2 mm and product size is 0.1 mm (about 200 mesh).
11. Energy consumption per unit mass of product is low due to coarse particle production.	11. Energy consumption per unit mass of product is high due to fine particle production.

Ultrafine Grinders

- Many commercial powders must contain particles averaging 1 to 20 μm in size. Mills which reduce solids to such fine particles are called as **ultrafine grinders**.

2.7 FLUID ENERGY MILL

- Grinding takes place by attrition.
- A fluid energy mill is a size reduction unit in which size reduction results from attrition between rapidly moving particles of the material being ground. A source of compressed air or gas or high pressure superheated steam that enters the grinding chamber through nozzles in the periphery at high speed provides energy to the particles to achieve high velocities.
- In fluid energy mill there is no moving parts and no grinding media.
- It consists of a flat horizontal cylindrical chamber provided with tangentially arranged jet nozzles in the inner wall. The energy for milling (grinding) is supplied by a compressed air or nitrogen gas. The compressed air/gas issuing through the nozzles forms a very high velocity tangential circle within the grinding chamber. The material to be ground is fed into the same tangential circle through a venturi feeder. The material in the circle gets rapidly accelerated, causing it to impact against itself, hence breaking the particles to the low micron range. The particles that are larger in size are held towards the outer periphery of the chamber by centrifugal force, while the particle smaller in size travel in a spiral movement towards the central outlet from where they exit into a cyclone below for bottom discharge.

- The fluid energy mill can handle powders having an initial size ranging from 150 microns and can grind materials upto one micron. Usually powders from pulverizers are handled in it. The materials that can be processed include food products, antibiotics, pigments, dyes, cosmetics, etc.

2.8 OPEN-CIRCUIT AND CLOSED-CIRCUIT GRINDING

- In many machines, the feed material is reduced to satisfactory size by passing it once through the machine.
- If the material is passed only once through the machine (crushing or grinding), and no attempt is made to return the oversize material to it for further reduction, the process is known as *open-circuit grinding.*
- If the partially ground material from the machine is sent to a size separation unit, from where the undersize is withdrawn as the product and the oversize material is returned to the machine for reground, the process is known as *closed-circuit grinding.*
- In case of coarse particles, the size separation unit is a screen or grizzly while it is some form of classifier in case of fine powders. Closed-circuit grinding though useful for any crusher, it is commonly employed to machines yielding a fine product.

2.8.1 Open Circuit Grinding

- Open circuit grinding consists of one or more grinding mills arranged in series or parallel without classification equipment. This method discharges a final ground as it comes from a mill and there is no return of coarse discharge back to the mill.
- Some examples of open circuit grinding are : (i) Ball mill, (ii) Rod mill and (iii) Combination of ball mill and rod mill.

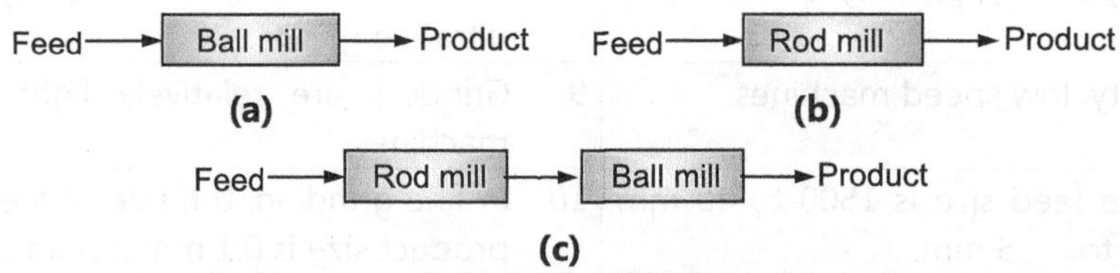

Fig. 2.10 : Open circuit grinding systems

- Some conditions that favour open circuit grinding are : (i) Small reduction ratios and (ii) Coarse reduction of particles.
- Advantages of open circuit grinding are : (i) Simplicity of operation, (ii) Minimum equipment requirements.

2.8.2 Closed Circuit Grinding

- Closed circuit grinding consists of one or more grinding mills with classification equipment. The mills discharge ground product to classifier which returns the coarse product from it to the mill for further grinding.
- Some examples of closed circuit grinding are : (i) Ball mill and classifier, (ii) Rod mill and classifier and (iii) Rod mill, ball mill and classifier.

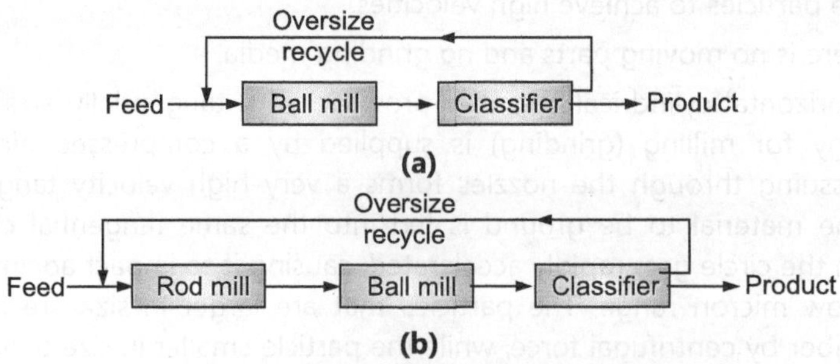

Fig. 2.11 : Closed circuit grinding systems

- Some advantages of closed circuit grinding are : (i) higher capacity, (ii) lower power consumption per ton of product, (iii) suitable for reduction to fine and ultrafine sizes, (iv) avoids coarse material in the final ground product by returning it to the mill, (v) eliminate overgrinding by removing fines early.
- Some conditions that favour closed circuit grinding are : (i) larger reduction ratios and (ii) fine reduction of particles.

The equipments which are used in industry are :

1. Jaw crusher in Cement Industry.
2. Ball mill in Paint Industry.
3. Ultrafine grinders in Cosmetic and Pharmaceutical Industries.
4. Cutters in Leather Tanning Industry.
5. Hammer mill in Food Industry.

Size reduction operation is carried out in coal washeries, ore processing, cement industry, paint industry, chemical industry and food processing industry.

SOLVED EXAMPLES

Example 2.1 : *A certain crusher accepts a feed material having a volume-surface mean diameter of 19 mm and gives a product of volume-surface mean diameter of 5 mm. The power required to crush 15 tonnes per hour is 7.5 kW. What will be the power consumption if the capacity is reduced to 12 tonnes per hour ?*

Solution : We have,

$$\frac{P}{\dot{m}} = K_r\left[\frac{1}{\bar{D}_{sb}} - \frac{1}{\bar{D}_{sa}}\right]$$

where P is the power consumption in kW, $\dot{m}$ is the feed rate in t/h and $\bar{D}_{sb}$, $\bar{D}_{sa}$ is the surface-volume mean diameters of product and feed respectively.

Case I : P = 7.5 kW, $\dot{m}$ = 15 t/h, $\bar{D}_{sb}$ = 5 mm = 0.005 m , $\bar{D}_{sa}$ = 19 mm = 0.019 m

$$\frac{7.5}{15} = K_r\left[\frac{1}{0.005} - \frac{1}{0.019}\right]$$

$$K_r = \mathbf{3.4 \times 10^{-3}} \qquad \text{... Ans.}$$

Case II : $\dot{m}$ = 12 t/h, P = ?, $K_r = 3.4 \times 10^{-3}$

$$\frac{P}{12} = 3.4 \times 10^{-3}\left[\frac{1}{0.005} - \frac{1}{0.019}\right]$$

$$P = \mathbf{6\ kW} \qquad \text{... Ans.}$$

Example 2.2 : *What will be the power required to crush 150 tonnes per hour of limestone if 80 percent of the feed passes 50 mm screen and 80 percent of the product a 3.125 mm screen ?*

Work index of limestone = 12.74.

Solution : We have,

$$\frac{P}{\dot{m}} = 0.3162\ W_i\left[\frac{1}{\sqrt{D_{pb}}} - \frac{1}{\sqrt{D_{pa}}}\right]$$

In this equation, P is in kW and D_p is in mm.

$\dot{m}$ = 150 t/h, W_i = 12.74

D_{pb} = product size = 3.125 mm

D_{pa} = feed size = 50 mm

$$\frac{P}{150} = 0.3162 \times 12.74\left[\frac{1}{\sqrt{3.125}} - \frac{1}{\sqrt{50}}\right]$$

$$P = \mathbf{256.4\ kW} \qquad \text{... Ans.}$$

Example 2.3 : *Find out the critical speed of the ball mill by using the following data :*

Diameter of ball mill = 450 mm

Diameter of ball = 25 mm

Solution : Data : Diameter of ball mill = 450 mm

Diameter of ball = 25 mm

The critical speed of a ball mill is given by $N_c = \frac{1}{2\pi}\sqrt{\frac{g}{R-r}}$, in r.p.s.

$$D = \text{diameter of ball mill in m}$$

$$D = 450 \text{ mm} = 0.45 \text{ m}$$

$$\therefore \quad R = 0.225 \text{ m}$$

$$\text{Diameter of ball} = 25 \text{ mm} = 0.025 \text{ m}$$

$$\therefore \quad r = 0.0125 \text{ m}$$

$$g = 9.81 \text{ m/s}^2$$

$$N_c = \frac{1}{2\pi}\sqrt{\frac{9.81}{0.225 - 0.0125}} = 1.08 \text{ r.p.s.}$$

$$= 1.08 \times 60 = 64.88 \text{ r.p.m.} \approx 65 \text{ r.p.m.}$$

Critical speed = **65 r.p.m.** **... Ans.**

Example 2.4 : *A pair of rolls is to take a feed equivalent to sphere 38 mm in diameter and crush them to sphere having a diameter of 12.7 mm. If the co-efficient of friction is 0.29, what should be the diameter of the rolls ?*

Solution : μ = co-efficient of friction = 0.29.

For particles to be drawn between the rolls and crushed $\mu > \tan \alpha$, where α is half the angle of nip. Hence α should be less than $\tan^{-1}$ (0.29) or 16° 17'. So for margin of safety, take angle of nip to be 16°.

For crushing rolls, we have

$$\cos \alpha = \frac{r+d}{r+R}$$

where

r = radius of roll

d = radius of largest possible particle in the product

$$d = \frac{12.7}{2} \text{ mm} = 6.35 \text{ mm}$$

$$R = \text{radius of feed particle} = 38/2 = 19 \text{ mm}$$

Substituting the values of R, d and α in above equation

$$\cos (16) = \frac{r + 6.35}{r + 19}$$

$$0.96126 (r + 19) = r + 6.35$$

$$\therefore \quad r = 307.5 \text{ mm}$$

$$\text{Diameter of rolls} = 2r = 2 \times 307.5 = 615 \text{ mm}$$

These odd size rolls are not made. Therefore, the **diameter of rolls is 600 mm.** **... Ans.**

Example 2.5 : *Calculate the operating speed of the ball mill from the following data :*

(i) Diameter of ball mill = 500 mm

(ii) Diameter of ball = 40 mm

(iii) Operating speed is 50% of the critical speed of the mill.

Solution : The critical speed of the ball mill in revolutions per second is given by,

$$N_c = \frac{1}{2\pi}\sqrt{\frac{g}{R-r}}$$

where $g = 9.81\ m/s^2$

Diameter of ball mill = 500 mm = 0.5 m

∴ R = radius of ball mill = 0.25 m

Diameter of ball = 40 mm = 0.04 m

∴ r = radius of ball = 0.02 m

$$N_c = \frac{1}{2\pi}\sqrt{\frac{9.81}{0.25-0.02}} = 1.04 \text{ r.p.s.} = 1.04 \times 60 = 62.4 \text{ r.p.m.} \cong 62 \text{ rpm}$$

The operating speed of ball mill = 0.5 N_c = 0.5 × 62 = **31 r.p.m.** **... Ans.**

Example 2.6 : *What rotational speed, in revolutions per minute, would you recommend for a ball mill 1200 mm in diameter charged with 75 mm balls ?*

Solution : The critical speed of ball mill is given by

$$N_c = \frac{1}{2\pi}\sqrt{\frac{g}{R-r}}, \text{ in r.p.s.}$$

where $g = 9.81\ m/s^2$

R = radius of the ball mill

Diameter of ball mill = 1200 mm

∴ R = 1200 / 2 = 600 mm = 0.60 m

r = radius of the ball

Diameter of the ball = 75 mm

∴ r = 75/2 = 37.5 mm = 0.0375 m

$$N_c = \frac{1}{2\pi}\sqrt{\frac{9.81}{0.60-0.0375}} = 0.665 \text{ r.p.m.} = 39.90 \text{ r.p.m.} \approx 40 \text{ r.p.m.}$$

Operating speed of the ball mill is 50 to 75% of the critical speed.

Operating speed = 50 to 75% of 40 r.p.m. = 20 to 30 r.p.m.

The rotational speed that can be recommended is between **20 to 30 r.p.m.** **... Ans.**

Example 2.7 : *A certain set of crushing rolls has rolls of 1000 mm diameter and 375 mm width face. They are set so that the crushing faces are 12.5 mm apart. The manufacturer recommends their speed to be 50 to 100 r.p.m. They are employed to crush a rock having specific gravity 2.35 and the angle of nip is 31° 30'. What is the maximum permissible size of the feed and maximum actual capacity of rolls in tonnes per hour if the actual capacity is 12% of the theoretical ?*

Theoretical capacity in t/h, Q = 4.352×10^{-7} N.D.w.d.s, where N in r.p.m., D (roll diameter) in mm, w (width) in mm, d (half the gap/width between roll surface) in mm and s (specific gravity).

Solution : For crushing rolls, we have :

$$\cos\alpha = \frac{r+d}{r+R}$$

where r = radius of roll = 1000/2 = 500 mm

d = (gap between the rolls)/2 = 12.5/2 = 6.25 mm

R = radius of feed particle in mm

For a margin of safety, take $2\alpha = 31°$

$\therefore \quad \alpha = 15°$

$\cos\alpha = 0.9659$

$$0.9659 = \frac{500 + 6.25}{500 + R}$$

$\therefore \quad R = 24.12$ mm

Diameter of the feed particle = 48.24 mm ≈ **48 mm.** ... **Ans.**

Theoretical capacity of the rolls is given by

for N = 50 r.p.m.

$Q = 4.352 \times 10^{-7} \times N \times D \times w \times d \times s$

where D = 1000 mm

w = 375 mm

d = 6.25 mm

s = 2.35

$Q = 4.352 \times 10^{-7} \times 50 \times 1000 \times 375 \times 6.25 \times 2.35 = 119.85$ t/h

Actual capacity at 50 r.p.m. = 12% of theoretical = 0.12 (119.85) = 14.38 t/h

For N = 100 r.p.m.

Q theoretical = $2 \times 119.85 = 239.7$ t/h

Actual capacity at 100 r.p.m. = 0.12 (239.7) = 28.76 t/h

Maximum capacity (at speed of 100 r.p.m.) = **28.76 t/h.** ... **Ans.**

Example 2.8 : *Calculate the operating speed of the ball mill from the data given below :*

Diameter of ball mill = 800 mm, diameter of ball = 60 mm

If (I) operating speed is 55% less than the critical speed.

(II) critical speed is 40% more than the operating speed.

Solution : The critical speed of a ball mill is given by

$$N_c = \frac{1}{2\pi}\sqrt{\frac{g}{R-r}}, \text{ in r.p.s.}$$

where R is radius of a ball mill and r is the radius of ball.

$g = 9.81 \text{ m/s}^2$

$R = 800/2 = 400$ mm = 0.40 m

$r = 60/2 = 30$ mm = 0.03 m

$$N_c = \frac{1}{2\pi}\sqrt{\frac{9.81}{0.40 - 0.03}} = 0.82 \text{ r.p.s.}$$

(I) Operating speed is 55% less than the critical speed.

55% of the critical speed = $0.55 \times 0.82 = 0.45$ r.p.s.

$\therefore$ Operating speed = 0.82 − 0.45 = **0.37 rps (22 r.p.m.)** ... **Ans.**

OR : Operating speed = (1 − 0.55) × Critical speed = (1 − 0.55) × 0.82) = **0.37 r.p.s.** ... **Ans.**

(II) Critical speed is 40% more than the operating speed.

$\therefore$ Critical speed = 1.40 (operating speed)

$\therefore$ Operating speed = 0.82/1.40 = **0.586 r.p.s. (35 r.p.m.)** ... **Ans.**

Example 2.9 : *A certain set of crushing rolls has rolls of 1000 mm diameter by 375 mm width of face. They are set so that the crushing surfaces are 12 mm apart at the narrowest point. The angle of nip is 30°. What is the maximum permissible size of feed ?*

Solution :

$$r = \text{radius of roll} = 1000/2 = 500 \text{ mm}$$

$$R = \text{radius of feed particle} = ?$$

$$d = \text{gap between the rolls} / 2 = 12/2 = 6 \text{ mm}$$

$$\alpha = \text{angle of nip}/2 = 30/2 = 15^\circ$$

For crushing rolls, we have :

$$\cos\alpha = \frac{r+d}{r+R}$$

$$\cos(15) = \frac{500+6}{500+R}$$

$$0.9659 = \frac{500+6}{500+R}$$

$$\therefore \quad R = 23.86 \text{ mm}$$

$$\text{Size of feed particle} = 2R = 2 \times 23.86 = 47.72 \text{ mm} \approx \mathbf{48\ mm}$$ ... **Ans.**

Example 2.10 : *What should be the diameter of a set of rolls to take feed of size equivalent to 38 mm spheres and crush to 12.7 mm ?*

The co-efficient of friction is 0.35.

Solution :

$$\mu = \tan\alpha$$

$$\therefore \quad \tan\alpha = 0.35$$

$$\therefore \quad \alpha = \tan^{-1} 0.35 = 19.29^\circ$$

$$\cos\alpha = 0.944$$

For crushing rolls, we have :

$$\cos\alpha = \frac{r+d}{r+R}$$

where

$$r = \text{radius of roll} = ?$$

$$R = \text{radius of feed particle} = \frac{38}{2} = 19 \text{ mm}$$

$$d = \text{gap between rolls} / 2 = \frac{12.7}{2} = 6.35 \text{ mm.}$$

$$0.944 = \frac{r+6.35}{r+19}$$

$$\therefore \quad r = 206.9 \text{ mm}$$

$$\text{Diameter of the rolls} = 2r = 2(206.9) = 413 \text{ mm}$$

$$\text{Diameter of the rolls} = 413 \text{ mm.}$$

Such odd sizes are not available, so 400 mm rolls should be used.

$$\text{Diameter of rolls} = \mathbf{400\ mm}$$ ... **Ans.**

For $\mu > \tan d$, the problem may be solved by taking a value of α slightly lower.

Important Points

- **Size reduction** refers to *an operation wherein particles of solids are cut or broken into smaller pieces.*
- **Crushing efficiency :** It is defined as the *ratio of the surface energy created by crushing to the energy absorbed by the solid.*
- **Rittinger's law :** It states that *the work required for reduction of particle size is directly proportional to the new surface created.*
- **Kick's law :** It states that *the work required for crushing a given material is proportional to the logarithm of the ratio between the initial and final diameters.*
- **Bond' law :** It states that *the work required to form particles of size D_p from very large feed is proportional to the square root of the surface-to-volume ratio of the product (S_p/v_p), $S_p/v_p = 6/\phi_s D_p$.*
- Crushers are slow-speed machines employed for coarse reduction of large quantities of solids.
- **Angle of nip** is the *angle formed by the tangents to the roll faces at a point of contact with a particle to be crushed.*
- *A revolving mill is a cylindrical shell slowly rotating on a horizontal axis and charged with a grinding medium to about half its volume.*
- The minimum speed at which centrifuging occurs in a ball mill is called the critical speed of the ball mill.
- Open-circuit grinding consists of one or more grinding mills arranged in series or parallel without classification equipment, whereas closed-circuit grinding consists of one or more grinding mills with classification equipment.

Practice Questions

1. Give the classification of size-reduction machines.
2. Name the four common ways of breaking solids in size-reduction machines.
3. Describe the construction of Blake jaw crusher.
4. Write in brief the construction and operation of ball mill.
5. What do you mean by closed-circuit grinding and open-circuit grinding ?
6. State why centrifuging is not desirable in a ball mill.
7. Define critical speed and give the formula for calculating the critical speed.
8. Define angle of nip and give the relationship between angle of nip, feed size, gap between rolls and diameter of rolls.
9. Draw a neat diagram of jaw crusher and name its parts.
10. State Rittinger's and Kick's law.
11. What should be the diameter of set of rolls which accepts feed equivalent to spheres of 50 mm in diameter and crush them to spheres having a diameter of 15 mm ? The co-efficient of friction is 0.30.

    (Ans. 750 mm)
12. Differentiate between crushing and grinding operation.
13. Differentiate between jaw crusher and gyratory crusher.
14. Draw a neat diagram of conical ball mill and name its parts.
15. Differentiate between Blake jaw crusher and Dodge jaw crusher.

❑❑❑

CHAPTER THREE

SIZE SEPARATION OF SOLIDS

INTRODUCTION

- Solids may be separated from solids in the dry state by methods such as screening (size separation-separation according to size), magnetic separation and electrostatic separation. In this chapter, we will deal with screening.
- In chemical industry, the problem of separating solid particles that is encountered is that of separation of a single solid material into a number of size fractions or to obtain an uniform material for incorporation in a system wherein a certain chemical reaction is occurring.

3.1 CONCEPT AND IMPORTANCE OF SCREENING OPERATION

Screening :

(i) A method of separating solid particles according to size alone is called screening.

(ii) Screening is an operation for separating solids on the basis of size alone.

(iii) It refers to the separation of solid materials on the basis of size using screens of known openings.

(iv) Screening is the separation of a mixture of solid particles of various sizes into two or more fractions by means of a screening surface.

- In screening, a mixture of solid particles of various sizes is dropped on a screening surface/screen (a surface provided with suitable openings) which acts as a multiple go and no-go gage. The material that passes through a given screen/screening surface is called the undersize or minus (–) material while the material that remains on the screen/screening surface is called the oversize or plus (+) material.
- A single screen can make a single separation of the material charged into two fractions. These are called unsized fractions as only the upper or lower limit of the particle sizes they contain is known and the other limit is not known. The material can be separated into sized fractions i.e. the fractions in which both the maximum and minimum particle sizes are known, by passing it through a series of screens of different sizes. Screening is much more commonly adopted for dry particulate solids and occasionally for wet particulate solids. In this topic, we will limit our discussion to screening of dry particulate solids.

Materials for Screens :

Industrial screens are made from metal bars, woven wire cloth, silk bolting cloth, perforated or slotted metal plates. Many varieties and types of screens are available for different purposes but we will discuss few representative types.

Necessity of or reasons for carrying out screening operation/importance of screening operation :

The screening operation is industrially carried out in order to

(i) remove the fines from a feed material before a reduction equipment such as jaw crusher, ball mill or rod mill,

(ii) prevent an incompletely crushed material (oversize) from entering into other unit operations,

(iii) produce a commercial or process-grade material to meet specific particle size limits,

(iv) remove the fines from a finished product prior to shipping.

3.2 CLASSIFICATION OF SCREENS ON THE BASIS OF PERFORMANCE

Comparison of Ideal and Actual screens

- The objective of a screen is to accept a mixture of various sizes of grains and separate it into two fractions, namely an underflow and overflow. The underflow is the one that is passed through the screen and the overflow is the one that is rejected by the screen.

- An **ideal screen** is *the one which sharply separates the feed mixture in such a way that the smallest particle in the overflow is just larger than the largest particle in the underflow.* The ideal separation defines a cut diameter D_{pc} (a typical particle dimension) which makes the point of separation between the undersize and oversize fractions and is nearly equal to the mesh opening of the screen.
- Fig. 3.1 (a) shows the performance of an ideal screen in terms of the screen analysis of the feed. The point 'C' in the curve is a cut point. The fraction A consists of all particles larger than cut diameter (D_{pc}) while the fraction B consists of all particles smaller than cut diameter.

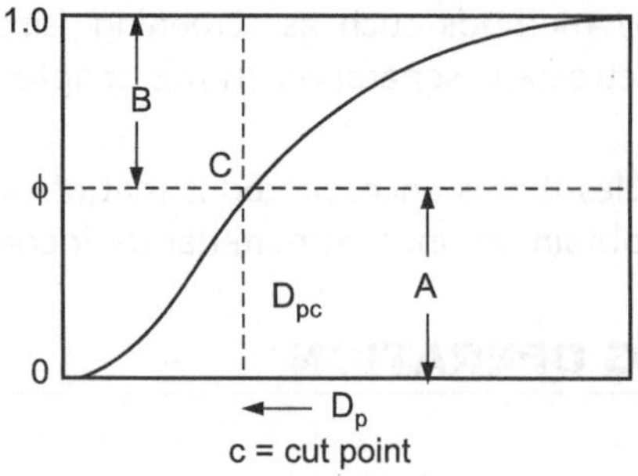

(a) Ideal screening

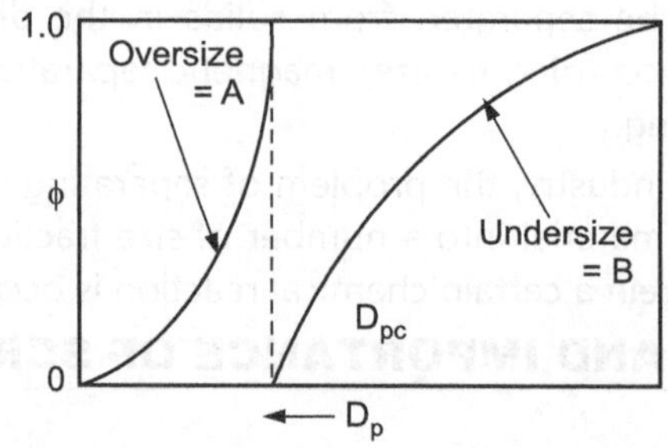

(b) Screen analysis of products from ideal screening

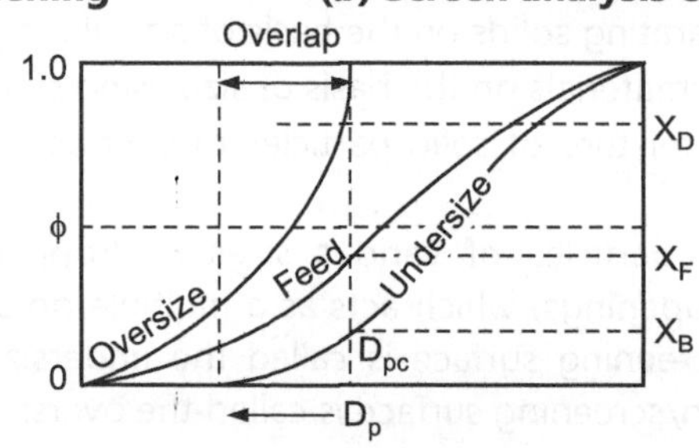

(c) Actual screening

Fig. 3.1 : Ideal versus actual screening

- Material A is the overflow and material B is the underflow. Fig. 3.1 (b) shows the screen analysis of the ideal fractions A and B. The first point on the curve for B and the last point on the curve for A have the same abscissa, and there is no overlap of these curves.
- Actual screens do not yield a sharp separation. With actual screens, the screen analysis of the overflow and underflow are similar to those shown in Fig. 3.1 (c). The overflow is found to contain the particles smaller than the cut diameter, and the underflow is found to contain particles larger than the cut diameter. The curves for A and B overlap.
- With standard testing screens, it is possible to get the closest separations in case of spherical particles. In this case also there is some overlap. In case of needlelike, fibrous particles or particles which have tendency to aggregate into clusters, the overlap is of a higher magnitude. Commercial screens usually give poorer performance than testing screens of the same mesh handling the same feed mixture.

Comparison of Ideal screen and Actual screen :

Ideal screen	Actual screen
1. Yields sharp separation.	1. Does not yield sharp separation.
2. Efficiency of the screen is 100%.	2. Efficiency of the screen is less than 100%.
3. Such screens do not found in practice / reality.	3. Such screens are available in practice.
4. The overflow will contain only particles larger than the cut diameter.	4. The overflow may also contain particles smaller than the cut diameter.
5. Underflow will contain only particles smaller than the cut diameter.	5. Underflow may also contain particles larger than the cut diameter.

3.3 TYPES OF STANDARD SCREEN SERIES

- Most particulate systems which are of practical interest consist of particles of a wide range of sizes. A number of methods of particle-size determination are available but most particle-size determinations are made by the screen analysis when the particles are within the size range that can be measured by screens.
- Usually, for carrying out the analysis, standard screens of either the Tyler standard screen series, U.S. sieve series or Indian standard sieves are used. The testing sieves with square openings are constructed of woven wire screens, the mesh and dimensions of which are standardised. Every screen is identified in meshes per inch.
- In coarse screens, the term mesh refers to the distance between adjacent wires or rods. While in fine screens, the **mesh** is *the number of openings per linear inch counting from the centre of any wire to a point exactly one inch distant* (e.g., a 200-mesh screen will have 200 openings per linear inch).
- *The minimum clear space between the edges of the opening in the screening surface* is termed as **screen aperture or screen-size opening.**
- The **Tyler standard screen series** is based on a 200-mesh screen with a wire 0.0053 mm (0.0021 in) in diameter, giving a clear opening of 0.074 mm (1/200 – 0.0021 = 0.0029 in).
- The screens coarser than a 200-mesh screen have their mesh and wire diameter so adjusted that the area of opening in any one screen is approximately twice the area of the opening in the next finer screen. This means that the ratio of the linear size of the openings in any screen to that in the next finer screen is $\sqrt{2}$ (1.41). Therefore, a 150-mesh screen will have an opening of 0.104 mm ($\sqrt{2} \times 0.0029 = 0.0041$ in) with a wire of 0.064 mm (1/150 – 0.0041 = 0.0026 in) diameter.
- Testing sieves of the Tyler standard sieve series are used to determine the efficiencies of screening equipments and work of crushing and grinding machinery.
- **The Indian Standard test sieves** satisfy requirements of IS : 460 (Part-I) for the wire cloth test sieves and IS : 460 (Part-II) for the perforated plate test sieves with respect to widths of aperture, wire diameter and screening areas. The sieves from 22 micron to 5.6 mm size have woven wire cloth fixed in spun brass frames. The sieves of size 5.6 mm to 125 mm are of woven wire cloth or perforated sheet in GI frames.

3.4 TYPES OF SCREEN ANALYSIS

There are two methods of reporting screen analysis : (1) Differential analysis, (2) Cumulative analysis.

- **Differential Analysis :** *The screen analysis in which the weight fraction of the material retained on each screen is reported in a tabular or a graphical form as a function of the mesh size/screen opening is called* ***differential analysis***.
- **Cumulative Analysis :** *The screen analysis in which the cumulative weight fraction of the material retained (cumulative oversize) or passing through (cumulative undersize) each screen is reported in a tabular or a graphical form as a function of the screen opening is called* ***cumulative analysis.***
- The fine particles are generally specified according to their screen analysis. A screen analysis of a material is carried out by using testing sieves. A set of standard screens is arranged serially in a stack in such a way that the coarsest of the screens is at the top and the finest of the screens is at the bottom.
- An analysis is carried out by placing the sample on the top screen and shaking the stack in a definite manner, either, manually or mechanically, for a definite length of time. The material retained on each screen is removed and weighed.
- For reporting the screen analysis, the amount of material retained on each screen is expressed as the weight fraction of the total sample.

- The screen analysis of a sample is reported either in a tabular form or as graphs. The results of a screen analysis can be reported in a tabular form to show the weight fraction of the material retained on each screen as a function of the mesh size. As the particles retained on any one screen are passed through the screen immediately above it, two numbers are needed to specify the size, one for the screen through which the fraction passes and the other for the screen on which that fraction is retained. Hence, the notation 10/14 means through 10 mesh and on 14 mesh (i.e., the material is such that it passes through the screen of mesh number 10 and collects on the screen of mesh number 14). An analysis reported in a tabular form in this manner is called a **differential analysis.**
- The material that is retained on the screen is the oversize or plus (+) of that screen and the material that passes it is the undersize or minus (–). Thus, a – 10 + 14 fraction means the fraction of the material that passes through a 10-mesh screen but is retained on a 14-mesh screen.

Table 3.1 shows a typical differential screen analysis.

Table 3.1

Mesh	Screen opening microns	Avg. particle size microns	Weight fraction retained
6/8	2362	2845	0.017
8/10	1651	2006	0.235
10/14	1168	1410	0.298
14/20	833	1000	0.217
20/28	589	711	0.105
28/35	417	503	0.062
35/48	295	356	0.028
48/65	208	252	0.017
65/100	147	178	0.010
100/150	104	126	0.005
150/200	74	89	0.002
Pan			0.004
			1.0

- The average particle size of the material retained on any particular screen (cited in Table 3.1) is calculated as the arithmetic mean of two screen openings used to obtain the fraction.
- The second method of reporting screen analysis is a **cumulative analysis**. The cumulative analysis is obtained from the differential analysis by adding cumulatively, the individual weight fractions of material retained on each screen, starting with that retained on the largest mesh, and tabulating or plotting the cumulative sums against the screen opening of the retaining screen under consideration.
- If we define ϕ by the equation of the form :

$$\phi = \Delta\phi_1 + \Delta\phi_2 + ... + \Delta\phi_{N_T} = \sum_{N=1}^{N_T} \Delta\phi_N \quad ... (3.1)$$

where $\Delta\phi_1$, $\Delta\phi_2$ ...... are the weight fractions of material retained on screens 1, 2, ... numbered serially from top of the deck.

- Then, the cumulative analysis is the relation between ϕ and screen opening. The quantity ϕ is the weight fraction of the sample that consists of particles larger than the screen opening. For the entire sample, the value of ϕ is unity. The cumulative analysis corresponding to the differential analysis of Table 3.1 is shown in Table 3.2.

Table 3.2 : Cumulative screen analysis

Mesh	Screen opening in micron	Cumulative fraction retained (φ) (oversize)
6	3327	0.0
8	2362	0.017
10	1651	0.252
		(0 + 0.017 + 0.235)
14	1168	0.55
		(0 + 0.07 + 0.235 + 0.298)
20	833	0.767
28	589	0.872
35	417	0.934
48	295	0.962
65	208	0.979
100	147	0.989
150	104	0.994
200	74	0.996
Pan	–	1.0

ϕ = cumulative fraction retained on the screen

$\therefore$ $1 - \phi$ = cumulative fraction passing through the screen.

- The cumulative analysis is also reported by incorporating a cumulative fraction passing through the screen as shown below :

Mesh	Screen opening	Cumulative fraction retained (oversize)	Cumulative fraction passing through screen (undersize)
6	3327	0.0	1.0
8	2367	0.017	0.983
:	:	:	:
:	:	:	:
Pan	–	1.0	0.00

- A graphical presentation of the screen analysis can be done on ordinary graph papers but more conveniently on semilogarithmic papers where the size is plotted on a logarithmic scale. A semilogarithmic plot avoids the crowding of most of the data into small section of the diagram. The results of screen analysis represented as fractional or cumulative distribution curves are shown in Fig. 3.2 (a) and (b).
- The fractional distribution curve is a plot of the weight fraction of material retained against the average particle size and the cumulative distribution curves shown in Fig. 3.2 (c) are obtained by plotting the fraction of the total weight of particles having a size greater than or less than a given screen opening against the screen opening.

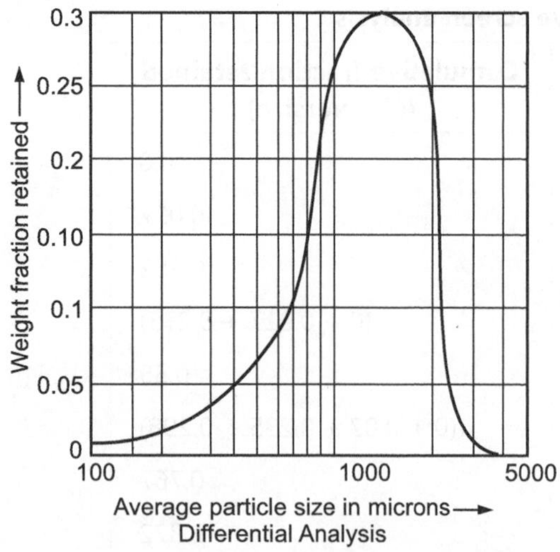

(a) Fractional distribution curve

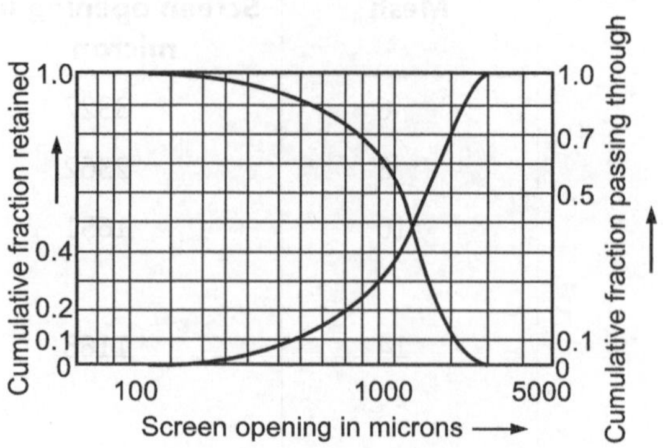

(b) Cumulative distribution curve

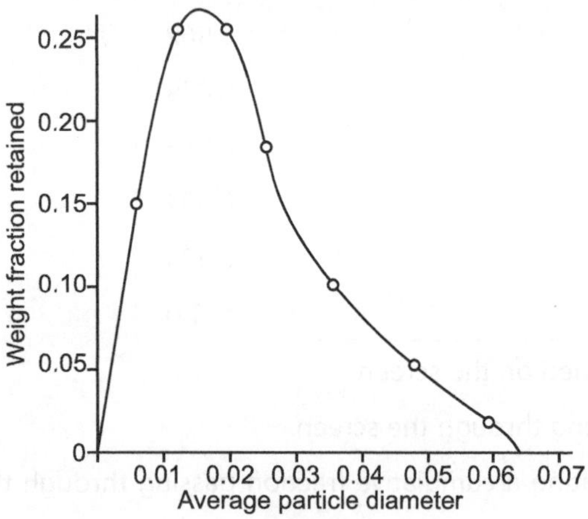

(c) Fractional distribution plot on ordinary graph paper

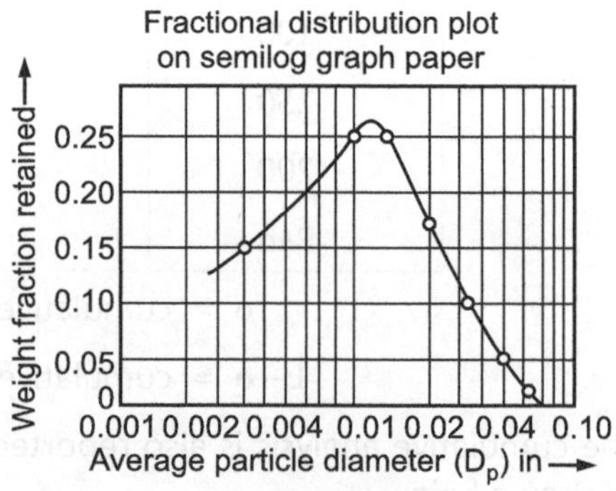

(d) Fractional distribution plot on semilog graph paper

Fig. 3.2 : Fractional and cumulative distribution curves

3.5 CAPACITY AND EFFECTIVENESS OF SCREENS

- The capacity and effectiveness are measures of the performance in industrial screening. The **capacity** of a screen is *the mass of material that can be fed per unit time to a unit area of the screen.* Capacity and effectiveness in screening operations are closely related. For obtaining maximum effectiveness the capacity must be small, and the large capacity is attainable only by reduction in the effectiveness. As the capacity and effectiveness are opposing factors, a reasonable balance must be done between them in actual practice.
- One can control the capacity of a screen simply by varying the mass flow rate of feed to the screening equipment. For a given capacity, the effectiveness that can be achieved depends upon the nature of the screening operation. For a given undersize particle, the overall chance of passage through the screen is a function of the number of contacts between the particle and the screen surface, and the probability of its passage in a single contact. When the screen is overloaded, the number of contacts is small and the probability of passage on contact is reduced due to the interference of the other particles. The effectiveness is improved by reducing the capacity since then there will be more contacts per particle and better chance for passage on each contact.

- The factors which tend to reduce the capacity and lower the effectiveness are : blinding, cohesion of particles to the screen surface, oblique direction of approach of the particles to the screen surface and interference of bed particles with the motion of any one.
- The moisture content of the feed adversely affects the screening operation as the damp particles are prone to stick to the screen surface and to each other.

Blinding of screen :

- It refers to the phenomenon wherein elongated, sticky, etc. particles become wedged into the openings during screening and thus prevent the other particles from passing through it. Thus, the blinding of a screen means the plugging of the screen with solid particles.
- Due to blinding an appreciable fraction of the screen becomes inactive. The blinding tendency is more pronounced with fine screens than with coarse screens.

Definition and Derivation of the Effectiveness of a Screen

Material Balances over a Screen :

- Consider that the feed to a screen consists of materials A and B, where A is the oversize and B is the undersize material. Out of the total materials fed to the screen, some part of it is removed as overflow and remaining part of it is collected as underflow.

Let F be the mass flow rate of feed, (kg/h)

D be the mass flow rate of overflow, (kg/h)

B be the mass flow rate of underflow, (kg/h)

x_F be the mass fraction of material A in feed.

x_D be the mass fraction of material A in overflow.

x_B be the mass fraction of material A in underflow.

The mass fractions of material B in feed, overflow and underflow are $1 - x_F$, $1 - x_D$, and $1 - x_B$ respectively.

Overall material balance over a screen is

$$F = D + B \quad \text{... (3.2)}$$

Material balance of 'A' over a screen is

$$x_F.F = x_D.D + x_B.B \quad \text{... (3.3)}$$

From Equation (3.2), we have

$$F - B = D \quad \text{... (3.4)}$$

Substituting the value of D from Equation (3.4) into Equation (3.3), we get

$$x_F.F = x_D (F - B) + x_B.B \quad \text{... (3.5)}$$

$$x_F.F = x_D.F - x_D.B + x_B.B$$

$$(x_D - x_F) F = (x_D - x_B) B \quad \text{... (3.6)}$$

$$\frac{B}{F} = \frac{x_D - x_F}{x_D - x_B} \quad \text{... (3.7)}$$

Similarly, elimination of B from Equations (3.2) and (3.3) gives

$$\frac{D}{F} = \frac{x_F - x_B}{x_D - x_B} \quad \text{... (3.8)}$$

- The **effectiveness of a screen (screen efficiency)** is *a measure of the success of the screen in closely separating undersize and oversize materials.* In the case of a perfectly functioned screen, all the oversize material 'A' would be in the overflow and all the undersize material would be in the underflow.
- The **screen effectiveness** based on the **oversize material** is *the ratio of the amount of oversize material A that is actually in the overflow to the amount of oversize material A in the feed.*

$$\begin{matrix}\text{Screen effectiveness} \\ \text{based on material A}\end{matrix} = E_A = \frac{D.x_D}{F.x_F} = \frac{\text{Quantity of oversize in the overflow}}{\text{Quantity of oversize in the feed}} \quad \text{... (3.9)}$$

Similarly, the screen effectiveness based on the undersize material is given by

$$E_B = \frac{B\,(1 - x_B)}{F\,(1 - x_F)} \quad \text{... (3.10)}$$

The overall effectiveness of a screen can be given by

$$E = E_A \cdot E_B \quad \text{... (3.11)}$$

where E is the overall effectiveness of a screen.

Substituting the values of E_A and E_B from Equations (3.9) and (3.10) into Equation (3.11), we get

$$E = \frac{D.B.\,x_D\,(1 - x_B)}{F^2\,x_F\,(1 - x_F)} \quad \text{... (3.12)}$$

Substituting the values of D/F and B/F from Equations (3.8) and (3.7) into Equation (3.12), we get

$$E = \frac{(x_F - x_B)\,(x_D - x_F)\,.\,x_D\,(1 - x_B)}{(x_D - x_B)^2\,(1 - x_F)\,x_F} \quad \text{... (3.13)}$$

3.6 FACTORS AFFECTING PERFORMANCE OF SCREENS

Variables in Screening Operations :

- There are many variables in a screening operation that can be very easily changed. The best combination of these variables results into the excellent performance of a screening equipment.

1. Method of feeding :

In order to obtain the maximum capacity and efficiency, the screening equipment must be fed properly. The material should be spread evenly over a full width of the screening surface and should approach the screening surface in a direction parallel to the longitudinal axis of the screen and must be fed at a low rate.

2. Screening surfaces :

Use of single-deck screens in series results into the most efficient operation as in case of multiple-deck screens lower decks are not fed so that their entire area is not used and each separation requires a different combination of angle, speed, and amplitude of vibration for the best performance.

3. Screen slope :

As the screen slope increases, the rate at which the materials travels over the screening surface increases and at the same time, it reduces the bed thickness. The increase of rate of travel means an increase in the quantity passing over the screen per unit time. Reduction in the bed thickness allows the fines to approach the screen surface and pass through it. However, the slope cannot be increased beyond a certain value because beyond that value (limit) the material will travel down the screen much faster without getting screened and the screening efficiency reduces drastically.

4. Vibration amplitude and frequency :

One has to select the proper amplitude of vibration to prevent blinding of the screening cloth and for long bearing life. The frequency of vibration affects the capacity of the screening equipment by regulating the number of contacts between the material and the screening surface.

5. Moisture in feed :

The moisture associated with the feed material adversely affects the screening operation and should be removed.

3.7 TYPES OF SCREENING EQUIPMENTS

- Screening equipments can be classified on the basis of size of material as the screens may be required to pass grains ranging from several mm in diameter down to 200-mesh.
  1. Grizzlies (fixed inclined screens) are used for the coarse screening of large lumps.
  2. Trommels (revolving screens) are generally used for fairly large particles.
  3. Shaking and vibrating screens are used in a coarse range and also for fine sizing.
- In the screening operation, coarse particles pass easily through the large openings in a stationary surface but for fine particles the screen must be agitated by shaking, gyrating or vibrating it mechanically or electrically. Fig. 3.3 shows typical screen motions.

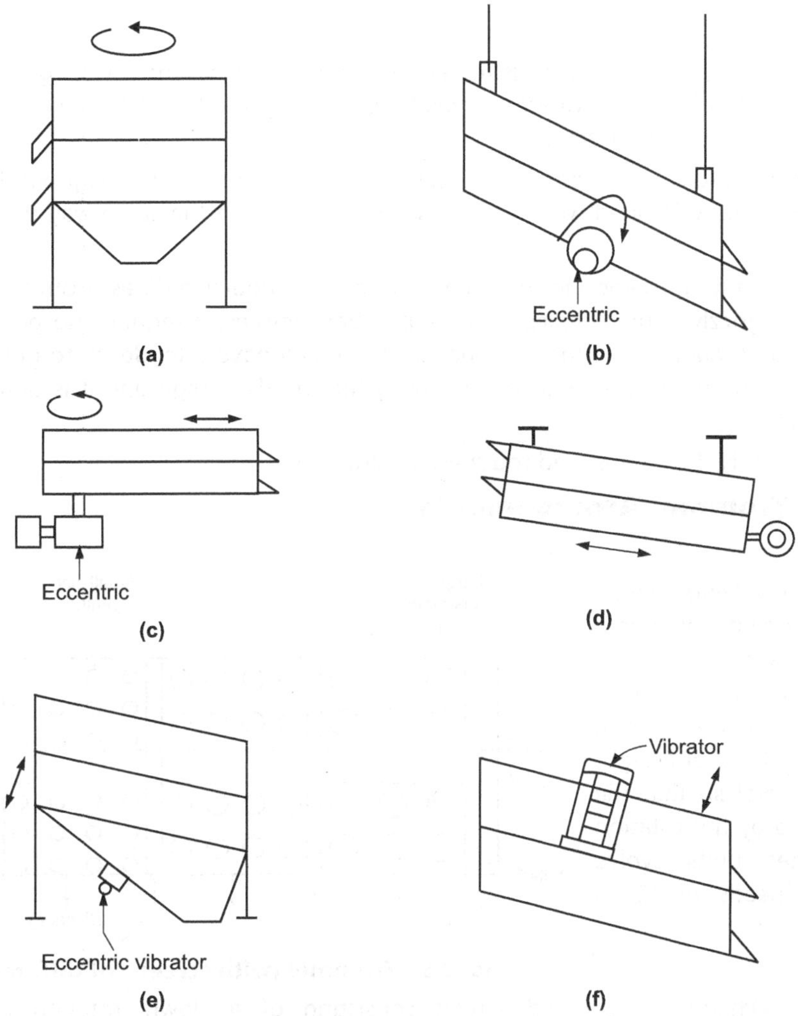

Fig. 3.3 : Motions of screens

(a) Gyrations in horizontal plane; (b) Gyrations in vertical plane; (c) Gyrations at one end and shaking at other; (d) Shaking; (e) Mechanically vibrated; (f) Electrically vibrated

3.7.1 Grizzlies / Grizzly Screens

Construction :

- A grizzly is a grid of parallel metal bars set in an inclined stationary frame, with a slope of 30 to 45°. The slope, and therefore the path of the material is parallel to the length of the bars. The length of the bar may be upto 3 m and the spacing between the bars is 50 to 200 mm. The material of construction of the bars is manganese steel to reduce wear. Usually, the bar is shaped in such a way that its top is wider than the bottom, and hence the bars can be made fairly deep for strength without being choked by material passing partway through them.
- A stationary grizzly is usually used for a dry free flowing material and is not satisfactory for a moist and sticky material.

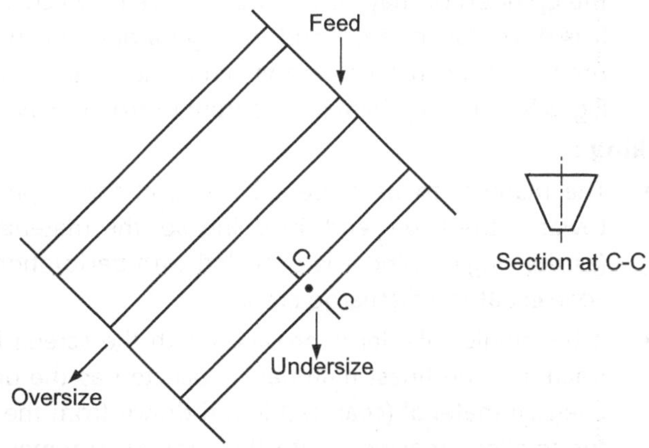

Fig. 3.4 : Grizzly

Working :

- A coarse feed (say from a primary crusher) is fed at the upper end of the grizzly. Large chunks roll and slide to the lower end (the tail discharge), whereas small lumps having size less than the opening in the bars fall through the grid into a separate collector.
- If the angle of inclination to the horizontal is greater, greater is the output (throughput) but the lower is the screen efficiency. Stationary inclined woven-metal screens operate in the same way that separate particles 12 to 100 mm in size.
- A grizzly finds its greatest application in the separation of the undersize (fines) from the feed to a primary crusher. A stationary grizzly is the simplest of all separating devices. It requires no power and is the least expensive to install and maintain. As the openings in the grizzly have a tendency to get blocked by wedge shaped particles, the labour requirement for operating the grizzly is high and it is difficult to change the openings in the bars.
- Grizzlies are used for only the coarest and roughest separations.

3.7.2 Trommel and Trommel Arrangements

Construction :

- A trommel is a revolving screen consisting of a cylindrical frame surrounded by wire cloth or perforated plate (which acts as a screening surface). It is open at one or both ends and inclined at a slight angle to the horizontal so that the material is advanced by the rotation of the cylinder. These units revolve at relatively low speeds of 15 to 20 rpm.

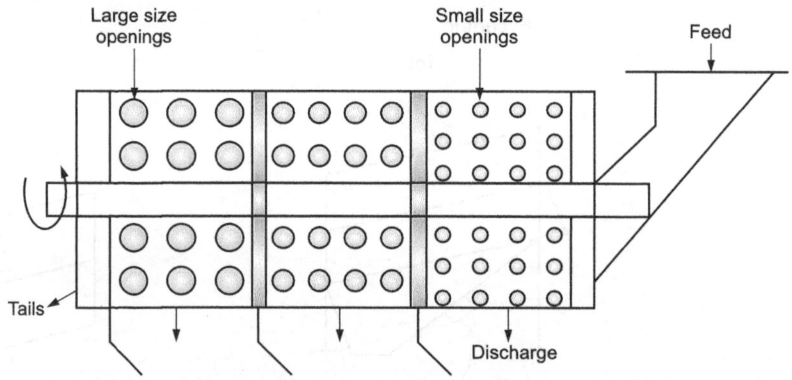

Fig. 3.5 : Trommel (with screens of different openings)

- A trommel is a mechanically operated screen consisting of a slowly rotating perforated cylinder (or cylindrical frame) surrounded by wire cloth (or perforated plate) with its axis at a slight angle to the horizontal.
- The perforations in the screening surface may be of the same size throughout (i.e., over the whole length of the cylinder) or may be of different size in which case the small size perforation section is near the feed end. It is driven at the feed end through a gear mechanism. It has a feed point at the upper end, an undersize product discharge below the screening surface and a oversize discharge at the opposite end (lower end). Fig. 3.5 shows a schematic diagram of trommel having sections of different size perforations.

Working :

- The material to be screened is fed at the upper end and gradually moves down the screening surface towards the lower end. In doing so, the material passes over the apertures of gradually increasing size (as the single cylinder is provided with perforations ranging from the finest desired at the feed end to the coarsest at the discharge end).
- If the single cylinder is provided with the screen having three different size perforations then we get four fractions. The finest material is collected as the underflow in the compartment near the feed end and the oversize material (coarsest) is withdrawn from the discharge end. Such type of arrangement is usually used for smaller capacities. With this type of trommel, there is a tendency of blockage of the apertures by the large material and the screen with the finest opening being the weakest it is subjected to the largest wear.
- The operating speed of a trommel is 30 to 50% of the critical speed (the critical speed is the one at which the material is carried completely round in contact with the screening surface).

Various trommel arrangements

- For separation of a given material into several size fractions, several trommels are operated in series. The first trommel of a series may have the coarsest perforations so that it produces the coarsest finished product which is delivered to the next trommel and so on. In such a case, it is most convenient to place the trommels one above the other as shown in Fig. 3.6 (a).
- When the first trommel of the series has the smallest perforations, the oversize material passes to the next trommel and so on. In such a case, it is most convenient to put the screens in line, end to end as shown in Fig. 3.6 (b).
- If the screening equipment of this type is put into service for finer separations then the cylinder may be covered with a fine wire or silk cloth instead of a perforated plate or coarse wire screen. Such an equipment is usually called a reel.
- Fig. 3.6 (d) shows several concentric cylinders. The innermost is the longest and has coarsest perforations. The outer ones are successively shorter and have finer perforations.
  In this arrangement, maximum load is given to the strongest screen but the construction is complicated and expensive.
- Trommels are well suited for relatively coarse materials (1/2 in or over).

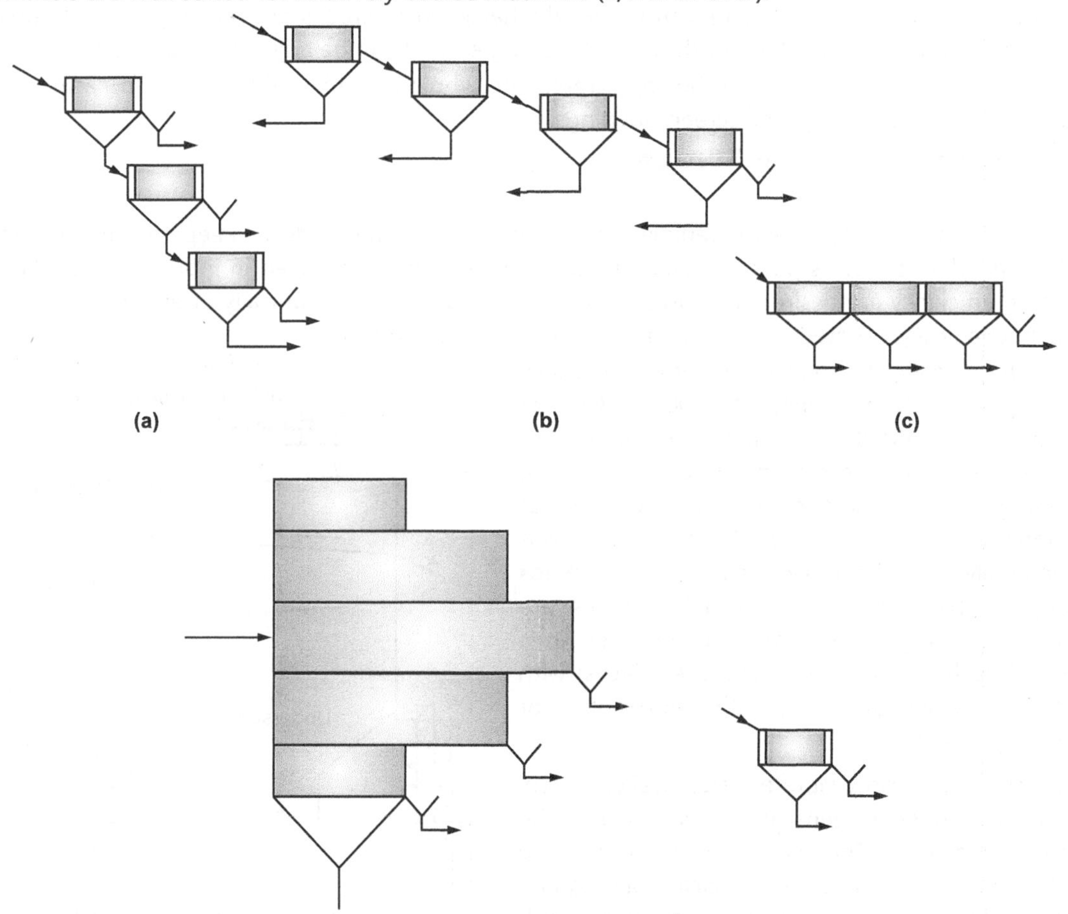

Fig. 3.6 : Trommel arrangements

(a) One-size screen to each trommel, coarsest trommel first; (b) One-size screen to each trommel, finest trommel-first; (c) Single trommel with different perforations; (d) Concentric trommels with coarsest trommel inside; (e) Trommel with single-size perforations.

Comparison of Grizzlies and Trommels :

Grizzlies	Trommels
1. These are stationary inclined screens.	1. These are revolving screens.
2. Usually, screen is a grid of metal bar.	2. Usually, screen is a perforated cylindrical member.
3. Openings in screen are large.	3. Openings in screen are small.
4. They handle large size feed.	4. They handle small size feed.
5. Capacity is large.	5. Capacity is relatively small.
6. Labour requirement is large.	6. Labour requirement is low.
7. Cheap construction.	7. Relatively expensive construction.

Trommels are well suited for relatively coarse material (12 mm and over).

3.7.3 Gyratory Screens

- Gyratory screens which are gyrated vertically contain several decks of screens arranged one above the other and held in a box or casing. The screens are arranged such that the coarsest screen is at the top and the finest at the bottom. Discharge ducts are provided for the screens to permit removal of the several fractions. The casing is inclined at an angle ranging from 16 and 30° with the horizontal. The gyrations are in a vertical plane about a horizontal axis and are produced by an eccentric shaft fixed in the floor of the casing halfway between the feed location and discharge. The screens are rectangular and fairly long. The speed of gyration, the amplitude of throw and the angle of tilt can be adjusted as per requirements.
- The mixture of particles is fed on the top screen. The whole assembly of screens and casing is gyrated to screen the particles through the screen openings.

3.7.4 Shaking and Vibrating Screen

Vibrating screens :

- In some situations, the screen is rapidly vibrated with small amplitude to keep the material moving and prevent blinding as far as possible. Vibrating screens are commonly used in industry where large capacity and high efficiency are desired. The vibrations may be produced mechanically or electrically, accordingly we have mechanically vibrated screens and electrically vibrated screens.
- The vibrations may be produced either mechanically or electrically with frequency of 1800 to 3600 or even more per minute. Mechanical vibrations are generally passed on from high speed eccentrics to the casing and from there to inclined screens so that the whole assembly is vibrated. Electrical vibrations are generally passed on from heavy duty solenoids directly to the screens so that only screens are vibrated. Vibrating screens may be mounted in a multideck fashion (not more than three decks) with the coarsest screen at the top, either horizontally or inclined upto 45°.
- Fig. 3.7 shows a directly vibrated (electrically) screen. Electrically vibrated screens are widely used in the chemical industry. The vibrating screens have accuracy of sizing, increased capacity per square meter and low maintenance cost per ton of material.

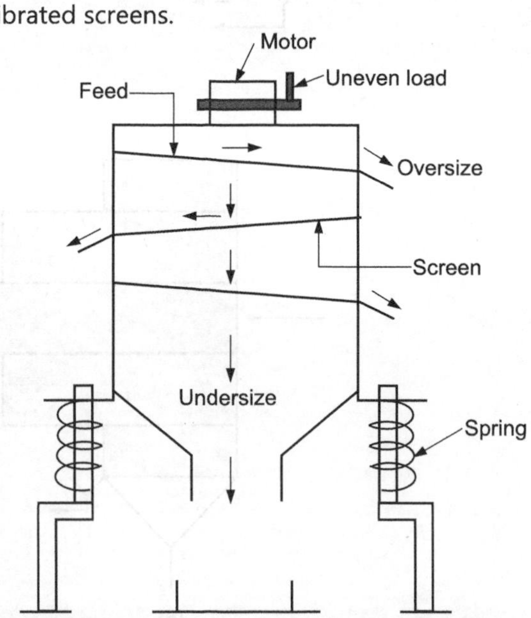

Fig. 3.7 : Vibrating screen (Lab. Model)

Operation of the vibrating screen :

- In case of a single screen, the vibrations are given to the screen to effect the separation of solid particles into two size fractions.

- In case of vibrating screen (generally consists of three decks), the material to be separated is fed to the top screen and simultaneously the screens are vibrated either electrically or mechanically at a frequency of 1000 to 3500 per minute (1000-3500 vibrations per minute). Due to vibrations the particles on the screen are kept moving and due to inclination given to the screens, the oversize material travels along the screen and is collected separately. The undersize material passes through the screen and is collected. Four fractions are obtained with a three deck screen.

SOLVED EXAMPLES

Example 3.1 : *A quartz mixture having a certain screen analysis is screened through a standard 10 mesh screen. Calculate : (a) the mass ratio of overflow and underflow to feed and (b) the effectiveness of the screen.*

Due to blinding an appreciable fraction of the screen surface becomes inactive. The blinding tendency is more pronounced with fine screens than with coarse screens.

Data : $D_p = D_{pc} = 1.651$ mm, $x_F = 0.47$, $x_D = 0.85$ and

$x_B = 0.195$ (cumulative mass fractions)

Solution : x_F = Mass fraction of material A in feed = 0.47

x_D = Mass fraction of material A in overflow = 0.85

x_B = Mass fraction of material A in underflow = 0.195

The mass ratio of overflow to feed is

$$\frac{D}{F} = \frac{x_F - x_B}{x_D - x_B} = \frac{0.47 - 0.195}{0.85 - 0.195} = \mathbf{0.4198 \approx 0.42} \quad \text{... Ans. (a)}$$

The mass ratio of underflow to feed is

$$\frac{B}{F} = \frac{x_D - x_F}{x_D - x_B} = \frac{0.85 - 0.47}{0.85 - 0.195} = \mathbf{0.58} \quad \text{... Ans. (a)}$$

OR :

$$\frac{B}{F} = \frac{F - D}{F} = 1 - \frac{D}{F} = 1 - 0.42 = \mathbf{0.58} \quad \text{... Ans. (a)}$$

The overall effectiveness of the screen is

$$E = \frac{(x_F - x_B)(x_D - x_F)\, x_D (1 - x_B)}{(x_D - x_B)^2 (1 - x_F)\, x_F} = \frac{(0.47 - 0.195)(0.85 - 0.47)(1 - 0.195)(0.85)}{(0.85 - 0.195)^2 (1 - 0.47)(0.47)}$$

$$= \mathbf{0.6691 \approx 0.67} \quad \text{... Ans. (b)}$$

OR :

$$E = \frac{DB\, x_D (1 - x_B)}{F^2 x_F (1 - x_F)} = \frac{D}{F} \cdot \frac{B}{F} \cdot \frac{x_D (1 - x_B)}{x_F (1 - x_F)} = 0.42 \times 0.58 \times \frac{0.85\,(1 - 0.195)}{0.47\,(1 - 0.47)}$$

$$= \mathbf{0.66914 \approx 0.67 \text{ or } 67\%} \quad \text{... Ans. (b)}$$

Example 3.2 : *A dolomite mixture having the following screen analysis is screened through a standard 100 mesh screen. Calculate the effectiveness of the screen and the mass ratios of overflow and underflow to feed.*

Screen analysis :

Mesh	*Feed*	*Oversize*	*Undersize (weight %)*
35	*7.07*	*13.67*	*0.00*
48	*16.60*	*32.09*	*0.00*
65	*14.02*	*27.12*	*0.00*
100	*11.82*	*20.70*	*2.32*
150	*9.07*	*4.35*	*14.32*
200	*7.62*	*2.07*	*13.34*
– 200	*33.80*	*0.00*	*70.02*
	100	*100*	*100*

Solution : From the screen analysis provided, the cumulative mass fractions for a 100-mesh screen are :

Feed : $$x_F = \frac{7.07 + 16.60 + 14.02 + 11.82}{100} = 0.4951$$

= mass fraction of A in feed

Overflow : $$x_D = \frac{13.67 + 32.09 + 27.12 + 20.70}{100} = 0.9358$$

= mass fraction of A in overflow

Underflow : $$x_B = \frac{0.00 + 0.00 + 0.00 + 2.32}{100} = 0.0232$$

The mass ratio of overflow to feed is

$$\frac{D}{F} = \frac{x_F - x_B}{x_D - x_B} = \frac{0.4951 - 0.0232}{0.9358 - 0.0232} = \mathbf{0.517} \quad \text{... Ans.}$$

The mass ratio of underflow to feed is

$$\frac{B}{F} = \frac{x_D - x_F}{x_D - x_B} = \frac{0.9358 - 0.4951}{0.9358 - 0.0232} = \mathbf{0.4829} \approx \mathbf{0.483} \quad \text{... Ans.}$$

OR : $$\frac{B}{F} = \frac{F - D}{F} = 1 - \frac{D}{F} = 1 - 0.517 = \mathbf{0.483} \quad \text{... Ans.}$$

The overall effectiveness of the screen is

$$E = \frac{DB\, x_D (1 - x_B)}{F^2 x_F (1 - x_F)} = \frac{D}{F} \cdot \frac{B}{F} \cdot \frac{x_D (1 - x_B)}{x_F (1 - x_F)}$$

$$= \frac{0.517 \times (0.483)\,(0.9358)\,(1 - 0.0232)}{0.4951\,(1 - 0.4951)} = \mathbf{0.9131} \text{ or } \mathbf{91.31\%} \quad \text{... Ans.}$$

OR : $$E = \frac{(x_F - x_B)\,(x_D - x_F)\, x_D (1 - x_B)}{(x_D - x_B)^2 (1 - x_F)\, x_F}$$

$$= \frac{(0.4951 - 0.0232)\,(0.9358 - 0.4951)\,(0.9358)\,(1 - 0.0232)}{(0.9358 - 0.0232)^2 (1 - 0.4951)\,(0.4951)} = \mathbf{0.9131} \text{ or } \mathbf{91.31\%} \quad \text{... Ans.}$$

Important Points

- A method of separating solid particles according to size alone is called screening.
- Grizzlies, trommels, vibrating screens and oscillating screens are the screening equipments used in practice.
- The screen analysis in which the weight fraction of the material retained on each screen is reported in a tabular or graphical form as a function of the mesh size/screen opening is called differential analysis.
- The screen analysis in which the cumulative weight fraction of the material retained or passing through each screen is reported in a tabular or graphical form as a function of the screen opening is called cumulative analysis.
- The variables in screening operations are : method of feeding, screening surfaces, screen slope, moisture in feed and vibration amplitude and frequency.
- The effectiveness of a screen is a measure of the success of the screen in closely separating oversize and undersize materials.
- The capacity of a screen is the mass of material that can be fed per unit time to a unit area of the screen.

Practice Questions

1. Define screening.
2. Define the terms : differential analysis and cumulative analysis.
3. Derive the equation for effectiveness of a screen.
4. Differentiate between actual screen and ideal screen.
5. Explain in brief the variables in screening operations.
6. Compare grizzlies with trommels.
7. Draw neat sketches of any two motions of screens.

CHAPTER FOUR

SEPARATION OF SOLIDS BASED ON SPECIFIC PROPERTIES

4.1 TYPES OF SIZE SEPARATION BASED ON PROPERTIES

- In this chapter, we will deal with the methods for separating solid particles based on specific properties.
- Generally, screening is the most satisfactory method for separating relatively coarse materials/coarse solids according to size, but with very fine particles (which would clog the fine apertures of the screen or for which it is not possible to make the openings sufficiently fine), the method is impracticable and in such cases a form of settling process is used. This method of separation depends on differences in the behaviour of particles in a moving fluid and separate materials/solid particles according to their terminal falling velocities which in turn depend on size and density.
- Other methods of separation depend on differences in the magnetic properties (magnetic separation), electrical properties (electrostatic separation) or surface properties (froth flotation) of the materials.

4.1.1 Classification

- *Classification is a method of separating solid particles into fractions based upon/according to their terminal falling/settling velocities.*
- Suppose, for example, that the solid particles to be separated are fed in suspension into a tank (containing water) of large cross-sectional area. When the feed stream enters the tank, the horizontal velocity component decreases and the particles start to settle. The faster-settling particles will reach the bottom of the tank before the slower-settling particles. Thus, the faster-settling particles will accumulate near the inlet/entrance, while the slower-settling particles which are carried farther (because of relatively slow-settling rates) and will concentrate nearer the exit/outlet.
- Another way to separate the particles would be to place two particles having different settling velocities (rates) in a rising stream of water. If the velocity of the water is so adjusted that it lies between the terminal settling velocities of the two particles then the slower-settling particle will be carried upward by the water and the faster-settling particle will simultaneously move downward against the water stream and settle out to the bottom, thus achieving a separation.
- A device that separates the solids into two fractions is called a classifier. The product streams that are obtained from any classifier are : (i) a partially drained fraction containing the coarse material, called the sand and (ii) a fine fraction together with the remaining liquid medium, called the overflow.
- In the classification operation, the coarse solids that are settled at the bottom of a pool of fluid pulp are removed by gravity, mechanical means or induced pressure, while the solids which do not settle are taken out as an overflow from the pool.
- All wet classifiers work upon the difference in rate of settling/settling rate between coarse and fine particles. The settling rate of a particle depends upon its size and density and the particle will settle under the conditions of free settling or hindered settling depending upon the concentration of solids.
- When the particle is at sufficient distance from the vessel walls and from other particles, so that its fall is not affected by them, the process is called free settling. In practice, the concentrations of suspensions (high concentrations of solids to liquid) used in the industry is usually high so that the particles are very close together and thus the collision between the particles is practically continuous.
- When the motion of the particle is impeded or affected by other particles (as they being very close to each other), the process is called hindered settling.

4.1.2 Gravity Settling Tank

- It is the simplest type of classifier. It consists of a large tank with provisions for a suitable inlet and outlet.
- A slurry feed enters the tank through an inlet connection. As soon as the slurry feed enters the tank, its linear velocity decreases as a result of the enlargement of cross-sectional area. Solid particles start to settle under the influence of gravity.
- The faster-settling particles (coarse particles) will be collected at the bottom of the tank near the inlet/entrance, while the slower-settling particles (small particles) will be carried farther into the tank before they reach the bottom of the tank. The very fine particles are carried away in the liquid overflow from the tank.
- Vertical baffles placed at various distances from the inlet within the tank allow for the collection of several fractions (different grades of particles) according to the terminal falling velocities. Because of the occurrence of considerable overlapping of size, no sharp separation is possible with this classifier.

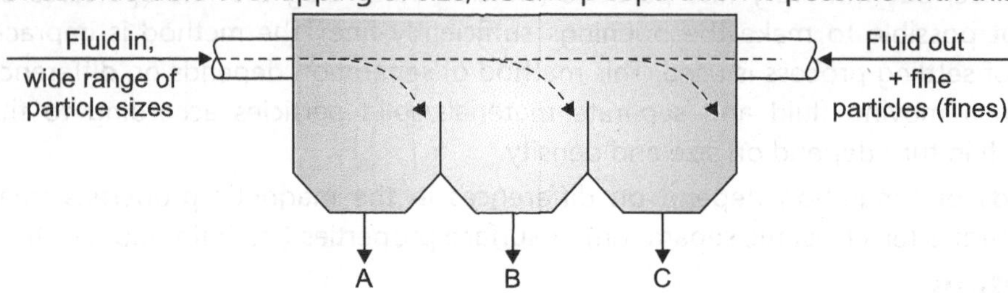

A = Coarse particles

B = Intermediate particles

C = Small particles

Fig. 4.1 : Gravity settling tank

4.1.3 Cone Classifier

- A cone classifier is simply a cone (conical vessel), installed point down, with a discharge launder around the top (of the cone).
- The feed is introduced in the form of a suspension through a fed inlet provided at the centre at the top. The coarse fraction (the partially drained fraction containing the coarse material) collects at the point of the cone (i.e., at the apex) and is withdrawn periodically or continuously. The fine fraction along with the remaining portion of the liquid is removed from the launder as an overflow. The separation achieved with this unit is only an approximate one.
- Cone classifiers are used for relatively crude work because of low cost of installation. They are used in ore-dressing plants.

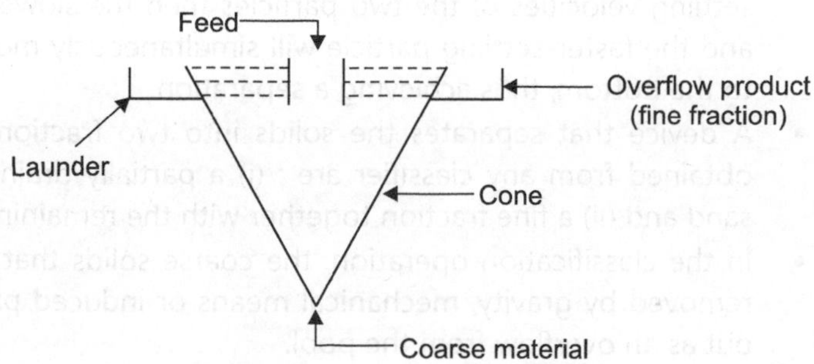

Fig. 4.2 : Cone classifier

4.1.4 Double-cone Classifier

- This classifier uses hydraulic water for classification (a stream of additional water supplied to a classifier is called hydraulic water).
- The double-cone classifier is shown in Fig. 4.3. It consists of a conical vessel incorporating a second hollow cone in it. The inner cone is slightly larger in angle, arranged apex downwards and is movable in a vertical direction. The bottom portion of the inner cone is cut away and its position (height) relative to the outer cone is regulated by a screw adjustment (not shown).

- The feed to be separated is fed in the form of a suspension to the centre of the inner cone. It flows downward through the inner cone and out at a baffle at the bottom of the inner cone. Hydraulic water is fed near the outlet for the coarse material. The solids from the inner cone and a rising stream of water are mixed below the inner cone. Then they flow through an annular space between the two cones. Classification occurs in the annular space, the small/fine particles are carried away in the overflow, whereas the large particles/coarse particles settle against the hydraulic water to the bottom and are removed periodically.

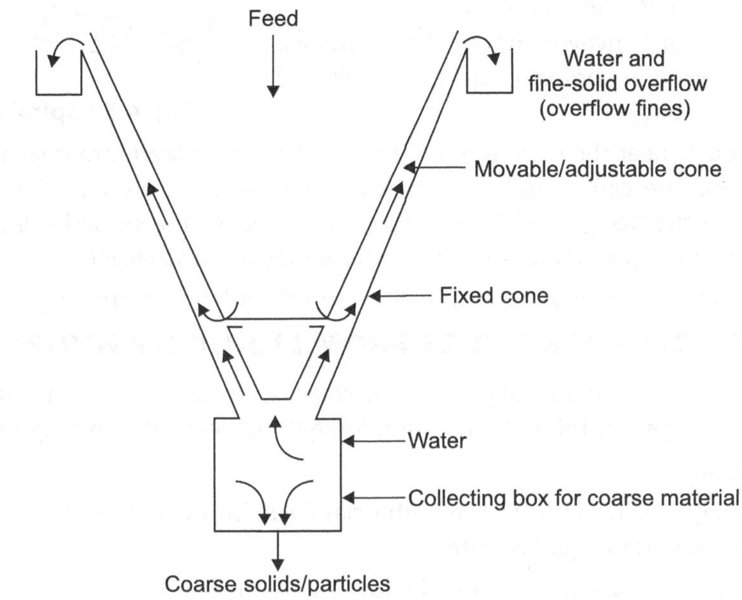

Fig. 4.3 : Double-cone classifier

4.1.5 Mechanical Classifiers

4.1.5.1 Rake Classifier

- The rake classifier such as the Dorr classifier consists of a rectangular tank with a sloping/inclined bottom. The tank is provided with movable rakes (reciprocating rakes). The feed in the form of suspension (slurry) is introduced continuously near the middle of the tank. The lower end of the tank has a weir overflow (discharge weir) from which the fines that are not settled leave with the overflow liquid.

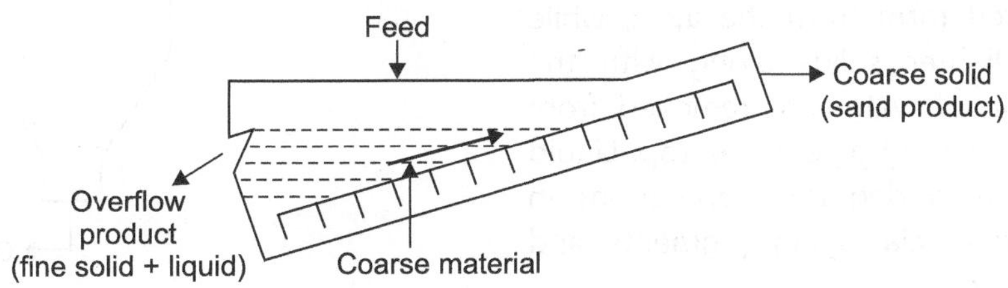

Fig. 4.4 : Rake classifier

- The heavy material (coarser particles) sink to the bottom of the tank. The rakes scrap the settled solids and move them upwards along the bottom of the tank towards the top/upper end of the tank from where they are discharged. The reciprocating rakes keep the slurry in continuous agitation. The time of raking stroke is so adjusted that fines do not have time to settle and so remain near the surface of the slurry, while the heavy particles have time to settle [they settle, scrapped upward and removed as a dense slurry (called the sand)].

4.1.5.2 Spiral Classifier

- It is a mechanical classifier. The spiral classifier such as the Akins classifier consists of a semicylindrical trough (a trough which is semicircular in cross-section) inclined to the horizontal. The trough is provided with a slow-rotating spiral conveyor and a liquid overflow at the lower end. The spiral conveyor moves the solids which settle to the bottom upward towards the top of the trough.

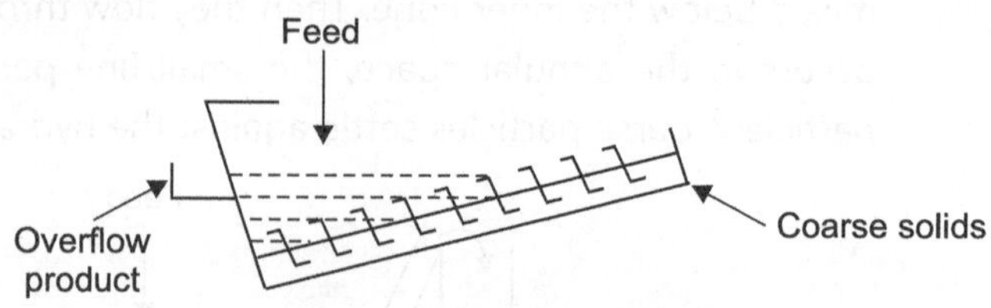

Fig. 4.5 : Spiral classifier

- Slurry is fed continuously near the middle of the trough. The slurry feed rate is so adjusted that fines do not have time to settle and are carried out with the overflow liquid. Heavy particles have time to settle, they settle to the bottom of the trough and the spiral conveyor moves the settled solids upward along the floor of the trough towards the top of the trough from where they are discharged.
- Rake and spiral classifiers are used along with ball mills in closed-circuit grinding.

4.2 SEPARATION OF SOLID PARTICLES FROM LIQUID BY HYDROCYCLONE

- A cyclone/cyclone separator is essentially a settling chamber in which the gravitational separating force is replaced by a much stronger centrifugal separating force (to increase the settling rate).

Liquid Cyclone (Hydrocyclone) :

- Cyclone separators may also be used to effect the classification of solid particles suspended in a liquid. In such cases, the commonly used liquid is water.
- Liquid cyclone has a top cylindrical section and a lower conical section terminating in an apex opening. The top vertical section is covered by a flat plate and is provided with a tangential inlet at the top. The cover has a downward-extending pipe to cut the vortex and remove the overflow product since the viscosity of water is much higher than that of a gas, the fluid resistance encountered in this cyclone is greater than that in the cyclone used for dust collection. Therefore, the diameter of this cyclone must be smaller in order to get a corresponding greater/larger centrifugal force. The pressure of a feed (induced by means of a pump) to the cyclone lie between 5 to 120 psi.
- The slurry feed is pumped into the cylindrical section tangentially. Coarse or heavy solids thrown out against the walls, travel down the sides of the cone section and are discharged in a partially dewatered form from the apex, while the smaller or lighter solids along with the remaining portion of water are removed from the downward extending pipe at the top. Liquid cyclones are used in degritting operations in alumina production, classifying pigments and ore-dressing practice.

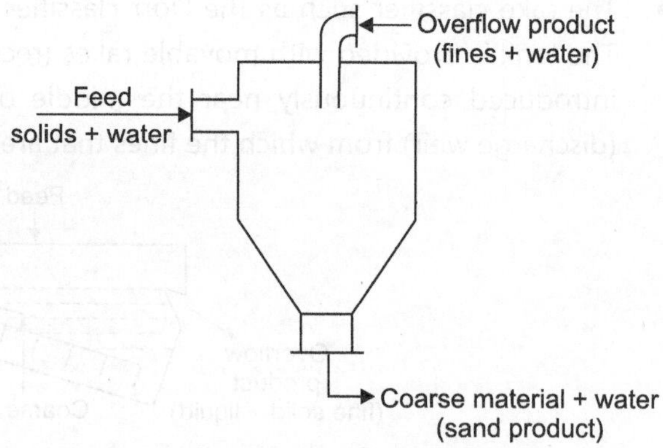

Fig. 4.6 : Liquid cyclone

4.3 JIGGING

- A **jig** is a mechanical device used for the separation of materials of different specific gravities by pulsating a stream of liquid (usually water) flowing through a bed of materials resting on a screen.
- Jigging is a method of separating materials of different specific gravities by the pulsation of a stream of liquid (water) flowing through a bed of materials resting on a screen.

  [pulsate ⇒ oscillate ⇒ move or swing back and forth at a regular rate.

  Jig ⇒ move up and down with a quick jerky motion.]

Principle of operation :

- Jigs separate solids by difference in density and size.
- Jigging is *a process of gravity concentration where solids are separated based upon the differences in the behaviour of particles through a moving fluid which in turn, depends upon densities/specific gravities.*
- Separation of solids of different specific gravities is achieved by the pulsation of a liquid stream flowing through a bed of solids on a screen. The liquid pulsates or jigs up and down and this action causes the heavy material to move towards the bottom of the bed and the lighter material to rise to the top. Each product is taken out separately.

Applications :

- Jigging is used for concentrating heavy minerals from the light minerals. It is commonly employed for coarse material having a size 20-mesh and above and where there is a sizeable/fairly large difference between the effective specific gravity (effective sp. gr. = sp. gr. of mineral – sp. gr. of water) of the valuable and the waste material.
- Jigs are simple in operation, consume very large quantities of water and have high tailings losses on metallic ores. They are used mostly to treat iron ores, few lead-zinc ores, etc.

Hydraulic Jig :

- It operates by providing very short periods for materials to settle due to which the particles do not attain their terminal falling velocities and initial velocities cause the separation. Thus, it is suitable for the separation of materials of a wide size range into various fractions.

Construction :

- Fig. 4.7 shows a hydraulic jig. It consists of a rectangular section tank with a tappered bottom. The tank is divided into two portions/compartments by a vertical baffle. In one compartment, a plunger is incorporated. It operates in a vertical direction giving a pulsating motion to the liquid. In the other compartment, a screen is incorporated. The separation of material is carried out over this screen. It is provided with a connection for feeding liquid during the upstroke. It is also provided with a bottom discharge connection for the removal of small particles of heavy material and gates at the side of jig for the removal of particles settled on the screen and for overflow.

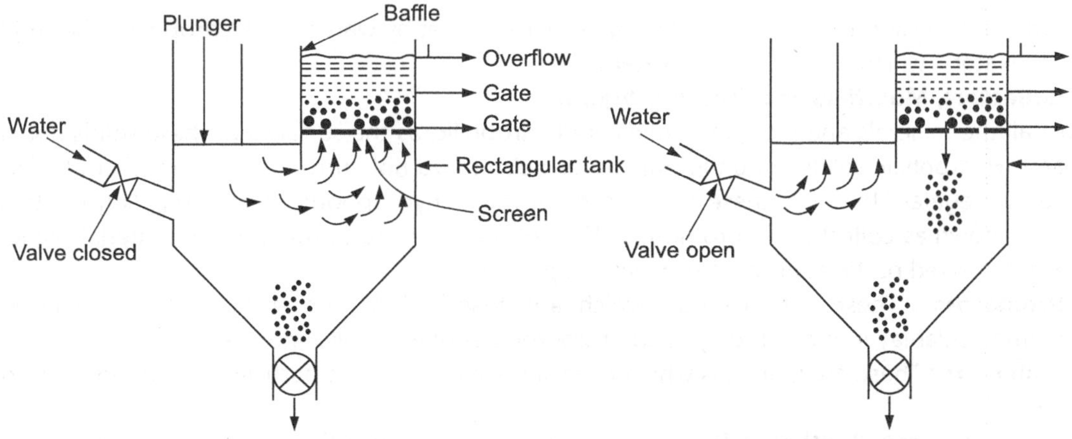

(a) Downward stroke **(b) Upward stroke**

Fig. 4.7 : Hydraulic jig

Working :

- The material to be separated is fed over a screen and is subjected to a pulsating action by oscillating liquid with the help of a reciprocating plunger. During the upward stroke of the plunger, input water is taken into the jig and there is no net flow through the bed of solids. During the downward stroke, water inlet is closed and particles on the screen are brought into suspension and they segregate according to their size and density such that the dense material is collected near the bottom of the screen.

- Very small particles of the dense material will pass through the screen and are collected at the bottom of the jig. Small particles of the less dense material (light material) carried by the liquid water are removed through an overflow. The material retained on the screen is removed through gates provided at the side.

The following four fractions are obtained from the jig :

1. Small and dense material passing through the screen collected at the bottom of the tank.
2. Small size less dense material in the liquid overflow.
3. Large size dense material segregated near a screen removed through a gate at the side.
4. Large size less dense material segregated above the dense material removed through a gate at the side.

4.4 FROTH FLOTATION

- Froth flotation processes are used for the separation of finely divided solids.
- Flotation refers to *an operation in which one solid is separated from another by floating one of them at or on the liquid surface*. In froth flotation, separation of a solid feed mixture depends upon differences in the surface properties of the materials involved. This technique is commonly used in mineral dressing. Mineral dressing refers to the method of treating ores at or near the mine site to produce one or more concentrates of valuable minerals and a tailings composed of waste or less valuable minerals.
- Froth flotation is used for treating the metallic ores that are finer than 48 to 65 mesh, or coal and certain non-metallics that are finer than 10 to 48 mesh. It is not possible to treat a coarser feed by froth flotation as the same cannot be suitably mixed and suspended by a floatation machine.

Principle of operation :

- Separation by froth flotation depends on differences in the surface properties of the materials.
- If the mixture is suspended in an aerated liquid (water), the gas bubbles will tend to adhere preferentially to the constituent which is more difficult to wet by the liquid (hydrophobic constituent) and so its effective density will be reduced to such an extent that it will rise to the surface (i.e., it will float on the surface of the liquid) and the material which has affinity for the liquid (hydrophilic material) gets surrounded by the liquid and it will simply sink, thus, achieving a separation. Frothing agents inducing the formation of a froth of sufficient stability are added to suspend or retain the particles in the froth on the surface before they are discharged.

  [Hydrophobic – failing to mix with water – which will not wet by water and Hydrophilic – having tendency to mix with water – which will get wet by water].

Promotors, Collectors, Modifiers and Frothing Agents :

- Almost all the minerals and inorganic solids are hydrophilic, as the surfaces of these solids get easily wetted by water. Hydrophilic solids are unflotable as air bubbles do not surround or cover them to form a particle bubble aggregate. However, these solids can be made hydrophobic (water repellent) with the help of reagents known as collectors or promotors. The collectors or promotors are the materials which selectively render the desired particles air-avid and water repellent.
  - **(i) Promotors :** These are materials which are adsorbed on the surface of the particles forming a unimolecular layer. A commonly used promotor is sodium ethyl xanthate.
  - **(ii) Collectors :** These are materials which form surface films on the particles. A commonly used collector is pine oil.
  - **(iii) Frothing agents/Frothers :** These are materials which induce the formation of a froth (which produce a froth) of sufficient stability in order to retain the particles of the constituent which is to be floated to be discharged as an overflow. Commonly used frothers include liquid soaps, pine oil, cresylic acid, methyl amyl alcohol and methylisobutylcarbinol.
- The valuable concentrates from froth flotation may be either the froth product which is collected and removed from the top, or the underflow product. In case of metallic sulfide ores of copper, nickel, etc. the valuable product collects in the froth and is removed from the top. In glass-sand flotation, iron bearing minerals collect in the froth, while the valuable product (high grade silica) is removed as an underflow product.

Flotation Machine/Flotation Cell

Construction :

- The mechanically agitated flotation cell consists of a tank having square or circular cross-section. It is provided with an agitator which violently agitates the pulp. A compressor / blower is used to introduce air into the system through a downpipe surrounding the impeller shaft. The bottom of the tank is conical and is provided with a discharge for tailings. An overflow is provided at the top for mineralised froth (or froth) removal.

Working :

- Water is taken into the cell, material is fed to the cell. The promotors and frothers are added. Agitations are given and air is bubbled in the form of fine bubbles. Air-avid particles due to reduction in their effective density, will rise to the surface and be held in the froth before they are discharged from the overflow. Hydrophilic particles will sink to the bottom and removed from the discharge for tailings.

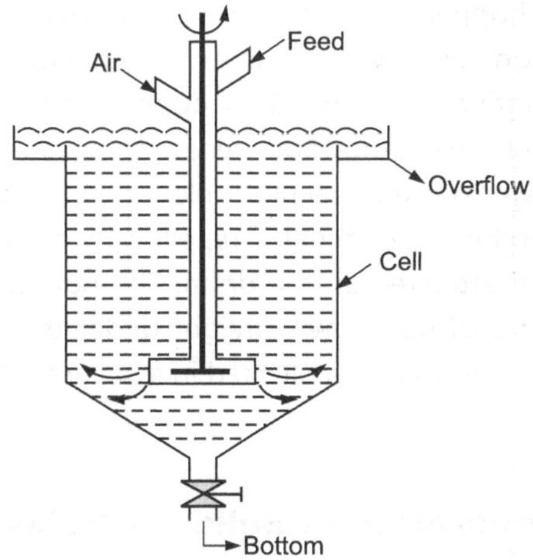

Fig. 4.8 : Froth flotation cell (Lab. model)

4.5 SEPARATION OF SOLID PARTICLES BASED ON ELECTRICAL AND MAGNETIC PROPERTIES

4.5.1 Electrostatic Separation

Principle : If one or more of the materials of a granular mixture can acquire a surface charge on or just before entering an electrostatic field, the grains/particles of that material will be attracted towards the active electrode or repelled from it depending upon the sign of the charge on the grains/particles.

- Electrostatic separation is a method of separation of solid particles based on the differential attraction or repulsion of charged particles under the influence of an electric field. Basically, the difference in electrical properties of different materials is responsible for such a separation.
- Charging of particles is an essential step in this separation. Solid particles can receive a surface charge by any one of the following methods :
  (i) Contact electrification.
  (ii) Electrification by conductive induction.
  (iii) Electrification by bombardment.
- Electrification by conductive induction : When an uncharged solid particle is placed on a grounded conductor in the presence of an electric field (i.e., when it comes in contact with a charged surface), the particle will rapidly acquire a surface charge by induction. A conductive particle acquires the same charge as the grounded conductor (it becomes charged to the same potential as the grounded conductor within a very short period of time) through its contact with the conductor while a dielectric particle is polarised and thus no net charge is generated on it. As a consequence of this induction, the conducting particle will be repelled by the surface/grounded conductor, while the dielectric particle will be unaffected. This method is used for making a finite separation between relative conductors and non-conductors.

4.5.1.1 Electrostatic Separator

- Electrostatic separation depends on differences in the electrical properties (conductivity) of the materials to be treated.
- The electrostatic separator shown in Fig. 4.9 consists of a grounded rotor/rotating drum, a hopper for feeding the solids, an active electrode, situated/placed at a small distance from the drum and collecting bins.

- The solids to be separated are fed on to a rotating drum, either charged or grounded, from a hopper. The conductive particles in a very short time will assume the potential of the rotating drum, which is opposite to that of an active electrode and hence, they get attracted towards the active electrode. The non-conductive material is repelled by the electrode and attracted by the drum. The non-conductive material falls down straight under the influence of gravity and is collected in a separate bin.

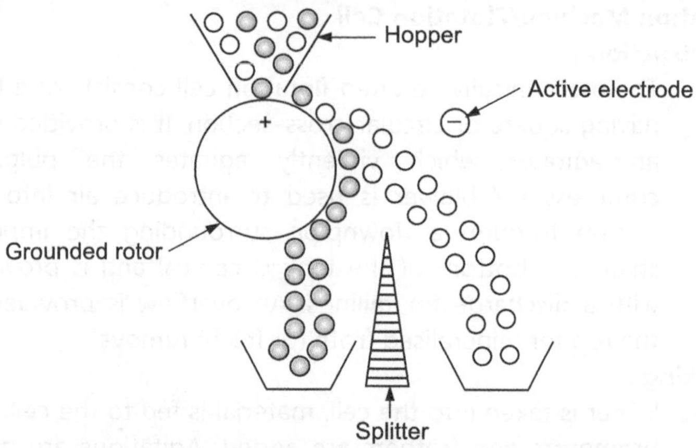

Fig. 4.9 : Electrostatic separator (separation by conductive induction)

4.5.2 Separation of Solid Particles Based on Magnetic Properties

- Magnetic separation is a method of separating solid particles by means of a magnetic field. In this method, materials having different magnetic attractability are separated by passing them through a magnetic field. The difference in magnetic properties of different materials is responsible for such a separation.
- Solids are classified as (i) diamagnetic solids – which when placed in a magnetic field are repelled by it and (ii) paramagnetic solids – which when placed in a magnetic field are attracted by the magnetic field. Therefore, when a mixture of above solids is subjected to/is passed through a magnetic field, magnetic solids are attracted towards it and non-magnetic solids are repelled and collected in separate bins.
- Magnetic separators are employed for tramp-iron removal (in this case they are called eliminators) and concentration (concentrators). (i) Magnetic pulleys and (ii) Magnetic drums.
- Tramp-iron magnetic separators are used for the removal of small quantities of magnetic material - tramp iron from the charge/feed to a size reduction machine (e.g. crusher or pulveriser) in order to protect the size reduction machine. Iron coarser than 1/8" (3.125 mm) is usually termed as tramp iron.

4.5.2.1 Magnetic Head Pulley

- A magnetic pulley/magnetised pulley is used for the removal of tramp-iron from the products handled on a belt conveyor.
- Magnetic pulleys (either electromagnetic or permanent-magnetic) having a diameter upto 1500 mm and a width upto 1500 mm are available. The belt speed ranges from 53 m/min for a pulley of diameter of 300 mm to 150 m/min for a pulley of 1500 mm diameter.
- A magnetic pulley is incorporated in a belt conveyor (carrying the charge/feed to a machine/equipment) at the discharge end. As the material is conveyed over this pulley, the magnetically inert material/non-magnetic material drops-off the belt (or is discharged from the belt) in a normal manner, whereas the magnetic material adheres to the belt and falls off from the underside where the belt loses contact with the pulley (i.e., when the belt leaves the magnetic field of the pulley).
- The material to be separated must be supplied in the form of a thin sheet/layer in order to subject all the particles to a magnetic field of the same intensity (power is applied to the magnetic pulley).

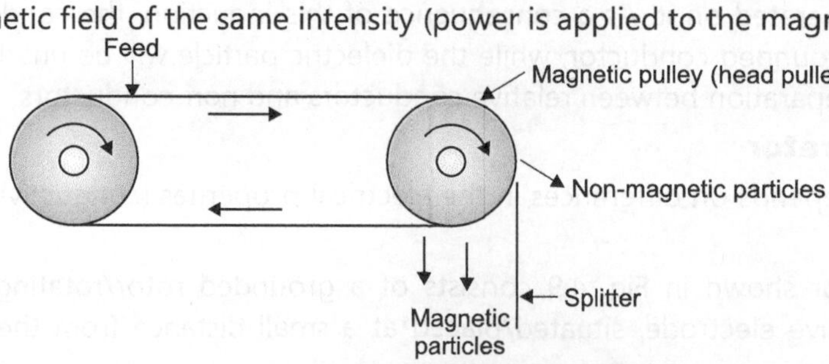

Fig. 4.10 : Magnetic head pulley

4.5.2.2 Magnetic Drum Separator

- A magnetic drum separator (Fig. 4.11) consists of a rotating drum incorporating stationary magnet assembly. The magnet arc covers approximately 165 degrees towards the discharge side of the drum. The feed is admitted at the top and is allowed to fall on the rotating drum. The non-magnetic material is discharged in the normal manner, while the magnetic material adheres to the drum and falls off the underside when the drum loses the contact of the magnet assembly.

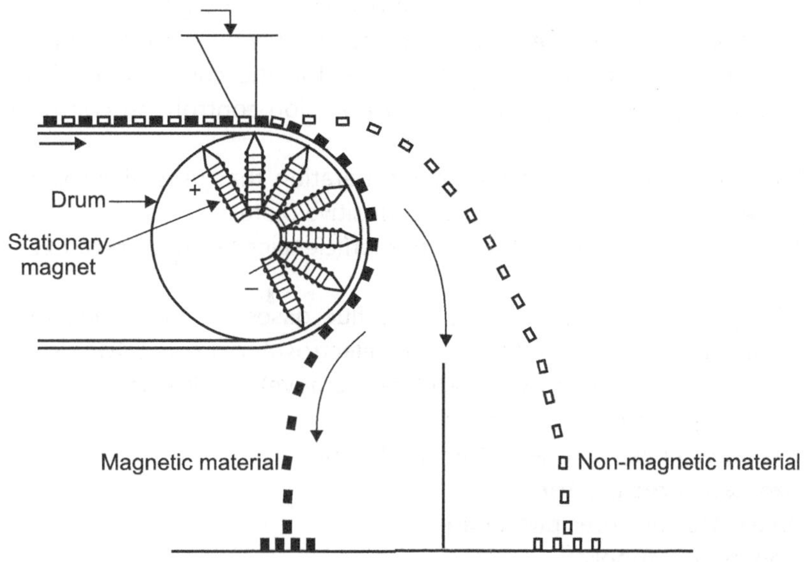

Fig. 4.11 : Magnetic drum separator

4.5.2.3 Ball-Norton Type Separator (Magnetic Separators as Concentrators)

- A typical concentrator used for separating magnetic ores from the associated mineral matter is the Ball-Norton machine. It consists of two horizontally staggered belt conveyors running parallel, one above the other as shown in Fig. 4.12. A hopper is provided for feeding the feed material to the lower belt and a stationary magnet assembly is incorporated in the upper belt conveyor near the discharge end.
- The material to be separated is fed to the lower belt in the form of a thin sheet and is conveyed under the second belt where it is subjected to a magnetic field. The non-magnetic material is discharged in the normal manner, whereas the magnetic material adheres to the lower side of the upper belt and thus carried some distance away from the discharge point of non-magnetic materials. It ultimately drops-off the belt in a separate compartment when it leaves the magnetic field, i.e., when the belt loses the contact of the magnet assembly. In this way the magnetic material is separated from the non-magnetic material.

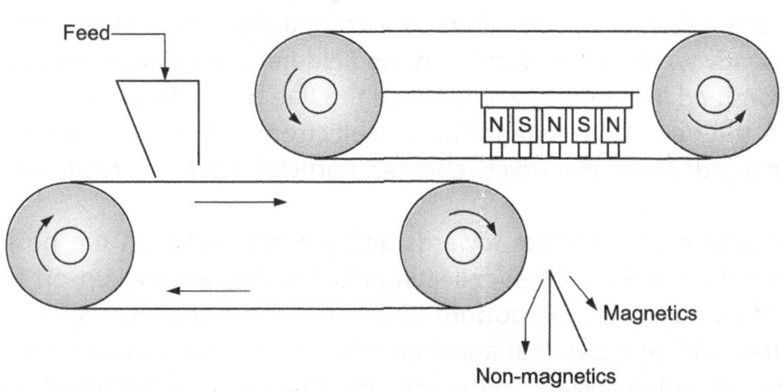

Fig. 4.12 : Ball-Norton magnetic separator

4.6 ELECTROSTATIC PRECIPITATOR

- An electrostatic precipitator is a device that uses static electricity/electrical energy for removing fine particles from a gas stream or air.
- An electrostatic precipitator is a device that removes dust and smoke particles from a flowing gas such as flue gases using the force of an induced electrostatic attraction (i.e., like charges repel, unlike charges attract each other).
- An electrostatic precipitator is a device to remove fine particles from a gas stream using an electrostatic force. The electrostatic force is the attractive or repulsive force between two electrically charged things.
- It is mainly used for the removal of the fine fly ash formed in the combustion of pulverised coal in thermal power plants/thermal power generating stations for air pollution control and controlling particulate emissions from industrial facilities.
- Electrostatic precipitation is a technique of solid gas separation that uses electrostatic forces and consists of discharge wires (negative plates) and collecting plates (positive plates).
- Electrostatic precipitators have relatively high collection efficiencies (around 99%) over a wide range of particle sizes (~ 0.05 - 5 μm).
- In order to remove dust and smoke particles from the flue gases produced due to the combustion of solid pulverised fuel in a furnace they are passed through the electrostatic precipitator installed in the way of the flue gases from the furnace to the chimney so that it will filter (remove) the flue gases before they enter the chimney from where they are discharged into the atmosphere.

  Principle : Like charges repel, unlike charges attract each other.

Construction of an Electrostatic Precipitator :

The components of an electrostatic precipitator are :

(i) 440 V, 50 Hz, 3-phase A.C. supply.
(ii) Transformer for step up or step down voltage.
(iii) Rectifier to convert A.C. supply to D.C. supply.
(iv) Discharge electrode (rod or wire mesh) that carries negatively charged voltage.
(v) Collecting electrode or collecting plate which carries positively charged high voltage.
(vi) Enclosure/casing with hopper and inlet/outlet gas flow connections.
(vii) Water spray arrangement.
(viii) Conveying system for discharged particles and mechanical vapours.

It has two sets of electrodes - one is positive and the other is negative. The negative electrodes are in the form of wire mesh or rod, whereas the positive electrodes are in the form of plates. The positive plates and negative electrodes are placed vertically in a precipitator casing alternatively one after the another.

The casing is provided with an inlet on one side for the flue gases and with an outlet on the opposite side for the clean gas. The casing is provided with an hopper at its bottom.

Working : The negative electrodes are connected to a negative terminal of a high voltage D.C. source, while positive plates are connected to the positive terminal of the D.C. source. In order to get stronger negativity in the negative electrodes, the positive terminal of the D.C source is grounded.

The medium between the electrodes is air. There is a corona discharge surrounding the negative electrodes owing to their very high negativity. The air molecules in the field between electrodes become ionized and so there will be plenty of free electrons and ions in the casing. When the flue gases stream enters into the unit, dust particles from the gases collide with the free electrons and the free electrons get attached to the dust particles and hence they become negatively charged. These negatively charged particles will be attracted due to electrostatic force of the positive plates.

Therefore, the charged dust particles move towards and get deposited on the positive plates. When enough particles are accumulated on the positive plates (collecting plates), they are shaken-off the collector by mechanical rappers. The particles fall into a hopper at the bottom of the unit and a conveyor system transports them away for disposal or recycling. The treated flue gases that are almost free from ash particles pass out of the precipitator and then through a stack to the atmosphere. Water spray from the top may be used for accelerating the removal of the dust from the collecting plates.

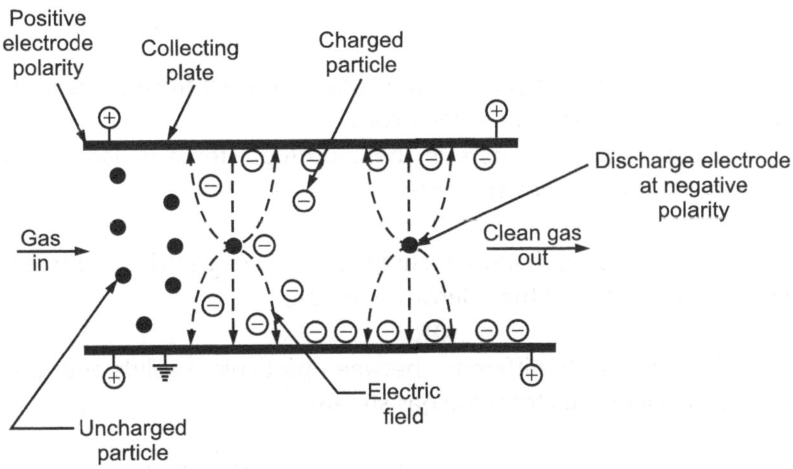

Fig. 4.13 : Electrostatic Precipitator (Top View)

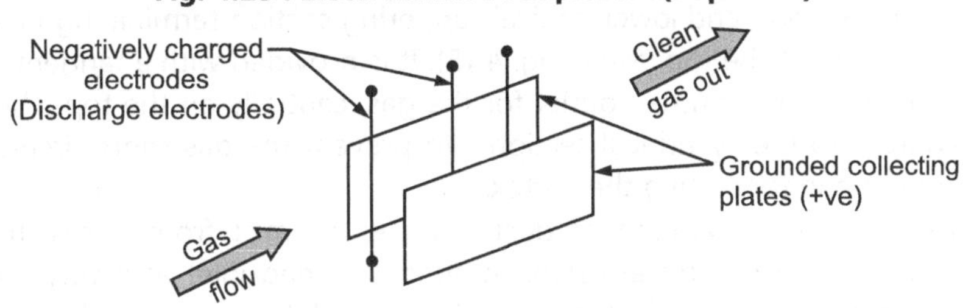

Fig. 4.14 : Electrostatic Precipitator

Typical applications of electrostatic precipitators are :

(i) Collection of fly ash from thermal power stations/plants/steam boilers.

(ii) Particle collection from furnace operations in metallurgical industries.

(iii) Particle collection from black liquor recovery furnaces in paper industries.

(iv) Cleaning of stack emissions in municipal incinerators.

(v) Particle collection from cement and gypsum manufacturing processes in cement industries.

It is commonly used device for air pollution control. These units are used in steel industries, thermal power plant, and chemical process industries.

Advantages :

- Used for collection of dry as well as wet pollutants.
- High collection efficiencies (high efficiency of removal of particles/pollutants).
- Low operating costs.
- High durability.
- Handles large gas volumes and heavy dust loads at low pressures.

Disadvantages :

- High capital costs.
- Requires large floor space.
- Non-flexible once installed.
- Not used to collect gas pollutants.

4.7 GAS - SOLID SEPARATION

The removal of particulate solids from a gas or air stream is of great industrial importance. The objectives of this separation operation are to control air pollution from various industrial plants, eliminate safety and health hazards from the workplace where in grinding, milling, etc. are carried out and recovery of valuable products, etc.

The solid-gas separation technique uses devices such as cyclone or cyclone separator, wet scrubber, electrostatic precipitators and fabric or bag filters. Cyclone separators are used for removal of large size particles and electrostatic precipitators are used to collect fine particles while the other two are used for intermediate sized particles.

4.7.1 Cyclone Separator

A cyclone separator is a device for removing particulate solids (particles) from a gas stream (usually flue gases or dirty air) by using centrifugal force (induced centrifugal force).

Cyclone separator is a centrifugal separator where in the centrifugal force is used for removal of particulate solids from gas streams (dirty or contaminated gas streams).

Principle:

(a) Centrifugation (Centrifugation is a mechanical technique that uses centrifugal force to separate particulate solids from a fluid stream according to their density and size).

OR

(b) The centrifugal force with the density difference between particulate solids and gas stream is responsible for the separation of particulate solids from the gas stream.

Construction and Working:

- A cyclone separator consists of a tappering cylindrical vessel, that is, it consists of a cylindrical vessel consisting of a top vertical section and lower conical/tappering section terminating in an apex opening. The vessel is closed by a flat plate at the top (Refer Fig. 4.15). It is provided with a tangential feed inlet nozzle in the cylindrical section near the top and an outlet for the gas, centrally on the top. The outlet provided is a pipe that extends inward into the cylindrical section - to prevent the gas short-circuiting directly from the feed inlet to the gas outlet and for cutting the vortex.
- In this separator, used for the separation of dust particles or mist from gases, the dust laden gas is introduced tangentially into the cylindrical vessel through the feed inlet at a very high velocity (30 m/s). Centrifugal force throws solid particles out against the wall of the vessel and they drop into a conical section of the cyclone and removed from the bottom/apex opening. The clean gas is taken out through the central outlet at the top.
- Usually cyclone separators are designed to remove particulate matter that is larger than or equal to 10 micrometers in size.

Uses:

- The cyclone separator is a device for separating solid particles from contaminated gas streams and thus used in industrial applications such as power generations, chemical industry, dairy industry, pharmaceutical industry. Usually cyclone separators are used in industries in pollution control applications at a first stage.
- These units are often used as a pre-treatment before the flue gases enter more effective pollution control devices.
- Their advantages and disadvantages include: Cyclone separators are not expensive to built, install and maintain and they have no moving parts. Due to this these units keep maintenance and operating cost low. They require very little space to install.
- They are not able to collect particulate matter/solids that is smaller than 10 micrometers in size. They can not handle tacky or sticky material. Their energy requirement is more.

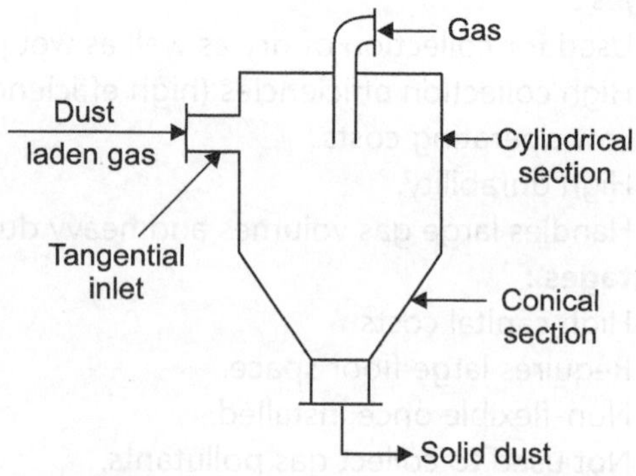

Fig. 4.15 : Cyclone separator/Cyclone

4.7.2 Fabric Filter

- A fabric filter is also known as a bag filter or bag house. The filter is usually in the form of cylindrical filter bags.
- A fabric filter is a device that removes particulate solids from a gas stream by passing the (contaminated) gas stream through a fabric (a filter medium) on which the particulated solids are deposited.
- A fabric filter is an air pollution control device that separates particulate matter from a dirty gas stream by passing the dirty gas through fabric filter elements.

Principle:

- The principle of fabric filter operation is filtration.
- A fabric filter utilizes fabric filtration of the removal of particulate matter from a gas stream.

Construction and Working:

- During the operation dust builds-up on the filter surface and to avoid excessive pressure drops on the gas side, the filter bags must be cleaned periodically. Fabric filters are classified into three types according to the method used to clean filter bags/filter material.
- The common methods of cleaning are shaking, reverse air and pulse jet and hence the types of fabric filters are:
  (i) Shaker fabric filter
  (ii) Reverse air fabric filter
  (iii) Pulse-jet fabric filter
- Gases contaminated with dust come in contact with filter bags inside a bag filter/fabric filter unit. Based on the type of fabric filter, dust comes in contact with filter bags (fabric filters in the form of bags) and collects either on the inside or outside of the bag.
- Here we will deal only with pulse-jet fabric filter.

4.7.2.1 Pulse-jet Fabric Filter

- Pulse-jet filters clean the filter bags using short duration pulses (or bursts) of compressed air.
- A pulse jet bag filter consists of a large array of filter bags for air/gas filtration and high velocity pulsed air is used in the reverse direction to the air or gas flow for cleaning of the filter bags.
- It consists of a cylindrical or rectangular casing provided with a hopper at the bottom. A large number of tubular bags of 12 to 40 cm diameter and 2 to 10 m long are incorporated in the casing. The bags are supported by metal or wire mesh cages and they hang in the casing from the holes in a tube sheet located at the top of the casing. The bags are closed at the bottom. The metal or wire mesh cage located inside the bags (or covered by bags) keep them from collapsing. It is provided with inlet/outlet connections for the gas streams, compressed air and dust discharge.
- In this filter, the contaminated gas stream enters through the side or through the hopper and flows into the bags and up through the tube sheet. The dust is collected on the external surface of the bags where it forms into a cake of dust. When the cake of an appropriate thickness is formed, a controller activates a sequential pulse of compressed air into the inside of the filter bags to one or more rows of the bags at a time.
- The dust cake dislodged by the compressed air pulse fall into the hopper and is removed from the filter.
- The pulse jet filters are commonly used as they do not contain moving parts, due to ease of operation, low energy requirement and minimal maintenance cost (advantages).

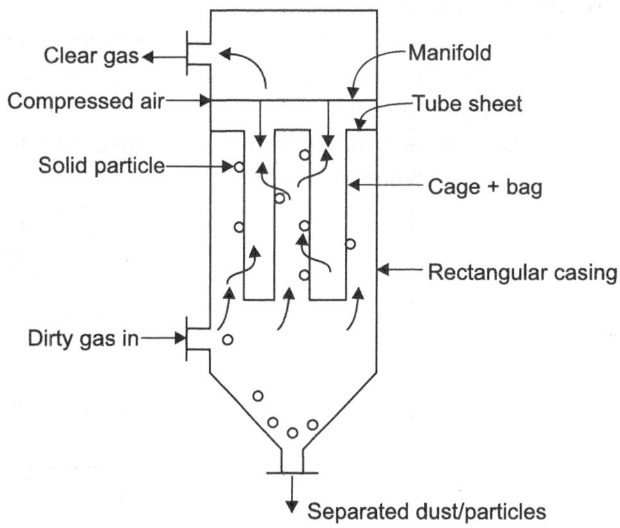

Fig. 4.16: Pulse-jet fabric filter

- They are commonly used in chemical and mineral processing plants, grain mills, asphalt plants, steel plants, food processing plants, etc. to maintain air quality and meet safety standards.

Application of fabric filters:

- Fabric filters are used in foundries, flour mills, metallurgical industries, cement industries, thermal power plants, food processing plants and chemical plants.

Filter Medium:

- Woven or felted materials are used to make bag filters. Woven filters are used with low energy cleaning methods such as shaking and reverse air and felted filters are used with a high energy cleaning method such as pulse-jet.
- While selecting the filter medium for bag houses, we have to consider the properties of the carrier gas and dust particles such as temperature, composition, flow rate, corrosive nature, etc. of the carrier gas and size, shape, proportion, etc. of the dust particles to be handled.
- The filter medium material used are: cotton, wool, nylon, polypropylene, dacron and fiber glass.

4.7.3 Wet Scrubbers

- In wet scrubbers, solid particles are removed from a gas stream by contacting the gas stream with a liquid and consequently transferring the particles into the liquid.
- The scrubbing liquid most commonly used is water.
- In a wet scrubber, dirty/contaminated gas stream is contacted with a scrubbing liquid and the particulate solids are collected in the scrubbing liquid.
- We scrubbers are most effective for removal of fine particles and mist, soluble particles, wet or sticky particulate matter, combustion dust, gases and vapours from contaminated gases.
- A wet scrubber is a pollution control device for removing particles and gases/vapours from industrial exhaust streams (e.g., flue gases, etc.) by bringing them in contact with a scrubbing liquid.
- These are used for reducing pollution in petroleum refineries, acid manufacturing plants, steel making plants, chemical process plants and large power plants.

4.7.3.1 Spray Tower Scrubber

- A spray scrubber/spray tower scrubber is a low energy wet scrubber used for particulate matter control.
- It is the simplest wet scrubber having a gas pressure drop of less than five inches of water and is capable of efficiently removing particulate solids/particles greater than about 5-10 micrometers (microns) in size.

Principle:

- Particulate solids are collected by impingement/impactation. The process of impingement is making solid particles impinge and adhere to water droplets (adsorb onto water droplets).

Construction and Working:

- A spray scrubber consists of an open vessel containing one or more sets of spray nozzles to distribute the scrubbing liquid. It is provided with a mist eliminator at the top to trap the entrained liquid droplets in the rising gas stream.
- It is provided with inlet/outlet connections for the dirty gas stream and the clean gas. Fig. 4.17 shows a counter current arrangement.

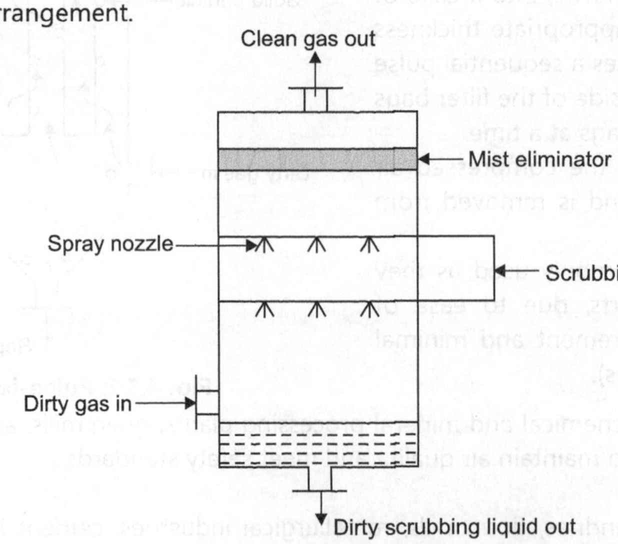

Fig. 4.17: Spray tower scrubber

- The contaminated gas stream enters at the bottom and passes upward through the liquid sprays that flow downward. The particles are collected when they impact the liquid droplets.
- The clean gas is removed from the top and the dirty scrubbing liquid is removed from the bottom.

4.7.3.2 Venturi Scrubbers

- A venturi scrubber is a wet scrubber that removes particulate matter from industrial exhaust streams by using a venturi.

Principle:

- It utilizes the energy from a high velocity gas stream to atomize a scrubbing liquid which then captures particulate solids from the gas stream due to impactation (solids will adhere to the liquid droplets).

Construction and Working:

- A venturi scrubber consists of a converging section (of a narrow length), a throat (the narrowest part of the venturi) and a diverging section (of a longer length). Spray nozzles are located in the throat or just at the upstream of the throat. The end of the diverging section is connected to a gas-liquid separator by an elbow. The gas-liquid separator is a cylindrical vessel provided with a mist eliminator, clean gas outlet and slurry discharge/drain.

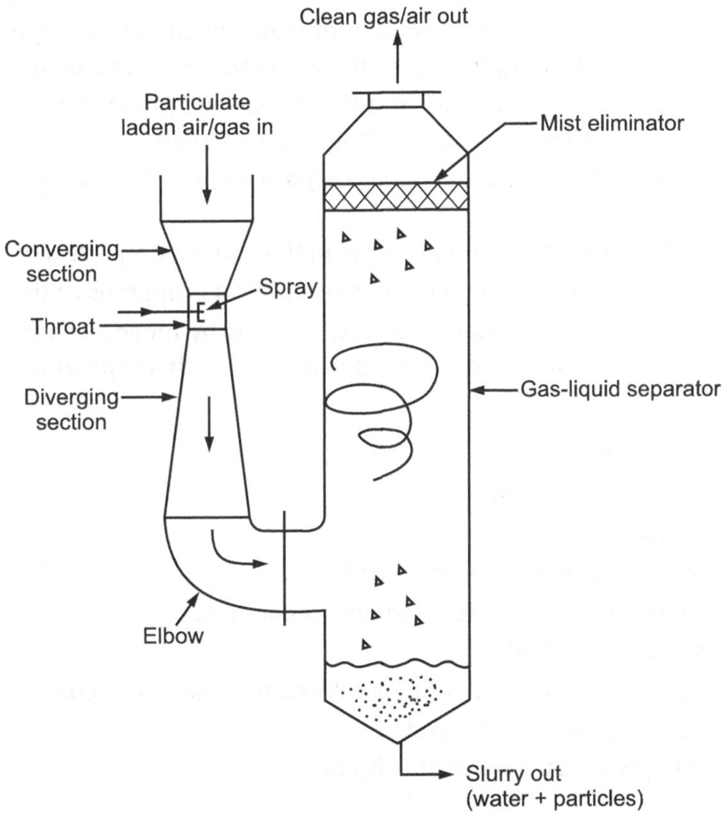

Fig. 4.18: Venturi scrubber

The dust laden gas stream is introduced from the upstream of the converging section and the scrubbing liquid is introduced through nozzles just at the upstream of the throat. The gas stream reaches a very high velocity at the throat section. The kinetic energy imparted by the gas stream results in shearing of the scrubbing liquid into tiny liquid droplets. Particulate solids get adhered/adsorbed on the surface of the droplets. The gas stream then enters the diverging section where it slows down to about its original velocity. Because of this, the droplets coalesce and thus become heavy to be carried out along the gas stream and consequently they settle out of the gas stream under the influence of gravity in a gas-liquid separator and finally drained from it. The clean gas leave the separator from the top via a mist eliminator.

- Venturi scrubbers are relatively inexpensive in terms of both initial cost and costs of operation but are relatively low-efficiency treatment devices.
- Venturi scrubbers are used to remove small particles (< 1 μm) from gas streams. However, these devices can also be used for larger particles.

Applications :

- Venturi scrubbers are used to control particulate matter control from industrial and commercial boilers fired with coal, oil or liquid waste. They are also used to control emission sources in the chemical, pulp and paper, iron and steel, mineral processing, food processing, etc.

Important Points

- *Classification is a method of separating solid particles into fractions based upon/according to their terminal falling/settling velocities.*
- Classification devices include gravity settling tank, cone classifier, rake classifier, double cone classifier and spiral classifier.
- A cyclone/cyclone separator is essentially a settling chamber in which the gravitational separating force is replaced by a much stronger centrifugal separating force (to increase the settling rate).
- A **jig** is a mechanical device used for the separation of materials of different specific gravities by pulsating a stream of liquid (usually water) flowing through a bed of materials resting on a screen.
- Jigging is a method of separating materials of different specific gravities by the pulsation of a stream of liquid (water) flowing through a bed of materials resting on a screen.
- Flotation refers to *an operation in which one solid is separated from another by floating one of them at or on the liquid surface.*
- Separation by froth flotation depends on differences in the surface properties of the materials treated.
- Electrostatic separation depends on differences in the electrical properties of the materials treated.
- Magnetic separation is a method of separating solid particles by means of a magnetic field. The difference in magnetic properties of different materials is responsible for such a separation.

Practice Questions

1. Define magnetic separation.
2. Define classification/wet classification.
3. Define electrostatic separation.
4. Draw a neat sketch of jig and explain its construction.
5. Draw a neat sketch of Drum separator for magnetic separation.
6. State the principle of froth flotation.
7. Explain in brief the construction and working of Ball-Norton magnetic separator.
8. Explain in brief rake classifier with a neat sketch.
9. Draw a neat sketch of cyclone used for dust collection.

☆☆☆

CHAPTER FIVE

FILTRATION AND SEDIMENTATION

5.1 CONCEPT OF FILTRATION

- *The separation of solids from a suspension in a liquid with the help of a porous medium or screen which retains the solids and allows the liquid to pass* is termed as **filtration**.
- The operation of separating a solid from a liquid by means of a porous medium (usually a wire or fabric filter cloth) is called as filtration. The medium retains the solid in the form of a porous cake, while the liquid passes through it.
- The mechanical separation of a solid from suspension in a liquid by passage through a porous medium which retains the solid and allows the liquid to pass is called as filtration.
- In filtration operation, the volume of the suspensions to be handled may vary from extremely large quantities (as in water purification) to relatively small quantities (as in the fine chemical industry), the suspensions may contain small or large proportions of solids and the valuable product may be the solid, the liquid, or both or sometimes none of them (e.g., the waste solids to be separated from a waste liquid prior to the disposal). The driving force required for a separation by filtration based upon the nature of a suspension may be divided into four categories, namely gravity, vacuum, pressure and centrifugal.
- Separation of suspended impurities from water (in water purification), separation of solid organic and inorganic materials from their slurry such as calcium carbonate, ammonium sulphate, sugar, paranitroaniline, etc. are some examples of filtration.

5.2 TYPES OF FILTRATION : CAKE FILTRATION AND DEEP BED FILTRATION

- Basically, there are two types of filtration : (i) Cake filtration, (ii) Deep bed filtration.
- In cake filtration, the proportion of solids in the suspension is large and most of the solid particles are collected in the cake which can later be detached from the filter medium.
- In deep bed filtration, the proportion of solids is very small and the particles of the solid being smaller than the pores of the filter medium will penetrate a considerable depth and ultimately get trapped inside the filter medium and usually no layer of solids will appear on the surface of the medium (e.g., water filtration).
- Thus, filters are divided into two main groups, **clarifying filters** and **cake filters**. Clarifying filters also called as deep bed filters are used to remove small amounts of solids to produce sparkling clear liquids, whereas cake filters separate large amounts of solids in the form of a cake of crystals. Clarifying filters find applications in water treatment. In this chapter, we will deal with cake filtration.

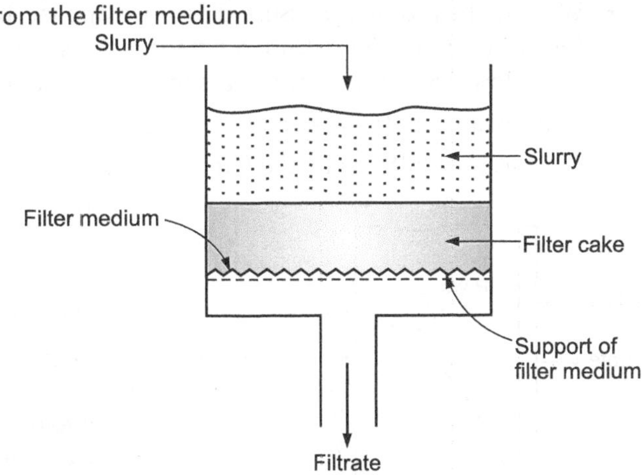

Fig. 5.1 : Principle of filtration

Principles of Cake Filtration :

- In cake filtration, the feed to be handled (two phase mixture) is called **slurry**, the bed of deposited solids on a porous membrane (filter medium) is called **cake** and the clear liquid leaving the filter medium is called **filtrate.**

- A typical cake filtration operation is shown in Fig. 5.1. During the initial period of flow, solid particles are trapped within the pores of a medium forming a true filter medium. The liquid passes through the bed of the solids and through the filter medium. In the early stage of filtration, the rate of filtration is high. As the cake thickness increases, the rate of filtration decreases for a given pressure differential across the filter medium. This is due to the fact that as the cake gradually builds upon the medium, the resistance to flow progressively increases.

5.3 CONSTANT RATE AND CONSTANT PRESSURE FILTRATION

Types of filtration (based on the pressure drop across a filtering medium) :

There are two types of filtration : 1. Constant pressure filtration, 2. Constant rate filtration.

- *The method in which the pressure drop over the filter is held constant throughout a run so that the rate of filtration is maximum at the start of filtration and decreases continuously towards the end of the run* is called **constant pressure filtration.** If the outlet pressure is constant, constant pressure filtration is carried out by applying a certain pressure at inlet and maintaining it constant throughout the run.
- *The method in which the pressure drop is varied usually from a minimum at the start of filtration to a maximum at the end of filtration so that the rate of filtration is constant throughout the run* is called **constant rate filtration.**
- In constant rate filtration, nearly constant rate of filtration is maintained by starting at low inlet pressure, and continuously increasing the pressure to overcome the resistance of cake, until the maximum pressure is reached towards the end of the run.
- In constant pressure filtration, application of high initial pressure results in a low rate of filtration as the first particles filtered will be compacted into a tight mass that largely fills the pores of cloth. In constant rate filtration, as the maximum pressure is reached towards the end of the run whole cycle is operated at less than the maximum capacity. To overcome the difficulties faced in the filtration types cited above, a practical solution is found out by carrying out the filtration at constant rate until the inlet pressure reaches a specified maximum and then to continue at constant pressure until the end of the run.

Distribution of overall pressure drop

- With the help of the pressure difference applied between the slurry inlet and the filtrate outlet, the filtrate is forced through a filter. During filtration, the solids are retained in the form of cake through which the filtrate must flow. The filtrate has to pass through three resistances in series –

  1. Resistance of the feed and filtrate channel, 2. Resistance of the cake, and
  3. Resistance offered by the filter medium.

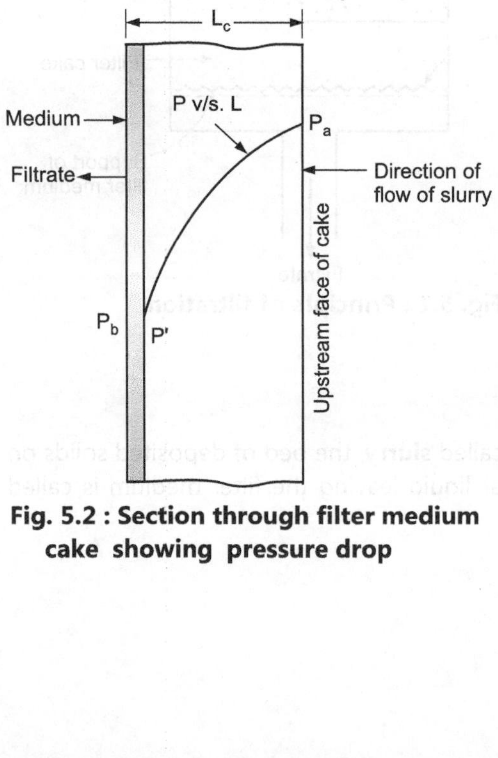

Fig. 5.2 : Section through filter medium cake showing pressure drop

- The overall or total pressure drop over the filter at any time is equal to the sum of the individual pressure drops over the medium and cake. Usually, the resistances offered by the inlet and outlet connections are small as compared with those offered by the cake and medium and thus, can be neglected.
- By the resistance of filter medium or filter medium resistance, we mean it is the entire resistance built up in the filter medium, including that from the trapped particles. Filter medium resistance is important in the early stages of filtration. The cake resistance is the one which is offered by all solids not associated with the filter medium. It is zero at the start of filtration and goes on increasing with time of filtration.

- If the resistance of the inlet and outlet channels is neglected then the overall pressure drop is the sum of pressure drops over the medium and cake.

$$\Delta P = P_a - P_b = (P_a - P') + (P' - P_b) = \Delta P_c + \Delta P_m \quad \text{... (5.1)}$$

where
P_a = inlet pressure
P_b = outlet pressure
P' = pressure at the interface between cake and medium
ΔP = overall pressure drop
ΔP_c = pressure drop over cake
ΔP_m = pressure drop over medium

Specific cake resistance :

- A specific cake resistance α can be defined by the equation

$$\alpha = \frac{\Delta P_c A}{\mu u m_c} \quad \text{... (5.2)}$$

where ΔP_c is the pressure drop over the cake, A is the filter area measured perpendicular to the direction of flow, u is the linear velocity of the filtrate based on the filter area, μ is the viscosity of the filtrate, and m_c is the total mass of solids in the cake. In the SI system, the units of α are m/kg and has dimensions of $M^{-1} L^1$. [ΔP_c in N/m^2, A in m^2, μ in $(N.s)/m^2$ and u in m/s]

Filter medium resistance :

A filter medium resistance can be defined by the equation

$$\frac{\Delta P_m}{R_m} = \mu u \quad \text{... (5.3)}$$

$$\therefore \quad R_m = \Delta P_m/\mu u \quad \text{... (5.4)}$$

where ΔP_m is the pressure drop over filter medium. In the SI system, R_m has the units of m^{-1}. The dimension of R_m is L^{-1}.

Rearranging Equation (5.2), we get

$$\Delta P_c = \mu u m_c \alpha/A \quad \text{... (5.5)}$$

Rearranging Equation (5.4), we get

$$\Delta P_m = \mu u R_m \quad \text{... (5.6)}$$

Substituting the values of ΔP_c from Equation (5.5) and ΔP_m from Equation (5.6) into Equation (5.1), we get

$$\Delta P = \Delta P_c + \Delta P_m = \frac{\mu u m_c \alpha}{A} + \mu u R_m$$

$$\therefore \quad \Delta P = \mu u \left[\frac{m_c \alpha}{A} + R\right] \quad \text{... (5.7)}$$

- In using Equation (5.7), it is convenient to replace u, the linear velocity of the filtrate, and m_c, the total mass of the solid in the cake, by functions of V, the total volume of filtrate collected in time t.
- If c is the mass of particles deposited in the filter per unit volume of filtrate, then the mass of solids in the filter at time t is given by

$$m_c = cV \quad \text{... (5.8)}$$

- The linear velocity of the filtrate is given by the equation

$$u = \frac{dV/dt}{A} \quad \text{... (5.9)}$$

where V is the volume of filtrate collected from the start of filtration to time t and A is the filter area normal to the direction of flow of filtrate.

- Substituting u from Equation (5.9) and m_c from Equation (5.8) in Equation (5.7), we get

$$\Delta P = \mu \frac{dV/dt}{A}\left[\frac{c\,V\,\alpha}{A} + R_m\right]$$

$$\frac{dt}{dV} = \frac{\mu}{A\,\Delta P}\left[\frac{c\,\alpha\,V}{A} + R_m\right] \qquad \text{... (5.10)}$$

or

$$\frac{dV}{A\,dt} = \frac{\Delta P}{\mu\left[\frac{c\,V\,\alpha}{A} + R_m\right]} \qquad \text{... (5.11)}$$

- Equation (5.11) expresses the differential or instantaneous rate of filtration per unit area of the filtering surface as the ratio of the pressure drop (driving force) to the product of the viscosity of the filtrate and the sum of the cake resistance and the filter medium resistance.

Constant Pressure filtration

When ΔP is constant, V and t are the only variables in Equation (5.10).

When t = 0, V = 0, and $\Delta P = \Delta P_m$. Therefore, Equation (5.10) becomes

$$\frac{\mu R_m}{A\,\Delta P} = \left(\frac{dt}{dV}\right)_o = \frac{1}{q_o} \qquad \text{... (5.12)}$$

Thus, Equation (5.10) may also be rewritten as

$$\frac{dt}{dV} = \frac{1}{q} = K_c\,V + \frac{1}{q_o} \qquad \text{... (5.13)}$$

$$\frac{dt}{dV} = \frac{\mu\,\alpha\,c\,V}{A^2\,\Delta P} + \frac{\mu\,R_m}{A\,\Delta P}$$

$$\frac{dt}{dV} = K_c\,V + \frac{1}{q_o} \qquad \text{... (5.14)}$$

where

$$K_c = \frac{\mu\,\alpha\,c}{A^2\,\Delta P} \qquad \text{... (5.15)}$$

Integrating Equation (5.14) between the limits : t = 0, V = 0, and t = t, V = V

$$\int_0^t dt = \int_0^V \left(K_c V + \frac{1}{q_o}\right) dV$$

$$t = \frac{K_c\,V^2}{2} + \frac{V}{q_o} \qquad \text{... (5.16)}$$

Substituting values of K_c and $1/q_o$ from Equations (5.15) and (5.12) in Equation (5.16), we get

$$t = \frac{\mu\,\alpha\,c}{2\,A^2\,\Delta P} V^2 + \frac{\mu\,R_m\,V}{A\,\Delta P}$$

$$t = \frac{\mu}{\Delta P}\left[\frac{c\,\alpha}{2}\left(\frac{V}{A}\right)^2 + R_m\left(\frac{V}{A}\right)\right] \qquad \text{... (5.17)}$$

Rearranging Equation (5.16), we get

$$\frac{t}{V} = \frac{K_c}{2}\,V + \frac{1}{q_o} \qquad \text{... (5.18)}$$

Hence, a plot of t/V v/s V is a straight line with a slope equal to $K_c / 2$ and an intercept equal to $1/q_o$. From this plot and Equations (5.12) and (5.15), the values of α and R_m can be calculated.

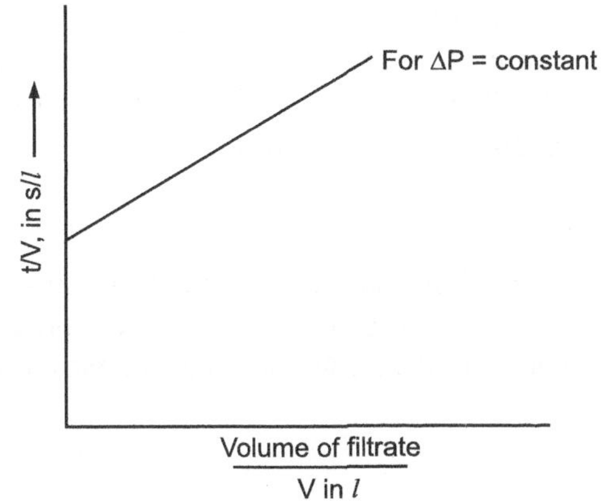

Fig. 5.3 : Plot of t/V v/s V

Empirical equations for cake resistance

- The variation of α with ΔP can be found out by conducting constant-pressure experiments at various values of ΔP.
- The specific cake resistance can be related to the pressure drop by the following empirical equation

$$\alpha = \alpha_o (\Delta P)^s \qquad \text{... (5.19)}$$

where α_o and s are empirical constants. The constant 's' is known as the compressibility coefficient of the cake. It is zero for incompressible sludges [i.e., α is independent of (ΔP) – the sludge is incompressible] and positive for compressible sludges [i.e., α increases with (ΔP) – the sludge is compressible]. For commercial slurries, the value of 's' usually lies between 0.2 and 0.8.

- For obtaining the value of s, we have to plot α as a function of (ΔP) on the logarithmic co-ordinates. The slope of the line obtained gives the value of 's' and intercept gives the value of α_o.
- In Equation (5.17), α is to be replaced by its value given by Equation (5.19) and then is directly applicable for use in the design of batch filters (constant pressure). The constants α_o, s, R_m must be evaluated experimentally, and the general equation can then be applied to conditions of varying A, ΔP, V, t, c and μ.
- If m_F is the mass of the wet cake and m_c is the mass of the dry cake after washing and drying, ρ is the density of the filtrate and c_s is the concentration of solids in the slurry in kg per m^3 of the liquid fed to the filter, then the mass of the particles that are deposited in the filter per unit volume of filtrate is given by

$$c = \frac{c_s}{[1 - (m_F / m_c - 1)]\, c_s / \rho} \qquad \text{... (5.20)}$$

where ρ in kg/m^3, c_s is in kg/m^3.

Constant-rate filtration :

- When filtrate flows at a constant rate, the linear velocity u is constant.

$$\therefore \qquad u = \frac{dV/dt}{A} = \frac{V}{A\,t} \qquad \text{... (5.21)}$$

We have :

$$\alpha = \frac{\Delta P_c\, A}{\mu\, u\, m_c} \qquad \text{... (5.22)}$$

and

$$m_c = cV \qquad \text{... (5.23)}$$

- Substituting u from Equation (5.21) and m_c from Equation (5.23) into Equation (5.22) and rearranging, we get

$$\frac{\Delta P_c}{\alpha} = \frac{\mu c}{t} \left(\frac{V}{A}\right)^2 \qquad \text{... (5.24)}$$

- If α is known as a function of ΔP_c, and if ΔP_m is estimated, then Equation (5.24) can be used for relating the overall pressure drop to time when the rate of filtration is constant. Equation (5.24) can be used directly if Equation (5.19) is used for relating α with ΔP_c. If α from Equation (5.19) is substituted in Equation (5.24) and if ΔP_c is substituted by $\Delta P - \Delta P_m$, then Equation (5.24) becomes

$$(\Delta P_c)^{1-s} = \alpha_o\, \mu\, c\, t \left(\frac{V}{A\,t}\right)^2 = (\Delta P - \Delta P_m)^{1-s} \quad \text{... (5.25)}$$

- The simplest method of correcting the overall pressure drop for the pressure drop through the medium is to assume the filter medium resistance to be constant during a constant rate filtration. Then by Equation (5.3), ΔP_m is also constant in Equation (5.25). As the only variables in Equation (5.25) are ΔP and t, Equation (5.25) can be written as

$$(\Delta P - \Delta P_m)^{1-s} = K_r\, t \quad \text{... (5.26)}$$

where

$$K_r = \mu\, u^2\, c\, \alpha_o \quad \text{... (5.27)}$$

Continuous Filtration

- In a continuous filter such as rotary drum filter, the feed, filtrate and cake move at steady constant rates. But for any particular element of the filter surface, conditions are unsteady. The process of filtration consists of several steps in series such as cake formation, washing, drying and discharging/scrapping and each of these steps involve gradual and continual change in conditions but the pressure drop across the filter during cake formation is held constant. Therefore, the equations obtained for discontinuous constant-pressure filtration may be applied to continuous filters with some changes.
- Let t be the actual filtering time (the time for which the filter element is immersed in the slurry). With this Equation (5.18) can be rewritten as

$$t = \frac{K_c V^2}{2} + \frac{1}{q_o} \quad \text{... (5.28)}$$

where V is the volume of filtrate collected in time t.

Equation (5.28) is a a quadratic equation in V which on solving yields

$$V = \frac{(1/q_o^2 + 2\, K_c t)^{1/2} - 1/q_o}{K_c} \quad \text{... (5.29)}$$

Substituting the values of q_o and K_c, we get

$$V = \frac{[(\mu R_m/A\Delta P)^2 + (2\,\mu\, c\, \alpha/A^2\, \Delta P)t]^{1/2} - \mu\, R_m/A\, \Delta P}{\mu\, c\, \alpha/A^2\, \Delta P}$$

Taking $\mu/A\, \Delta P$ common, we get

$$V = \frac{A\,[R_m^2 + 2\, \Delta P\, c\, \alpha . t/\mu]^{1/2} - R_m}{c\, \alpha}$$

$$\therefore \quad \frac{V}{A} = \frac{[R_m^2 + 2\, \Delta P\, c\, \alpha\, t/\mu]^{1/2} - R_m}{c\, \alpha}$$

Dividing both sides by t, we get

$$\frac{V}{At} = \frac{[(R_m/t)^2 + 2\, \Delta P\, c\, \alpha/\mu\, t]^{1/2} - R_m/t}{c\, \alpha} \quad \text{... (5.30)}$$

where V/t is the rate of filtrate collection and A is the submerged area of the filter.

If $\dot{m}_c$ is the rate of solids production, t_c is the cycle time, A_T is the total filter area, n is the drum speed and f is the fraction of the drum submerged, then

$$t = f\, t_c = f/n \quad \text{... (5.31)}$$

The rate of solids production using Equation (5.8) becomes

$$\dot{m}_c = \frac{m_c}{t} = \frac{cV}{t} = c\left(\frac{V}{t}\right) \quad \text{...(5.32)}$$

As A/A_T = f, the rate of solids production divided by the total area of the filter using Equations (5.30), (5.31) and (5.32) thus becomes

$$\frac{\dot{m}_c}{A_T} = \frac{[2c\,\alpha\,\Delta P\,fn/\mu + (n\,R_m)^2]^{1/2} - n\,R_m}{\alpha} \quad \text{... (5.33)}$$

The filter-medium resistance R_m includes the resistance offered by the cake which is not removed by the discharge mechanism provided and carried through the next cycle. If the filter medium is washed after the cake discharge, R_m is usually negligible. Therefore, neglecting R_m, Equation (5.33) becomes

$$\frac{\dot{m}_c}{A_T} = \left(\frac{2c\,\Delta P\,fn}{\alpha\,\mu}\right)^{1/2} \quad \text{... (5.34)}$$

If the specific cake resistance (α) varies with overall pressure drop as per Equation (5.19), then Equation (5.34) becomes

$$\frac{\dot{m}_c}{A_T} = \left(\frac{2\,c\,(\Delta P)^{1-s}\,f\,n}{\alpha_o\,\mu}\right)^{1/2} \quad \text{... (5.35)}$$

- Equations (5.33) and (5.34) are applicable to continuous vacuum filters as well as to continuous pressure filters.
- When R_m is negligible, Equation (5.34) predicts that the rate of filtrate flow varies inversely with the square root of the viscosity and of the cycle time. This is true in practice for thick cakes and long cycle times. But for short cycle times, it is not true and Equation (5.33) must be used.

5.4 FILTER MEDIUM (CHARACTERISTICS OF FILTER MEDIUM)

- In case of cake filtration, the choice of a filter medium is often the most important consideration in assuring satisfactory operation of a filter. The filter medium in any filter must meet the following requirements :
  1. It should retain the solids to be filtered, giving a reasonably clear filtrate.
  2. It should not plug or blind (low rate of entrapment of solids within its interstices).
  3. It should be mechanically strong to withstand the process conditions.
  4. It should be resistant to the corrosive action of fluid.
  5. It should offer as little resistance as possible to the flow of filtrate.
  6. It should possess ability to discharge cake easily and cleanly.
  7. It should have acceptable resistance to mechanical wear.
  8. It should be cheap.
  9. It should have long life.
- In cake filtration, the filter medium is frequently a textile fabric.
- Canvas cloth, woolen cloth, metal cloth of monel or stainless steel, glass cloth and synthetic fibre cloth - nylon, polypropylene, etc., are commonly used as filter media in industrial filtration practice depending upon the process conditions.
- For an alkaline slurry, nylon cloths are used while for an acidic slurry, polypropylene cloths are used as a filter medium.

5.5 FILTER AIDS

- Filtration of slurries containing very finely divided solids or slimy, deformable flocs is very difficult due to formation of a dense, impermeable cake that quickly plug the filter media. In such cases the porosity of the cake must be increased to allow passage of the filtrate at a reasonable rate. This is achieved by adding a filter aid to the slurry before filtration.

- A filter aid is a granular or fibrous material which packs to form a bed of very high voidage. Because of this, they are capable of increasing the porosity of the filter cake. A filter aid should be of *low bulk density*, should be *porous*, should be *capable of forming a porous cake*, and must be *chemically inert to the filtrate*.
- The commercial filter aids are diatomaceous earth - almost pure silica prepared from deposits of diatom (marine organisms) skeletons, expanded pearlite, and asbestos fibres. The filter aids are used for sludges that are difficult to filter and the use of filter aids is normally restricted to filtration technique in which the filtrate is valuable and the cake is the waste product.
- Methods of using filter aids :

  (i) adding a filter aid to the slurry before filtration, and

  (ii) precoating, i.e., by depositing a layer of a filter aid on the filter medium before filtration.
- Precoats prevent gelatinous solids from plugging the filter medium and give a clear filtrate. The precoat is a part of the medium rather than that of the cake. When the filter aid is directly added to the slurry before filtration, the presence of it increases the porosity of the sludge, decreases its compressibility and reduces the resistance of cake during the filtration operation.

5.6 FACTORS AFFECTING THE RATE OF FILTRATION

The rate at which the filtrate is obtained in a filtration operation, i.e., the rate of filtration depends upon the following factors :

1. Pressure drop across the feed inlet and far side of the filter medium.
2. Area of the filtering surface.
3. Viscosity of the filtrate.
4. Resistance of the filter medium and initial layers of cake.
5. Resistance of the filter cake.

- The rate of filtration is directly proportional to the pressure difference across a filter medium. Therefore, higher the pressure difference across a filter medium, higher will be the rate of filtration.
- The rate of filtration is directly proportional to the square of the area of a filtering surface. Therefore, higher the area of a filtering surface, higher will be the rate of filtration.
- The rate of filtration is inversely proportional to the viscosity of the filtrate. Therefore, higher the viscosity of the filtrate, lower will be the rate of filtration.
- The rate of filtration is inversely proportional to the resistance of a cake or filter medium. Therefore, higher the resistance of a cake or filter medium, lower will be the rate of filtration.

5.7 TYPES OF FILTRATION EQUIPMENTS

(a) Filters are generally divided into two major groups based on the function or goal of filtration (i.e., based on whether to produce a cake or sparkling liquid).

(i) Cake filters. (ii) Clarifying filters.

- *Filters that retain appreciable quantities of filtered solids on the surface of the filter medium* are referred to as cake filters.
- *Filters that remove small amounts of solids to produce sparkling clear liquids* are referred to as clarifying filters or deep bed filters. These filters are commonly employed in water treatment.

(b) Filters may be classified according to the method of operation or operating cycle as

(i) Batch filters, (ii) Continuous filters

(c) Filters may be classified based on the driving force used for separation, e.g., gravity, pressure, vacuum or centrifugal.

- In filtration operation, the filtrate is forced to flow through a filter medium by virtue of a pressure difference across the medium. The pressure difference may be created by gravity, superatmospheric pressure on the upstream of the filter medium, sub-atmospheric pressure on the downstream of the filter medium or centrifugal force across the medium. Therefore, filters may be classified as

  (i) gravity filters, (ii) pressure filters, (iii) vacuum filters, (iv) centrifugal filters.

Industrial cake filters are usually classified as follows :

1. Batch (discontinuous) pressure filters

   e.g., filter press - plate and frame press, pressure leaf filters.

2. Continuous pressure filters

   e.g., pressure filter-thickener, continuous rotary pressure filters.

3. Batch vacuum filters

   e.g., vacuum nutsches, vacuum leaf filters.

4. Continuous vacuum filters

   e.g., rotary drum filters, vacuum precoat filters.

5. Centrifugal filters (batch and continuous)

   e.g., suspended basket centrifuge - top driven or bottom driven, continuous filtering centrifugals.

- The most important **cake filters** which will be referred to are : plate and frame filter press, rotary drum filter, and basket centrifuge.
- In many cases, in the chemical industry, it is the solids that are wanted.
- The factors to be considered while selecting equipment for filtration and operating conditions are :
  1. Properties of the fluid such as viscosity, density and corrosiveness/chemical reactivity.
  2. Nature of the solid which includes particle size, size distribution, particle shape and packing characteristics of solid particles.
  3. Concentration of solids in slurry, i.e., feed slurry concentration.
  4. Quantity of slurry to be handled and its value.
  5. Valuable product of operation.
  6. Necessity of washing the solids.
  7. Initial investment.
  8. Necessity of pretreatment of the slurry for ease in filtration.
  9. Cost of labour and power.

5.7.1 Primary Filter - Sand Filters

- Sand filters (clarifying filters) are used for water treatment and water purification. The medium of this filter is sand of varying grades. When we have to remove taste and odour, the sand filter may include a layer of activated carbon. There are several kinds of sand filters : rapid (gravity) sand filters, slow sand filters, pressure sand filters and upflow sand filters.

5.7.1.1 Rapid Sand Filter

- It is a gravity filter and is widely used filter in the treatment of water. It consists of an open water tight tank 3 to 3.5 m deep, containing a filter bed, a layer of coarse sand 0.6 to 0.75 m thick. The size of sand particles ranges from 0.4 to 1 mm. The sand bed is supported by a layer of graded gravel (of the size range of 1 to 50 mm) 0.45 m thick. Below the gravel there is an under drainage system consisting of a central longitudinal conduit or manifold with strainers mounted on the top and pipes of small diameter called laternals that carry perforations on the sides and bottom. In the operation, water to be filtered is introduced from the top, it passes downward through the filter bed. During its flow the suspended impurities get trapped in the bed and almost clear water leaves the filter from the bottom. The filter bed is periodically cleaned by backwashing. During backwashing with water, the upward flow carries the deposited floc with it. The essential characteristics of rapid sand filters are :

  (i) rate of filtration is high

  (ii) cleaning is done through backwashing and

  (iii) careful pre-treatment of water is necessary.

5.7.1.2 Pressure Sand Filters

- Pressure sand filters are essentially same as rapid sand filters, except that the water is filtered through the filter bed under a suitable pressure and the filter medium is contained in a steel tank. These filters are commonly used for the treatment of boiler feed water. The water, instead of gravity fed, is pumped through the bed under pressure. Such units are built as vertical or horizontal units. The former being used for a relatively small amount of water and the latter for greater volumes. Pressure sand filtration is often carried after coagulation and sedimentation and if not, the coagulants are introduced to the filtered water pipe ahead of the filters. These filters are operated with a feed pressure of 2 to 5 bar.
- Vertical pressure filters range in diameter from 0.5 to 2.5 m and height from 2 to 2.5 m, while horizontal units are usually 2.5 m in diameter and are of any desired length upto 7.5 m. The pressure filters occupy less space than the gravity filters of the same capacity.

5.7.2 Pressure Filters

- *Filters which operate with super-atmospheric pressure on the upstream side of the filter medium and atmospheric or greater pressure at the downstream side of the filter medium* are termed as **pressure filters**. In these devices, the filtering pressure is applied on the upstream side by a liquid pump or by a compressed gas. Hence, pressure filters are fed by plunger, screw, diaphragm or centrifugal pumps. Since the cake discharge from a pressure environment is difficult, continuous filters are in limited use and most of the pressure filters are batch operated.

Advantages of Pressure filters :

(i) Use of high filtration pressure results in relatively rapid filtrations.

(ii) These filters are compact so they provide a large filtration area per unit of floor space occupied by the filter.

(iii) Batch pressure filters offer greater flexibility than any other filter at relatively low initial investment.

Disadvantages of Pressure filters :

(i) Difficult to adapt to continuous processes and the operating cost is high in many applications.

(ii) Continuous pressure filters are inflexible to some extent and are expensive.

- A filter press is the simplest and the most commonly used filtration equipment. Two main forms in which this press is made are : the plate and frame press, and the recessed plate press/chamber press.

5.7.2.1 Plate and Frame Filter Press

Construction :

- It consists of plates and frames arranged alternately and supported on a pair of rails. The plate is a solid piece having a ribbed surface. The frame is hollow and provides the space for the filter cake. The alternate arrangement of plates and frames results in the formation of chambers. The plates and frames are square or rectangular in shape and can be made of cast iron, stainless steel, nickel, aluminium, monel, hard rubber or plastics (polypropylene). Coated materials are also used (rubber or lead or epoxy resin covered).
- Filter cloths are placed over each plate to cover the plate surface on both sides so that hollow frame is separated from the plate by the filter cloth. The plates and frames have circular holes on the corners for feed and discharge as shown in Fig. 5.5. The filter cloths are also having holes that match the holes on the plates and frames. The filter cloths themselves act as gaskets.
- When the press is closed by means of a hand screw or hydraulically, a continuous channel is formed along the whole length of the press out of the corner holes in the plates, cloths, and frames. The frames have openings in the interior from the corner holes so that the slurry channel opens into the interior of frame (i.e., in the chamber formed between each pair of successive plates). At the bottom of the plates, holes are cored which connect the faces of the plates to the outlet cocks.

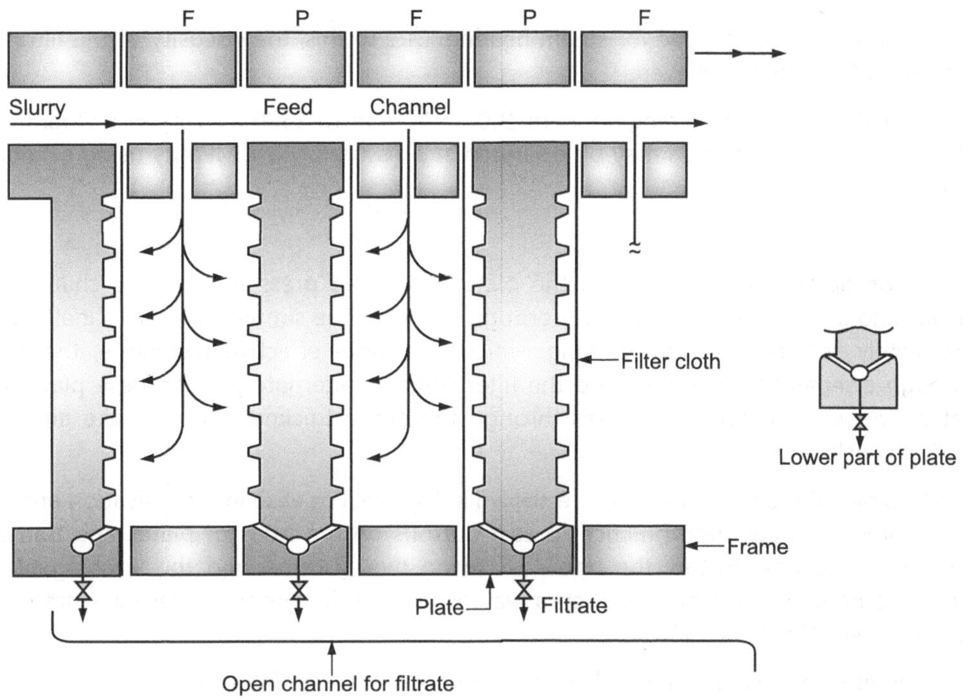

Fig. 5.4 : Plate and frame filter press (sectional view)

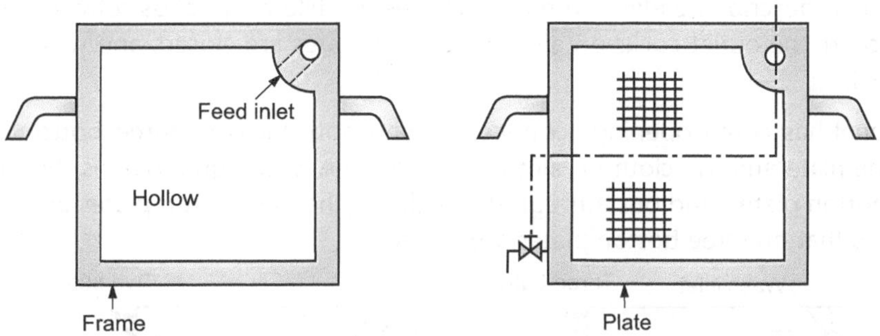

Fig. 5.5 : Plate and frame

Working :

- Slurry to be filtered is pumped through the feed channel. It runs into the chambers formed and fills the chamber completely (i.e., frames). As the feed pump continues to supply the slurry to be filtered, the pressure goes on increasing. Because of this, the filtrate passes through the cloth, runs down the faces of plates and finally leaves the filter through discharge cocks. (Fig. 5.4). The solids are deposited on the filter cloth. The two cakes are formed simultaneously in each chamber and these join when the frame is full and no more slurry can enter into it. The press is then said to be jammed. Wash liquid may be introduced in the press to remove soluble impurities from the solids and the cake is then blown with air to remove the residual liquid from it. The press is then dismantled, and the cake of solids scrapped off from each plate.
- In simple washing, the wash liquor is introduced through the feed channel and leaves the filter through the outlet cocks (i.e., it follows the same path as the slurry and filtrate). It is suited when the cake is uniform and permeable.
- In open discharge type, the filtrate is discharged through cocks into an open launder, so that the filtrate from each plate can be inspected and any plate can be isolated from the service by shutting off the cock if it is not giving a clear filtrate. Hence, it is used when absolutely clear filtrate is required. In closed discharge, the filtrate channel runs the entire length of press into a discharge pipe at one end. The closed technique is used when toxic or volatile materials are to be filtered.

- In many presses, arrangement is done for steam heating. Due to this, the viscosity of the filtrate is reduced and a higher rate of filtration is achieved.
- These units are made in plate sizes ranging from 100 × 100 mm to 1500 × 1750 mm. Operating pressures upto 700 kPa are common. The press may be operated at pressure upto 7 MPa by using a suitable material of construction.

Washing Press

- Washing of the precipitate is more easy in the plate and frame press than in the chamber press. Two methods of washing are simple washing and thorough washing. The simple washing is ineffective when the frame is completely full. In thorough washing, which is a more effective technique, the wash liquor is admitted through a separate channel behind the filter cloth on alternate plates. These plates are called as washing plates. The wash liquor thus passes through the entire thickness of the cake and is discharged through the drain cocks.
- The plates and frames of the washing press are shown in Fig. 5.6. For ease in identification and quick proper assembling the press, it is common practice to cast buttons on the sides of plates and frames. The non-washing plate is having one button, the frame is having two buttons, and the washing plate is having three buttons. The press is assembled in such a way as to give the order of plate and frame in the form 1 – 2 – 3 – 2 – 1 ...... etc. (Refer Fig. 5.7).
- The various channels lead to connections on the fixed head. During the filtration run,
a wash channel is closed by a valve on the head of the press. Filtration is carried out as in the non-washing plate and frame press described earlier. When the frames are filled (by solids retained on filter cloths), the feed channel is closed, the outlet cocks on all three button plates are closed, and wash liquor is introduced into a wash channel.
- As the wash channel has cored openings connecting with both faces of three botton plates, wash liquid enters between the plate and the cloth on all these plates. The wash liquid passes through the cake, down the faces of one botton plates, and out through the cocks on the one button plates as cocks on one bottom plates are open and that on three botton plates are closed.

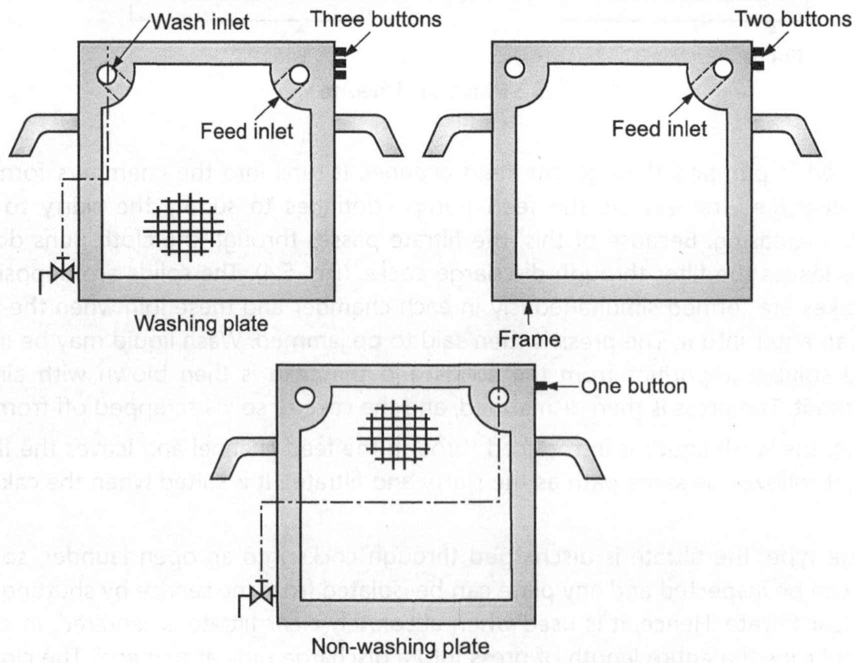

Fig. 5.6 : Plates and frames of washing press

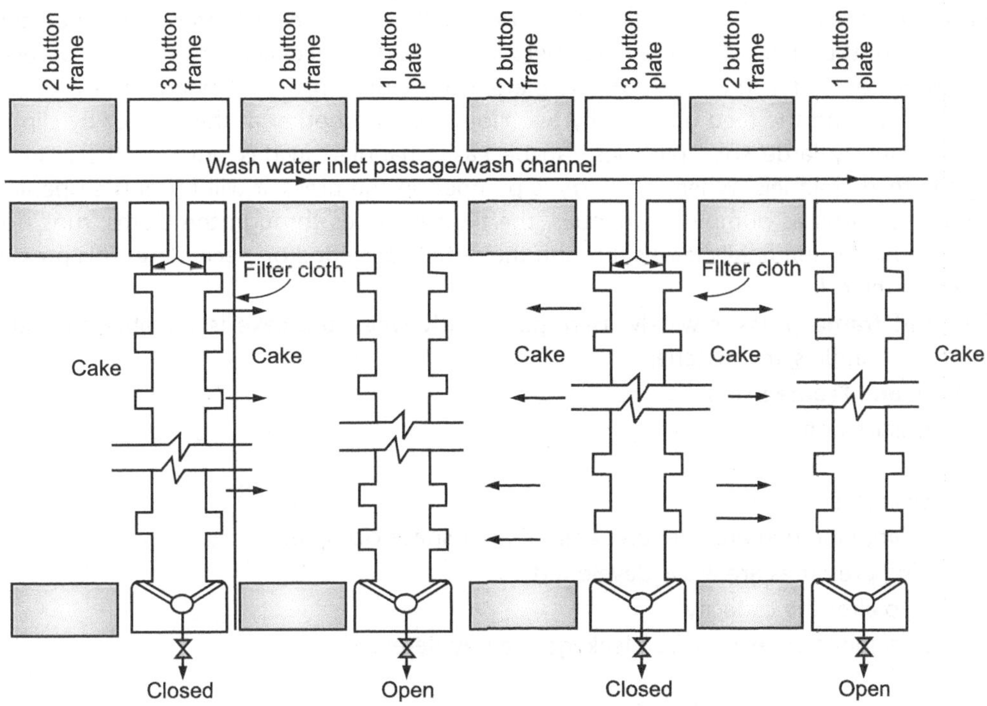

Fig. 5.7 : Section through Washing Press (Washing cycle)

- After washing, the excess liquid from the cake is removed by compressed air for easy discharge of the cake.
- In this press, the wash liquid passes through the whole thickness of the cake whereas the filtrate (during filtration) passes through only half the thickness of cake. The added resistance of the cake causes the liquid to distribute itself uniformly over the faces of three botton plates and thus, to pass through the cake uniformly.

Recessed Plate or Chamber Press :

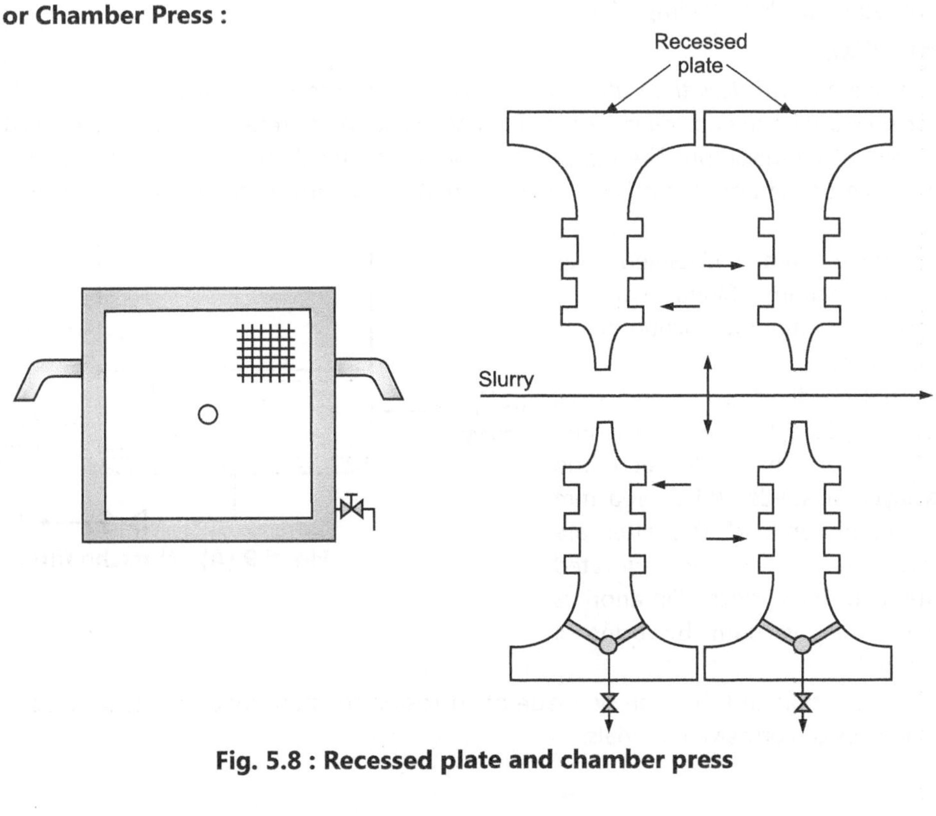

Fig. 5.8 : Recessed plate and chamber press

- It is similar to the plate and frame filter press except that the use of frames is avoided by recessing the ribbed surface of the plates. The filter chambers are formed in recesses between the successive plates. The fed is generally located in the centre of the plate. Filter cloth on the recessed surface of each side of the plate is sealed around the feed opening by two cloths sewn together at the hole or by clip nuts. The slurry containing relatively large solid particles can be easily handled in this press as there are no chances of blocking the feed channels. When the slurry is pumped in the press, it will fill all the opening between the cloths and afterwards as pumping continues, the filtrate passes through the cloths, runs down the ribbed surface of the plates and finally leaves through the outlets provided on each plate. This press is not adopted for washing of the cake.
- The **plate and frame press** is widely used, particularly when the cake is valuable and relatively small in quantity. It can handle slimy material.

Advantages of Plate and Frame press :

1. Simple in construction.
2. Low first cost.
3. Maintenance cost is low.
4. It provides a large filtering area per unit area of floor space occupied.
5. High operating pressures are easily developed.
6. It is possible to alter the capacity.
7. The majority of joints are external, so leakage is easily detected.
8. Flexibility.

Disadvantages of Plate and Frame press :

1. Labour requirement is very high.
2. Discontinuous in operation. Periodic manual dismantling results in high wear on the cloths. So the filter cloth life is relatively short.
3. Not suitable for high throughputs.
4. Presses frequently drip and leak, making housekeeping in the area a problem.
5. Washing of cake is likely to be imperfect.

5.7.3 Vacuum Filters

- *Filters which operate with less than atmospheric pressure on the downstream side of the filter medium and atmospheric pressure on the upstream side of the filter medium* are referred to as **vacuum filters**. Thus, these filters are limited to a maximum filtering pressure of one atmosphere. Vacuum filters need a vacuum pump which is a source of the filtration driving force (it creates vacuum on the downstream side) and is costly to operate.
- Vacuum filters are classified as discontinuous vacuum filters (vacuum nutsch filter) and continuous vacuum filters (rotary drum filter).
- A vacuum nutsch filter is an industrial version of a laboratory scale Buchner funnel, 0.90 m to 3 m in diameter and forming a layer of solids 100 to 300 mm thick. The components of this filter are vessel, woven mesh screen or perforated metal plate and filter cloth. Filtration is carried out under vacuum by using a vacuum pump.

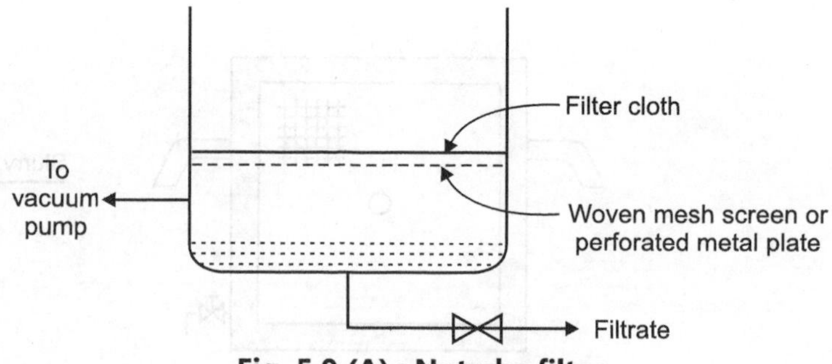

Fig. 5.9 (A) : Nutsche filter

- It is simple in construction and thus can be made of corrosion resistant materials. It is especially useful to filter experimental batches of corrosive materials.

Advantages of Vacuum filters :

(i) These filters can be designed as effective continuous filters.

(ii) Low labour requirement.

(iii) The filtering surface is easily accessible for inspection and repair as it can open to the atmosphere.

(iv) Low maintenance costs.

Disadvantages of Vacuum filters :

(i) We have to maintain a vacuum system.

(ii) Not suitable with filtrates that are volatile.

(iii) These units cannot handle difficulty filterable compressible solids.

(iv) Continuous vacuum filters are inflexible.

5.7.3.1 Rotary Drum Filter

- A rotary drum filter is the most common type of continuous vacuum filter. In this filter filtration, washing, partial drying and discharge of cake all take place automatically.

Construction :

- A rotary drum filter is shown in Fig. 5.9 (B). It consists essentially of a cylindrical sheet metal drum mounted horizontally. It may be from 50 to 400 cm in diameter and 50 to 800 cm long.
- The outer surface of the drum is formed of perforated plate. A filter medium such as canvas covers the outer surface of the drum which turns at 0.1 to 2 r/min in an agitated slurry trough. Inside the outer drum, there is a smaller drum with a solid surface.
- The annular space between the two drums is divided into number of compartments/sectors by radial partitions and separate connection is made between the compartments and a special type of rotary valve. As the drum rotates, vacuum and air are alternately applied to each compartment.
- Apart from cast iron, the other materials of construction of this filter include stainless steel, titanium, plastics such as PVC, etc. These materials give much improved corrosion resistance for many slurries.

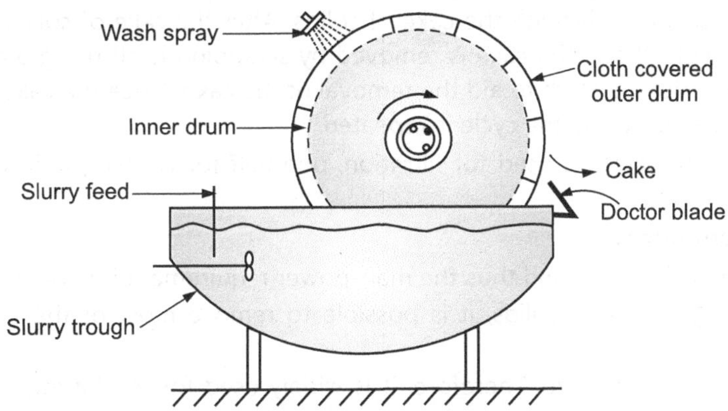

Fig. 5.9 (B) : Rotary drum filter

Working :

- The drum is immersed to the desired depth in the slurry which is mildly agitated to prevent the settling of the solids. Vacuum is then applied to the portion of drum which is submerged in the slurry through the rotary valve. Because of this, the liquid (filtrate) is sucked into the compartment and solids get deposited on the cloth to form a cake of the desired thickness which can be regulated by adjusting the speed of the drum. With higher speeds, thinner cake will be formed and consequently, high rate of filtration will be achieved. The filtrate from the compartment then goes to a filtrate collecting tank through the internal pipe and rotary valve.

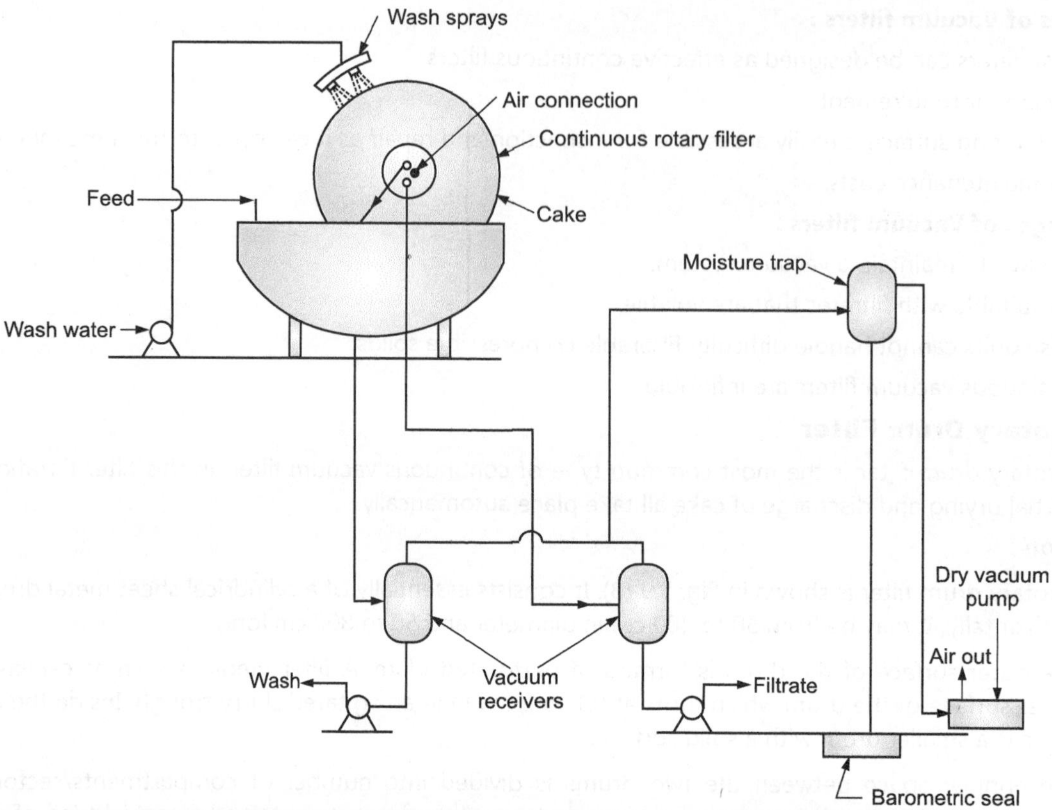

Fig. 5.10 : Flow sheet for continuous vacuum filtration

- As the portion of the drum on which the cake is formed comes out of the slurry, the cake is washed by spraying wash liquid. The wash liquid leaves the filter through the rotary valve and is collected separately in a separate tank. After washing, the cake enters into a drying zone as the drum rotates where the cake is partially dried by sucking air through the cake of solids. After the cake of solids has been sucked as dry as possible, vacuum is cut off and the cake is removed by scrapping it off using a adjustable doctor's knife. A little air is blown in under the cloth to aid the removal of the cake. Once the cake is removed from the drum sector, it re-enters the slurry and the cycle is repeated.
- Usually, one-third of the cycle is used for filtration, one-half for washing and air drying and one-sixth for cake removal.

Advantages of Rotary drum filter :

1. It is entirely automatic in action and thus the man-power requirement is very low.
2. With cake consisting of coarse solids, it is possible to remove most of the liquid from the cake before discharging.
3. It has a large capacity for its size. Therefore, it is widely used for the filtration of large quantities of free filtering material.
4. By changing the speed, it is possible to built up cakes of varying thickness. With fine solids, the thickness of cake is small and is large with coarse solids.

Disadvantages of Rotary drum filter :

1. The maximum available pressure difference is limited to less than one atmosphere.
2. As it being a vacuum filter, a difficulty is encountered in the filtration of hot liquids due to their tendency to boil.
3. It cannot be employed for materials forming relatively impermeable cakes or cakes that cannot be easily removed from cloth.
4. Initial cost of the filter and vacuum equipment is high.

5.8 CENTRIFUGAL FILTRATION

- In case of slurries containing coarse granular or crystalline solids forming a porous cake, the filtration operation can be carried by using centrifugal force rather than the pressure force.
- Centrifugal filters can be operated batchwise or in a continuous fashion. In these filters, the slurry is fed centrally to a rotating basket. The perforations in the walls of the basket are covered by a filter medium. The slurry is forced against the basket sides by pressure resulting from the centrifugal action, i.e., by centrifugal force. The liquor passes through the filter medium and the solids are retained by the medium. After building the cake to a predecided thickness, the feed is stopped and the cake of solids is spun for a short period to remove residual liquid from the cake.
- The principles of centrifugal separation and filtration are illustrated in Fig. 5.11.
- In Fig. 5.11 (a), a stationary cylindrical bowl contains a slurry (liquid + particulate solids of greater density than liquid). Since the bowl is not rotating, solids will settle at the bottom with a horizontal liquid surface above the solids.
- Fig. 5.11 (b) shows that the bowl is rotating about its vertical axis. In this case, the liquid and solids are acted upon by two forces – the gravity force acting downward and the centrifugal force acting horizontally. Normally, the centrifugal force is very large as compared to the gravity force and hence, the same may be neglected in comparison with the centrifugal force. Under the action of the centrifugal force, the solid particles are tightly pressed against the vertical bowl wall and the liquid layer assume the equilibrium position with an almost vertical inner surface as shown in Fig. 5.11 (b).
- If the wall of the bowl is perforated and perforations are covered with a filter medium such as a fine wire screen as shown in Fig. 5.11 (c), the liquid is free to flow outward but the solids are not. Almost all the liquid quickly flows out of the bowl, leaving behind the cake of filtered solids.

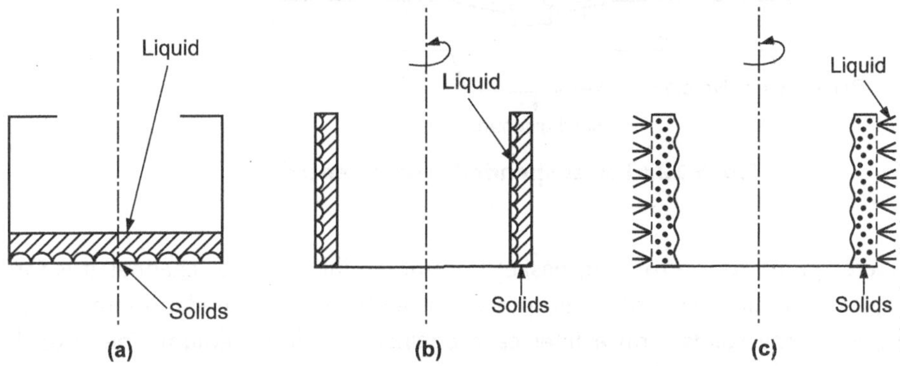

Fig. 5.11 : Principles of centrifugal separation and filtration

(a) Bowl stationary, (b) Sedimentation in rotating imperforated bowl,

(c) Filtration in rotating perforated basket

5.8.1 Centrifugal Filters

- A centrifuge or centrifugal is any rotating machine that utilises a centrifugal force for the separation of liquid from solids as well as for the separation of immiscible liquids of different densities. The essential components of a centrifuge machine are :
  1. a rotor or bowl in which centrifugal force is applied to the contents of bowl,
  2. a drive shaft,
  3. a drive mechanism e.g. electric motor,
  4. a frame for support, and align these and
  5. a casing.

Suspended batch centrifugal – Batch centrifuge

Construction :

- A batch centrifuge which is commonly used in industrial processing is the top-suspended centrifuge (See Fig. 5.12). It consists of a basket with perforated sides. The diameter of the basket ranges from 750 to 1200 mm and depth from 450 to 750 mm. The basket rotates at speeds between 600 to 1800 rpm. The basket is held at the lower end of a free swinging vertical shaft. The shaft is driven from above by an electric motor. The perforated sides (walls) of the basket are covered with a filter medium on the inside. The basket is surrounded by a casing provided with a filtrate discharge connection at the bottom. The basket and other parts may be constructed of mild steel, monel and stainless. In case of mild steel, they may be lined with lead, rubber, etc.

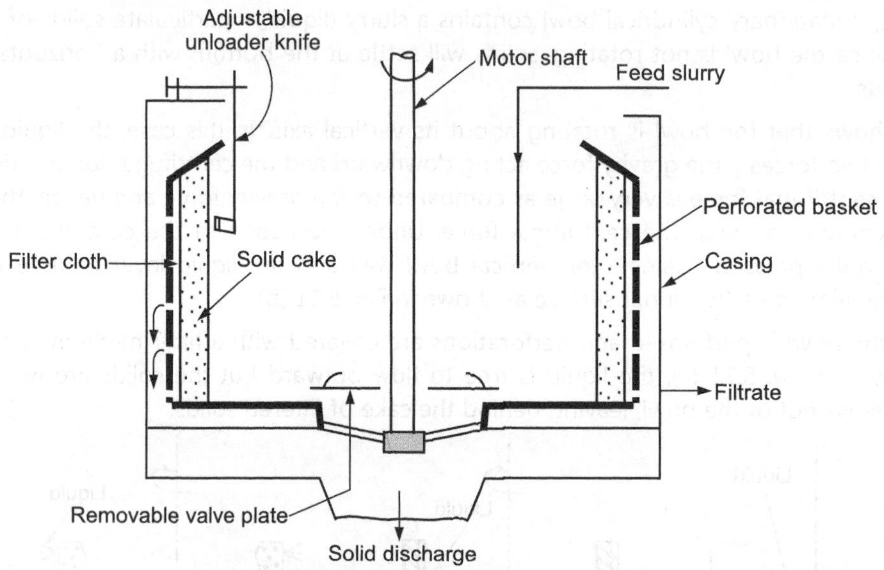

Fig. 5.12 : Top suspended basket centrifugal

Working :

- Slurry to be filtered is fed to the rotating basket through an inlet pipe or channel. It is forced against the basket sides by centrifugal force. The liquid passes through the filter medium into the casing and out a discharge pipe, while the solids form a filter cake against the filter medium. The cake thickness usually varies from 50 to 150 mm. The cake is washed by spraying wash liquid to remove the soluble material. It leaves the centrifuge through the discharge pipe. After washing is complete, the cake is spun as dry as possible, usually at a speed higher than that during the charging and washing steps. The motor is then turned off and the basket speed is reduced by the application of a brake. At the basket speed of 30 - 50 rpm, the cake is discharged by cutting it out with an unloader knife. The knife peels the cake off the filter medium and drops it through an opening in the basket floor. The valve which forms part of the bottom is opened to allow cake discharge into a receiver placed below. After unloading, the filter medium is rinsed clean and the cycle is repeated.

- These machines are widely used in sugar refining. They operate in sugar refining on short cycles of 2 to 3 minutes per load.

- Another design of basket centrifuge is the one which is driven from the bottom (under driven). In this machine, the drive motor, basket, and casing are all suspended from vertical legs mounted on a base plate. It may be top discharge or bottom discharge type.

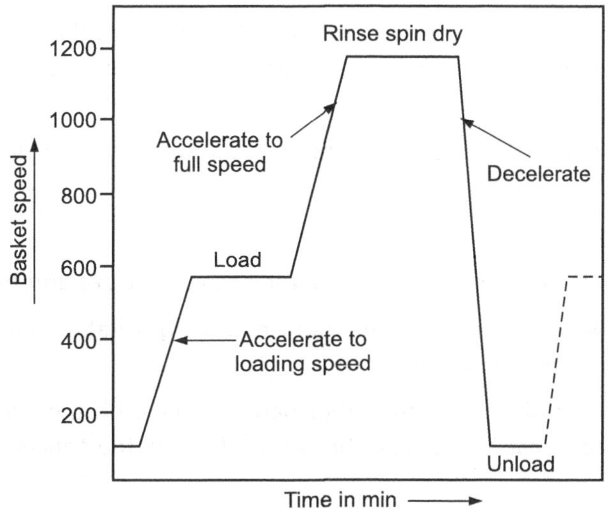

Fig. 5.13 : Typical operating cycle of a batch centrifuge machine

In case of bottom discharge under driven machines, the solids are plowed out through openings in the floor of the basket as in top-suspended machines. A typical operating cycle of a batch centrifuge is composed of various steps such as

(a) accelerate to loading speed,
(b) rinse screen,
(c) load,
(d) accelerate to full speed,
(e) wash cake,
(f) spin to dryness,
(g) decelerate to unloading speed, and
(h) unload.

- The drive may be variable-speed electric motor, either direct or through V-belts. A typical operating cycle for a variable speed automatic basket centrifuge is shown in Fig. 5.13. Centrifuge machines are also called as hydroextractor.

5.9 CONCEPT AND PRINCIPLE OF SEDIMENTATION

- *The separation of solids from a suspension in a liquid by gravity settling* is called ***sedimentation.*** The gravity force is responsible for the motion of solids through the liquid. In this operation, a dilute slurry is separated into a clear liquid and a slurry of higher solids content. The Dorr thickener is a common piece of equipment used for sedimentation.
- Removal of solids liquid sewage wastes (waste water treatment) and removal of suspended impurities from water to be used for domestic and industrial purposes (water treatment) are examples (application) of sedimentation.
- Sedimentation is one of the most widely used processes in the treatment of water. The simplest method of removing the suspended impurities is by plain sedimentation. The water is allowed to stand quiescent or move very slowly through a basin until the suspended impurities settle to the bottom of the basin and relatively clear water is drawn off from the top. The degree of removal of suspended impurities depends upon the length of retention period, the size of the suspended impurities and the temperature of water.

5.9.1 Types of Settling

- There are two types of settling processes by which particulates (solid particles) settle to the bottom of a liquid. 1. Free settling. 2.Hindered settling.

5.9.1.1 Free Settling

- It is the settling of the particles unaffeced by the other particles and the wall of the container.
- It refers to the process wherein the fall of the particle in the gravitational field through a stationary fluid is not affected by the other particles and the wall of the container.

- In this process, the individual particle does not collide with the other particles or with the wall of the container. This requires that the particles be at a sufficient distance from the wall of the container and also from each other.
- This type of settling process is possible only if the concentration of particulate solids in a suspension is very low.

5.9.1.2 Hindered Settling

- It is the settling of the particle impeded/affected by the other particles and the wall of the container.
- It refers to the process wherein the fall of the particle in the gravitational field through a stationary fluid is affected by the other particles and the wall of the container.
- In this process the particle collide with the other particles and with the wall of the container. This require that the particles be close to each other and this is turn demand the concentration of solids in a suspension to be high.
- Hindered settling is encountered when the concentration of solids in a suspension is large.
- For hindered settling, the settling velocity is considerably less than the terminal falling velocity under free settling condition.

5.9.2 Concept of Terminal Falling Velocity

- If a particle is allowed to settle in a fluid under the influence of gravity, it will increase in velocity until the accelerating force (force of gravity) is exactly balanced by the resisting force (drag force). When this happens there is no further change in the particles velocity and the particle will settle at a definite constant velocity. This velocity is known as the terminal falling or terminal settling velocity of the particle.
- The terminal falling velocity of the particle is affected by size, shape and density of the particles as well as the density and viscosity of the fluid.
- The terminal falling velocity of a particle freely falling in a fluid is the velocity of the particle when the drag force equals the downward force of gravity acting on the particle.

5.10 DIFFERENCE BETWEEN SEDIMENTATION AND FILTRATION

Sedimentation	Filtration
1. Sedimentation is defined as the removal of solid particles from a suspension by settling under gravity.	1. Filtration is defined as the separation of solid particles from a suspension by using a porous medium which retains the solid particles and allows the liquid to pass (through it).
2. The gravitation force – force due to gravity is responsible for separation (by sedimentation).	2. The pressure difference across the filter medium is responsible for separation (by filtration).
3. Filter medium is not required.	3. Filter medium is required.
4. The concentration of solids is very low in the suspension to be handled.	4. The concentration of solids is very large in the suspension to be handled in cake filtration.
5. In sedimentation, a clear liquid is the product of operation.	5. In cake filtration, wet cake of solids is the product of operation.
6. Sedimentation basins and thickeners are the equipments used for sedimentation.	6. Filters press, rotary drum filter etc. are the equipments used for filtration.
7. Usually a sludge is discarded from sedimentation.	7. Usually a filtrate is discarded from filtration.

5.11 DIFFERENCE BETWEEN SEDIMENTATION AND CENTRIFUGATION

Sedimentation	Centrifugation
1. The separation of solids from a suspension in a liquid by gravity settling is called sedimentation.	1. The separation of immiscible liquids or solids from liquids by the application of centrifugal force is called centrifugation.
2. The gravitation force is responsible for separation.	2. The centrifugal force is responsible for separation.
3. The force of gravity is comparatively very small and thus separation proceed slowly.	3. The centrifugal force is comparatively very grate/high and thus separation proceed very/ enormously fast.
4. Sedimentation basins and thickeners are used for sedimentation.	4. Various types of centrifuges are used for centrifugation.

5.12 DIFFERENCE BETWEEN SEDIMENTATION AND CLASSIFICATION

Sedimentation	Classification
1. The separation of solids from a suspension in a liquid by gravity settling is called sedimentation.	1. The separation of solid particles into fractions according to their terminal falling velocities is called classification.
2. The two products resulting by sedimentation are a clear liquid and a slurry of high solids content (sludge).	2. The two products resulting by classification are a partially drained fraction containing the coarse material and a fine fraction along with the remaining portion of the liquid medium.
3. Liquid medium is not required.	3. Liquid medium is required to effect separation.

5.13 LABORATORY BATCH SEDIMENTATION TEST AND SETTLING VELOCITY CURVE

- The mechanism of settling may be best described by batch settling test in a glass cylinder. Fig. 5.14 shows a series of observations of batch settling test.

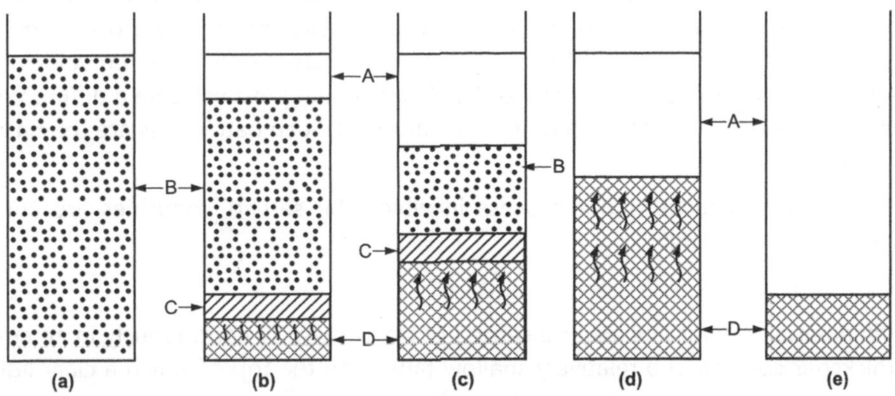

Fig. 5.14 : Laboratory Batch Settling test (Batch sedimentation)

- Fig. 5.14 (a) shows a cylinder containing a newly prepared slurry of uniform concentration of uniform solid particles throughout. As soon as the process starts, all the particles begin to settle and are believed to approach rapidly terminal settling velocities under hindered settling conditions. Various zones of concentration then are established as shown in Fig. 5.14 (b). The heavier faster settling particles settled at the bottom of a glass cylinder are indicated by zone D. Above zone D forms another layer, called zone C, a region of variable size distribution and non-uniform concentration. The boundary between C and D is usually obscure and is marked by vertical channels through which fluid is rising from the lower zone D as it compresses. Above zone C is zone B, which is a zone of uniform concentration, of approximately, the same concentration as that of the original pulp (suspension of solids is referred to as pulp in metallurgical work). Above zone B is zone A, which is a zone of clear liquid. If the original slurry is closely sized with respect to the smallest particles, the boundary between A and B is sharp.

- As sedimentation continues, the heights of each zone vary as shown in Fig. 5.14 (b), (c), (d). The heights of zones D and A increase at the expense of that of zone B while that of C remains constant. After further settling, zones B and C disappear, all the solids appear in zone D, but zone D may shrink further because of compression. During compression, the liquid associated with the solids in zone D is expelled in a clear zone.
- In a batch sedimentation operation as discussed, depths (heights) of various zones vary with time. The same zones will be present in continuous thickeners, but in a continuous sedimentation process, once the steady state is set up, the heights of each zone will be constant. Fig. 5.15 shows how the zones of Fig. 5.14 may be arranged in a continuously operating equipment such as a thickener.

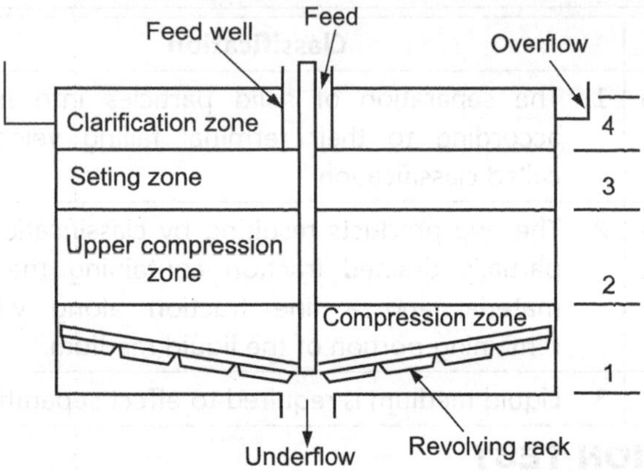

Fig. 5.15 : Settling zones in continuous thickener

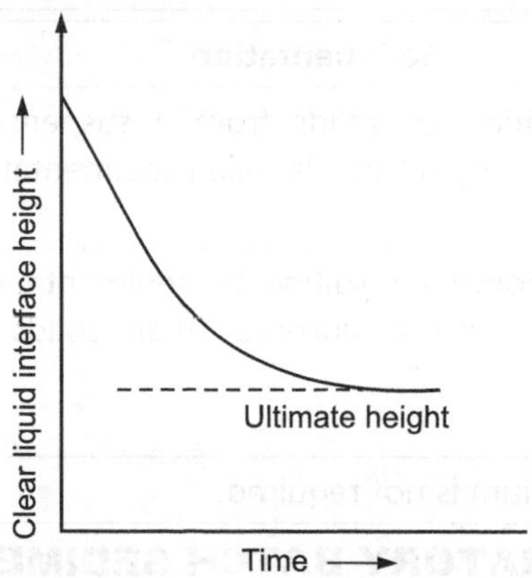

Fig. 5.16 : Batch-settling results

- In batch settling test carried out in the laboratory, the height of the liquid-solid interface (between zones A and B) is measured as a function of time. When the experimental data of height of interface v/s time are plotted, we get the curve as shown in Fig. 5.16. The slope of this curve at any point of time represents the settling velocity of suspension at that instant. During the early stage of settling process, the rate of settling is constant, as shown by the first portion of the curve. As time increases, the settling velocity decreases and steadily drops until the ultimate height is reached. The batch settling test will give a different curve for every sludge and somewhat different one for different concentrations. Such batch tests are the basis for design of continuous thickener.
- Thickening in sedimentation tanks is the process in which the settled impurities are concentrated and compacted on the floor of the tank.

Thickener

- Industrially, sedimentation operations may be carried out batchwise or continuously in an equipment called a *thickener*. A thickener consists of a relatively shallow tank from the top of which a clear liquid is taken off and the thickened liquid is withdrawn/removed from the bottom.
- In majority cases, the concentration of the suspension is high and hindered settling takes place. The rate of sedimentation can be artificially increased by the addition of coagulating agents such as alum, etc. which causes the precipitation of colloidal particles and the formulation of flocks. The suspension is also frequently heated which causes reduction in the viscosity of the liquid. Further, the thickener is frequently provided with a slow stirrer which helps in the consolidation of the sediment and also reduces the apparent viscosity of the suspension.

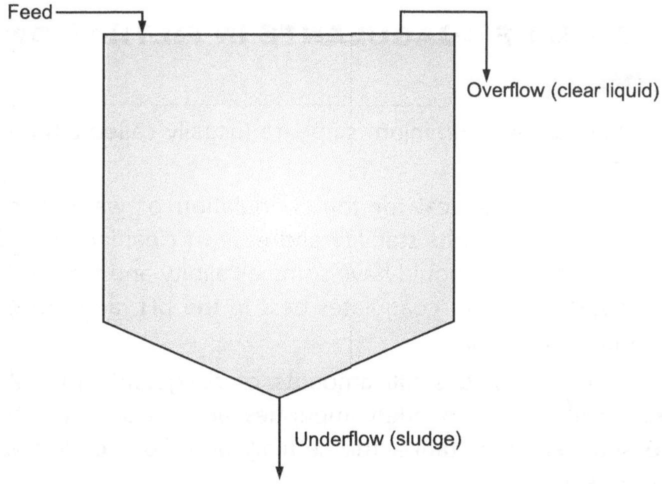

Fig. 5.17 : Schematic diagram of a thickener

5.14 TYPES OF THICKENERS

5.14.1 Batch Thickener

- A batch thickener usually consists of a cylindrical tank provided with openings for a slurry feed and product discharge. The bottom of the cylindrical tank is conical. The tank is filled with a dilute slurry, and the slurry is allowed to settle. After the sedimentation has proceeded for an adequate time, the clear liquid is decanted until sludge appears in the draw-off and the thickened liquid (sludge) is withdrawn from the bottom opening as indicated in Fig. 5.17.

5.14.2 Continuous Thickener

- A continuous thickener, such as the Dorr thickener consists of a flat bottomed, large diameter shallow-depth tank. It is provided with slow-moving radial rakes driven from a central shaft for removing the sludge. The slurry is fed at the centre of tank at a depth of 0.3 m to 1 m below the surface of the liquid, with a very little disturbance. The clarified liquid is continuously removed from an overflow which runs around the top edge of the tank (a launder) and the thickened liquor is continuously withdrawn from the outlet at the bottom. The slowly revolving rakes scrape the sludge towards centre of the bottom for discharge and remove water from the sludge as it stirs only the sludge layer. Thus, the solids are continuously moving downwards, and then inwards to the sludge outlet, whereas the liquid is moving upwards, and then rapidly outwards (Refer Fig. 5.18).

  The two functions of the thickener are :

  1. To produce a clear liquid, and 2. To produce a given degree of thickening of the suspension.

- For the production of clear liquid the upward velocity of the liquid must always be less than the settling velocity of particles. Thus, for a given throughput, the diameter of the tank determines the clarifying capacity of the thickener.

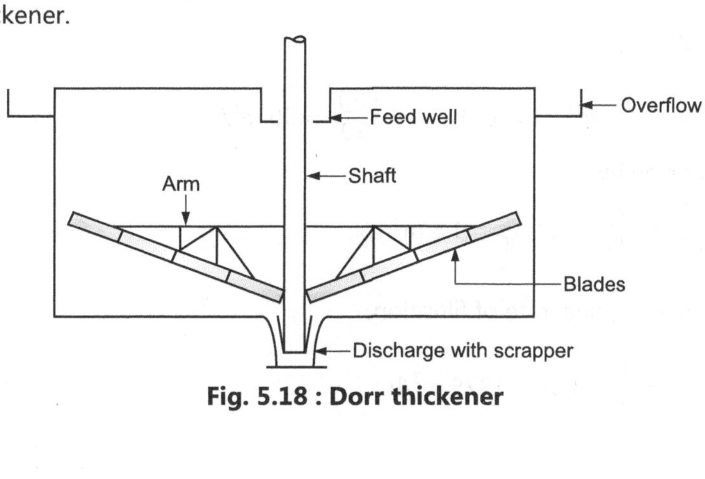

Fig. 5.18 : Dorr thickener

5.15 COAGULANTS AND ROLE OF COAGULANTS IN FILTRATION AND SEDIMENTATION

- The most widely used coagulants are : aluminium sulphate (usually called alum or filter alum) and ferrous sulphate (also known as copperas).
- Alum is the most commonly used chemical for the coagulation of water because of its excellent floc formation tendency, its relative economy, its stability and ease of cleaning. In order to react alum to form precipitate, it is necessary that the water should have some alkalinity and for this it is necessary to add soda ash or lime to water. It is found that alum coagulates best in the pH range of 6 to 8. It may be added in powdered form or in the form of solution.
- Chemical coagulation consists of adding small amounts of coagulants to water which form flocculant precipitates which coalesce with the suspended impurities and cause them to sink rapidly. When the coagulants are added to sedimentation tanks, the settling of solid particles will occur rapidly and the supernant liquid will be very clear.
- It is not possible to remove, as such, finely divided and colloidal particles, micro-organisms and colour producing compounds from water by use of sand filters. In order to remove these from water, coagulants are added to the water prior to filtration. When alum is added to water, it gets hydrolysed by natural or artificially created alkalinity of water with formation of the flocculant precipitate of aluminium hydroxide. The finely divided suspended matter, etc. gets adhered to this precipitate and are removed in sand filters.

SOLVED EXAMPLES

Example 5.1 : *For a sludge filtered in a washing plate and frame the filtration equation $V^2 = Kt$ holds good, where V is the volume of the filtrate obtained in time t. When the pressure is constant, 30 m^3 of filtrate is obtained in 10 h.*

(i) Calculate the washing time if 3 m^3 of wash water is forced to the cake at the end of filtration.

(ii) If the filtering area/surface is doubled keeping all other things constant, how long would it take to obtain 30 m^3 of filtrate ?

***Given :** The rate of washing is one-fourth the final rate of filtration.*

Solution : The filtration equation provided/given for a constant pressure filtration is

$$V^2 = Kt$$

where V is the volume of the filtrate obtained in time t.

Differentiating the above equation, we get

$$2V\,dV = K\,dt$$

$$\frac{dV}{dt} = \frac{K}{2V}$$

We have : $V^2 = Kt$

$$\therefore \quad K = \frac{V^2}{t}$$

Given : $V = 30\ m^3,\quad t = 10\ h,\quad K = \frac{(30)^2}{10} = 90\ m^6/h$

The final rate of filtration is given by

$$\left(\frac{dV}{dt}\right)_f = \frac{K}{2V} = \frac{90}{2 \times 30} = 1.5\ m^3/h$$

Given : Rate of washing $= \frac{1}{4}$ (final rate of filtration)

$$= \frac{1}{4} \times (1.5) = 0.375\ m^3/h$$

(i) Volume of wash water used = 30 m^3.

$$\text{Rate of washing} = \frac{\text{Volume of wash water}}{\text{Washing time}}$$

$$\text{Washing time} = \frac{\text{Volume of wash water}}{\text{Rate of washing}} = \frac{3}{0.375} = \mathbf{8\ h}$$ ... **Ans. (i)**

(ii) Filtering surface is doubled.

We know : Rate of filtration $\propto A^2$

$$\text{Rate} \propto A^2$$

$$\therefore \quad \text{Rate} = cA^2$$

$$(\text{Rate})_1 = c\,A_1^2$$ $(\text{Rate})_1$ is the rate with surface A_1

$$(\text{Rate})_2 = c\,A_2^2$$ ... with surface A_2

Given : $A_2 = 2A_1$

$$(\text{Rate})_1 = \frac{\text{Volume of filtrate}}{\text{Time}} = \frac{30}{10} = 3\ m^3/h$$

$$\frac{(\text{Rate})_2}{(\text{Rate})_1} = \frac{c\,A_1^2}{c\,A_2^2} = \frac{A_1^2}{A_2^2}$$

$$(\text{Rate})_2 = (\text{Rate})_1 \cdot \frac{A_1^2}{A_2^2} = 3 \times \frac{A_1^2}{(2A_1)^2} = 12\ m^3/h$$

We have to find the time required to obtain 30 m^3 filtrate when the surface is doubled.

$$\text{Time required} = \frac{\text{Volume of filtrate}}{(\text{Rate})_2} = \frac{30}{12} = \mathbf{2.5\ h}$$ ... **Ans. (ii)**

Example 5.2 : *A filter press is used to filter a sludge forming a nonuniform compressible cake. At a constant pressure difference, 6000 l of filtrate is obtained in 1 h. Washing is done with 1200 l of water, it proceeds exactly as filtration. The filtrate has the same properties as the wash water. Neglecting the resistance of filter cloth, calculate the washing time required.*

Given : *Rate of washing = 1/4 (final rate of filtration) for a filter press.*

Solution : For a constant rate filtration, neglecting the cloth resistance, we have

$$\frac{dt}{dV} = K_c V$$

which on integration gives $t = K_c \frac{V^2}{2}$

$$\therefore \quad K_c = 2t/V^2$$

Given : t = 1 h = 60 min ... V = 6000 l

$$K_c = \frac{2 \times 40}{(6000)^2} = 3.33 \times 10^{-6}\ \text{min}/l^2$$

We have : $$\frac{dV}{dt} = \frac{1}{K_c V}$$

The final rate of filtration is

$$\left(\frac{dV}{dt}\right)_f = \frac{1}{K_c V_f}, \quad \text{where } V_f = 6000\ l$$

$$= \frac{1}{3.33 \times 10^{-6} \times 6000} = 50\ l/\text{min}$$

Given : Rate of washing $= \frac{1}{4}$ (final rate of filtration)

$$= \frac{1}{4} \times 50 = 12.5\ l/\text{min}$$

Volume of wash water used $= 1200\ l$

$$\text{Rate of washing} = \frac{\text{Volume of wash water used}}{\text{Washing time required}}$$

$$\therefore \quad \text{Washing time} = \frac{\text{Volume of wash water}}{\text{Rate of washing}} = \frac{1200}{50} = \mathbf{24\ min} \quad \text{... Ans.}$$

Example 5.3 : *A plate and frame filter press when filtering a sludge gave 8 m^3 of filtrate in 1800 s and 11 m^3 of filtrate in 3600 s when filtration was stopped. Calculate the wahsing time if 3 m^3 of wash water is used to wash the cake. Neglect the resistance of a filter cloth and assume a constant pressure filtration.*

Given : Rate of washing = 1/4 (final rate of filtration)

Solution : For a constant pressure filtration, neglecting the filter cloth resistance, the filtration equation is

$$\frac{dt}{dV} = K_c V$$

$$\frac{dV}{dt} = \frac{1}{K_c V}$$

Integrating, we get

$$\frac{V^2}{2} = \frac{t}{K_c}$$

$$\therefore \quad K_c = \frac{2t}{V^2}$$

Let us find K_c.

Given : $V = 8\ m^3$ and $t = 1800$ s, $K_c = \frac{2 \times 1800}{(8)^2} = 56.25\ s/m^6$

The filtration was complete at t = 3600 s during which 11 m^3 of the filtrate was collected/obtained. Thus, the final rate of filtration is given by

$$\left(\frac{dV}{dt}\right)_f = \frac{1}{K_c V_f}, \quad V_f = 11\ m^3$$

$$= \frac{1}{56.25 \times 11} = 1.62 \times 10^{-3}\ m^3/s$$

Given : Rate of washing $= \frac{1}{4}$ [final rate of filtration]

$$= \frac{1}{4} \times 1.62 \times 10^{-3} = 4.04 \times 10^{-4}\ m^3/s$$

$$\text{Rate of washing} = \frac{\text{Volume of wash water}}{\text{Washing time}}$$

$$\text{Washing time} = \frac{\text{Volume of wash water}}{\text{Rate of washing}} = \frac{3\ m^3}{4.04 \times 10^{-4}\ m^3/s} = \mathbf{7425\ s} \quad \text{... Ans.}$$

Important Points

- *The separation of solids from a suspension in a liquid with the help of a porous medium or screen which retains the solids and allows the liquid to pass* is termed as **filtration**.
- In cake filtration, the feed to be handled (two phase mixture) is called **slurry**, the bed of deposited solids on a porous membrane (filter medium) is called **cake** and the clear liquid leaving the filter medium is called **filtrate.**
- *The method in which the pressure drop over the filter is held constant throughout a run so that the rate of filtration is maximum at the start of filtration and decreases continuously towards the end of the run* is called **constant pressure filtration.**
- *The method in which the pressure drop is varied usually from minimum at the start of filtration to a maximum at the end of filtration so that the rate of filtration is constant throughout the run* is called **constant rate filtration.**
- A filter aid is a granular or fibrous material which packs to form bed of very high voidage and therefore are capable of increasing the porosity of the filter cake.
- *Filters that retain appreciable visible quantities of filtered solids on the surface of the filter medium* are referred to as cake filters.
- *Filters that remove small amounts of solids to produce sparkling clear liquid* are referred to as clarifying filters or deep bed filters.
- *Filters which operate with super-atmospheric pressure on the upstream side of the filter medium and atmospheric or greater pressure at the downstream side of the filter medium* are termed as **pressure filters**.
- *Filters which operate with less than atmospheric pressure on the downstream side of the filter medium and atmospheric pressure on the upstream side of the filter medium* are referred to as **vacuum filters**.
- Sand filters (clarifying filters) are used for water treatment and water purification.
- A centrifuge is any rotating machine that utilises a centrifugal force for the separation of liquid from solids.
- *The separation of solids from a suspension in a liquid by gravity settling* is called ***sedimentation.***
- Sedimentation is defined as the removal of solid particles from a suspension by settling under gravity.
- **Free settling :** It is the settling of the particles unaffected by the other particles and the wall of the container.
- **Hindered settling :** It is the settling of the particles impeded by the other particles and the wall of the container.
- The terminal falling velocity of a particle freely falling in a fluid is the velocity of the particle when the drag force equals the downward force acting on the particle.
- Industrially, sedimentation operations may be carried out batchwise or continuously in an equipment called a *thickener*.
- Alum is the most commonly used chemical for the coagulation of water because of its excellent floc formation, its relative economy, its stability and ease of cleaning.

Practice Questions

1. Define filtration and state factors affecting the rate of filtration.
2. Classify industrial cake filters.
3. Draw a neat sketch of plate and frame filter press.
4. Write construction of plate and frame filter press.
5. Draw a sketch of rotary drum filter and explain its construction.
6. State the characteristics of a filter medium.
7. Draw a neat sketch of basket centrifuge and explain its construction.

8. State advantages and disadvantages of
   (i) plate and frame filter press and (ii) rotary drum filters.
9. Explain briefly the terms : constant rate filtration and constant pressure filtration.
10. Define sedimentation and state its applications.
11. Define the terms free settling and hindered settling.
12. Differentiate between sedimentation and filtration.
13. Explain in brief coagulants and their role in sedimentation and filtration.
14. Draw a neat sketch of Dorr thickener.
15. Differentiate between sedimentation and centrifugation.
16. Explain the term terminal falling/settling velocity.

❑❑❑

CHAPTER SIX

MIXING

6.1 CONCEPT OF MIXING

- Mixing is a process in which at least two separate materials (which may be present in the same or different phases) are taken and forced them to be randomly distributed through one another by some mechanical means.
- It is a physical process of reducing non-uniformities in fluids by eliminating gradients of concentration, temperature and other properties.
- The term mixing implies taking atleast two separate phases and causing them to distribute randomly through one another.
- A substance which is uniform throughout in physical state and chemical composition is called a homogeneous substance or a phase. Phases may be liquid, solid or gaseous. Therefore, mixing may involve gases, liquids or solids in any possible combination of two or more components - two different liquids, a liquid and a gas, a liquid and a powdered solid or two different or same solids.

6.2 HOMOGENEOUS AND HETEROGENEOUS MIXTURES

- A mixture in which its components/constituents are present in a single phase is called a homogeneous mixture. For example, a liquid mixture of methanol and water, a mixture containing CO_2, N_2 and O_2 gas.
- A mixture in which its components/constituents are present in distinct phases is called a heterogeneous mixture.

  For example, a liquid mixture of benzene and water forms a heterogeneous mixture made up of two immiscible liquid phases.

6.3 IMPORTANCE OF MIXING AND AGITATION

- In this chapter, we will deal with mixing of liquids with liquids, gases with liquids, liquids with solids and solids with solids.
- When the ratio of liquid to solid is large, mixing of solids with liquids can be performed in the same fashion as mixing of liquids with liquids. On the other hand, if the ratio of liquid to solid is small, solid-liquid mixing becomes similar to mixing of solids with solids (solid-solid mixing).
- It should be noted that agitation and mixing are not synonymous. Agitation refers to the induced motion of material in a circulatory pattern inside a tank or vessel, while mixing is the random distribution into and through one another, of two or more initially separate phases.
- The practical aims of mixing are :
  1. To promote a chemical reaction. It is the most important use of mixing in the chemical industry, since intimate contact between reacting phases/substances is necessary for a reaction to proceed properly.
  2. To produce simple physical mixtures - of two or more uniformly divided solids, two or more miscible liquids, etc.
  3. To carry out physical change - formation of crystals from a supersaturated solution.
  4. To accomplish dispersion in which a quasi-homogeneous material is produced from two or more immiscible fluids and from one or more fluids with finely divided solids.

- Based upon the objectives of the processing step, liquids are agitated for the following purposes :
  (i) Blending miscible liquids
  (ii) Dispersing a gas in the liquid
  (iii) Suspending or dispersing relatively lighter solid particles in the liquid to produce uniformity required for promoting mass transfer and assisting chemical reactions.
  (iv) Dispersing or contacting immiscible liquids
  (v) Promoting heat transfer between the liquid in the container and a coil or jacket surrounding the container.

6.4 MIXING LIQUIDS WITH LIQUIDS

- A propeller or a turbine in a tank is the most commonly used equipment for operations involving liquid-liquid and to some extent liquid-solid mixing.
- In liquid-liquid mixing, a system may contain liquids with or without solids that are not viscous (e.g., light oils) liquids with or without solids that are viscous but pourable (e.g. paints, heavy oils) and liquids with solids that form stiff pastes (oil-bond distempers).
- The usual form of equipment is a vertical vessel fitted with an agitator (i.e., an agitated vessel). The height of the vessel ranges from 1.5 to 2 times the diameter. The impeller diameter is usually one-third of the tank diameter.

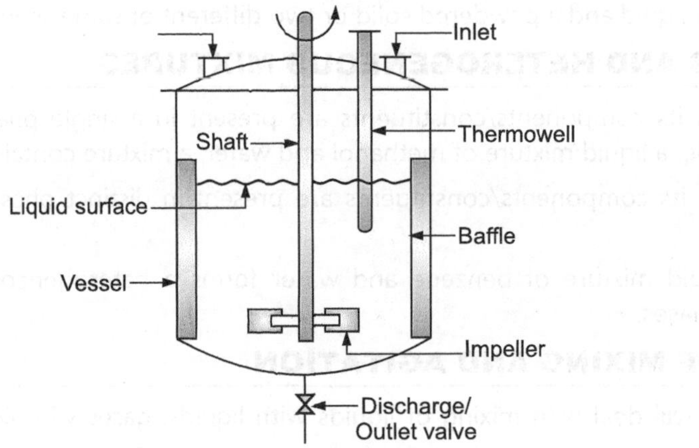

Fig. 6.1 : Typical Agitated Vessel

- An agitated vessel is a vertical, cylindrical vessel fitted with an agitator. The agitator is driven by an electrical motor directly or through a speed reducing gear box. It is provided with inlet and outlet connections, coil, jacket, etc.
- In the agitated vessel, the impeller creates a flow pattern, causing the liquid to circulate through the vessel and return ultimately to the impeller.
- An agitator is a combination of the impeller and shaft, i.e., impeller attached to the shaft. There are various types of impellers and so the agitator types. When we say turbine impeller, it is also termed as turbine agitator. The terms Impeller and Agitator are used interchangeably.

6.4.1 Construction and Flow Patterns of Impellers

- There are two types of impellers : 1. Axial flow impellers and 2. Radial flow impellers.
- Axial flow impellers make an angle of less than 90° with the shaft. They generate flow currents parallel to the axis of shaft.
- Radial flow impellers have blades parallel to the axis of the shaft. They generate flow currents in tangential (tangential to the circular path) or radial directions (perpendicular to the shaft).

- Impellers are further classified into three sub-types as :
  1. Propellers, 2. Paddles and 3. Turbines.
- Propellers and pitched blade turbines are axial flow impetters, whereas paddles, flat blade, curved blade, disc flat blade turbines are radial flow impellers.
- Axial impellers are used at high speeds to promote rapid dispersion and used at low speeds to keep solids in suspension. Radial flow impellers are used for large scale mixing of solid/liquid suspension.

6.4.1.1 Propellers

- A propeller is an axial-flow, high speed impeller commonly used for low viscosity liquids. It may be mounted centrally, off-centre or at an angle to the vessel. It is simple and portable. The diameter of propeller usually lies between 15 to 30% of the diameter of the vessel.
- A typical propeller is shown in Fig. 6.2. The most common propeller is a standard three bladed marine propeller. A propeller is shaped with a tapering blade to minimise the effect of centrifugal force and produce maximum axial flow.

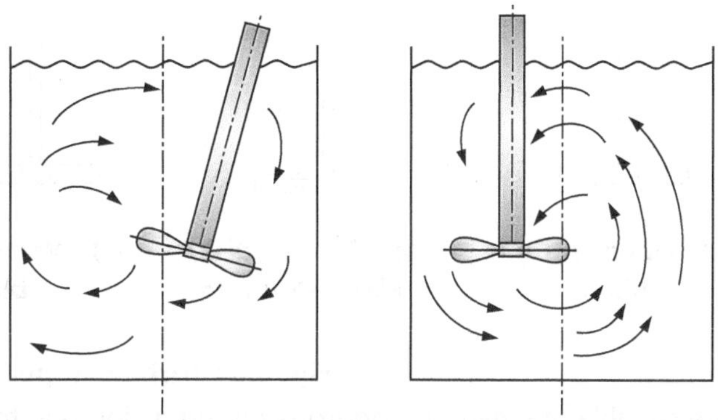

Fig. 6.2 : Standard three blade propeller Fig. 6.3 : Propeller, off centre and angular (unbaffled)

- Small propellers rotate at full motor speeds, whereas large ones rotate at a speed of 400 to 800 r.p.m.
- Propellers may also be mounted near the bottom of the cylindrical wall of a vessel as shown in Fig. 6.4 for blending low viscosity fluids or suspending slow settling sediments in very large tanks.

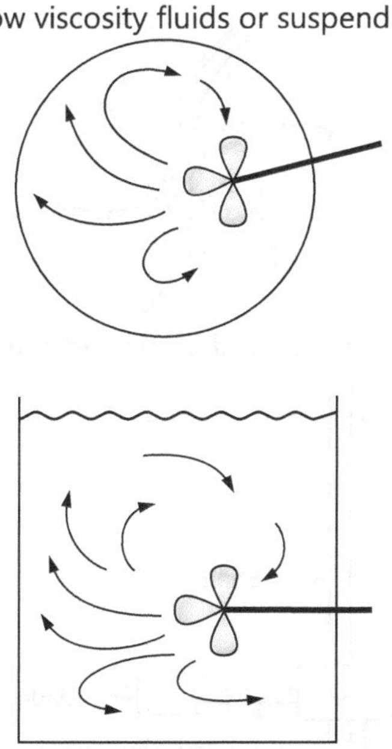

Fig. 6.4 : Side entering propeller

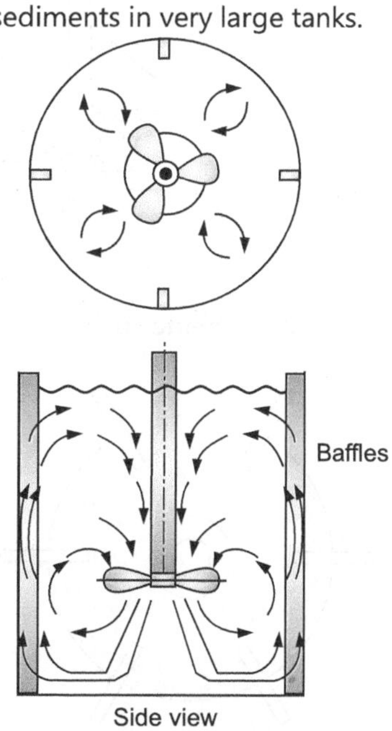

Fig. 6.5 : Flow pattern in a baffled vessel with centrally, mounted propeller or axial flow turbine

- Propeller drives the liquid straight down to the bottom of the vessel, at the bottom the stream spreads radially in all directions towards the wall, then the liquid flows upward along the wall, and finally returns to the suction of impeller from the shaft. Such a flow pattern is shown in Fig. 6.4. These agitators are used in situations where strong vertical currents are desired, e.g., for suspending heavy particles.

6.4.1.2 Turbines

- Various types of turbine impellers are shown in Fig. 6.6. Pitched blade turbine is an axial flow impeller while curved blade and flat blade turbines are radial flow impellers.

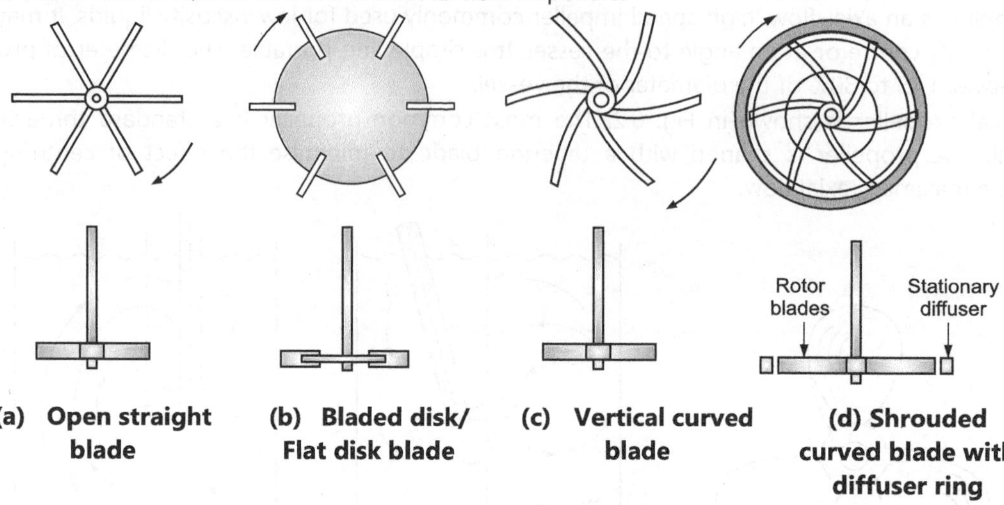

Fig. 6.6 : Turbine impellers

- They are capable of creating a vigorous mixing action due to centrifugal and rotational motions generated by them. A stator ring surrounding this impeller gives an efficient mixing action.

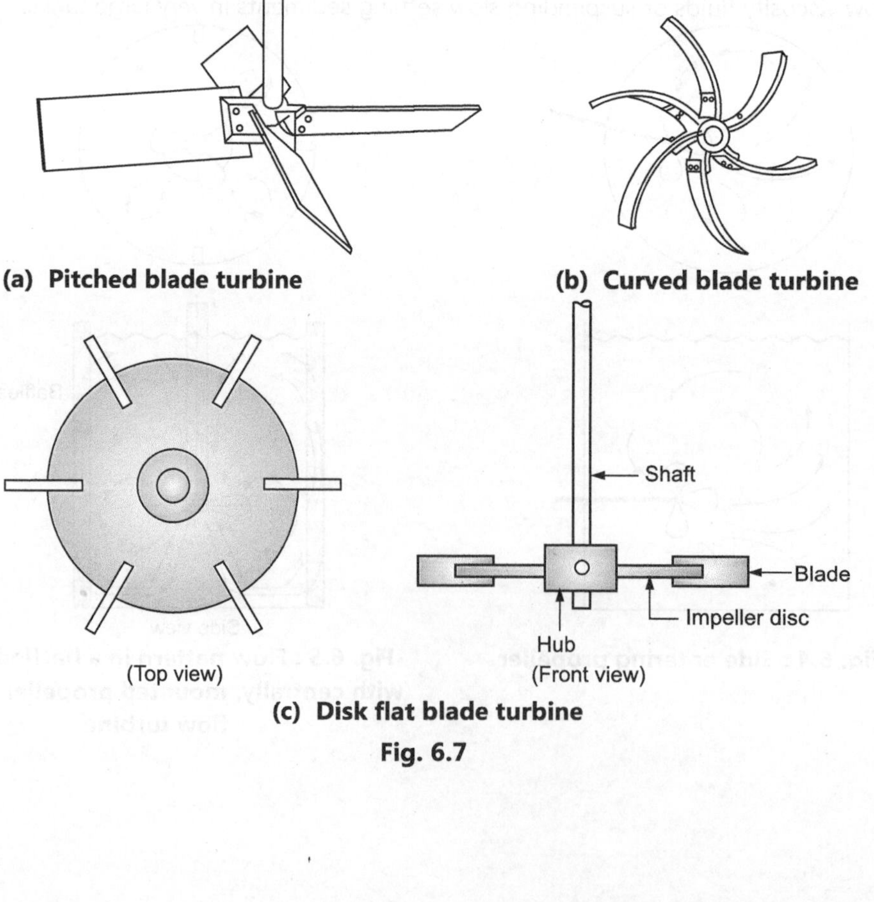

(c) Disk flat blade turbine

Fig. 6.7

- The blades of the impeller may be attached to a central hub or to a central disc. The diameter of the impeller is kept between one-third and one-sixth of the vessel diameter. The blade length is one-fourth of the impeller diameter. With a central disc, it is 1/8th of the impeller diameter. The blade angle of curved blade turbine may be between 30 to 60°. The impeller speed usually ranges from 50 to 250 r.p.m.
- Turbines are very effective over a wide range of viscosities (upto 10^4 cP).
- Turbine impellers drive the liquid radially against the wall, where the stream divides into two portions. One of the portions flows downward to the bottom and then returns to the centre of impeller from below, while other flows upward towards the surface and finally returns to the impeller from the above (Refer Fig. 6.8).

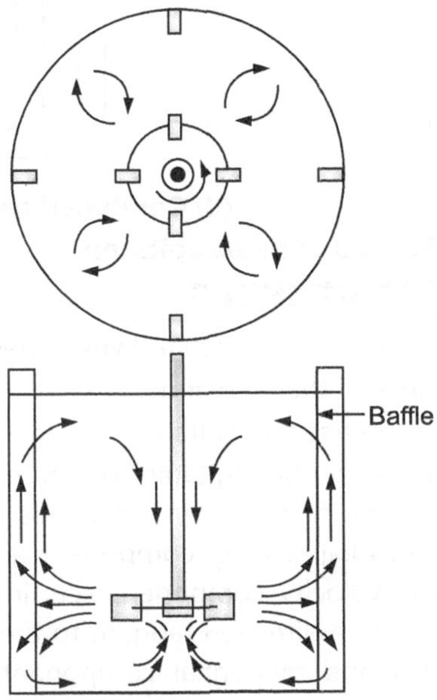

Fig. 6.8 : Flow pattern with turbine impeller in a baffled vessel

- Turbines are especially effective in developing radial currents, but with a baffled vessel they also induce vertical flow currents. To avoid vortexing and swirling with turbines, baffles or a diffuser ring can be used
- It is a common practice to locate the agitator at a height not less than one agitator / impeller diameter length from the bottom of a vessel and it should be submerged in the liquid at a depth equal to twice the diameter of agitator / impeller at low speeds and four times at high speeds. When the depth of the liquid is more than twice the agitator diameter, it may be advisable to use two impellers on the same shaft.

6.4.1.3 Paddles

- Paddle agitators with two or four flat blades are very common. The blades of these agitators are usually vertical and extend close to the vessel wall. They are simply pushers and cause the mass to rotate in laminar swirling motion with practically no radial flow along the paddle blades or any axial flow (vertical motion). The circulation is poor and the mixing action is insufficient. These rotate with a speed ranging from 20 to 150 r.p.m. The total length of this impeller lies between 50 to 80% of the inside diameter of the vessel (commonly 80% of the diameter). The width of the blade is 1/4 to 1/10th of the paddle diameter.
- In some designs, the shape of blades is similar to the shape of the bottom of a vessel so that they scrap the surface or pass over it with a close clearance. Such a type of paddle is known as an anchor agitator. Anchors are very useful for preventing deposits on a heat transfer surface as in reaction vessels and are commonly employed for obtaining improved heat transfer in high viscosity fluids but are poor mixers.
- Gate and anchor are used to sweep the entire peripheral area of the vessel, both walls and bottom.

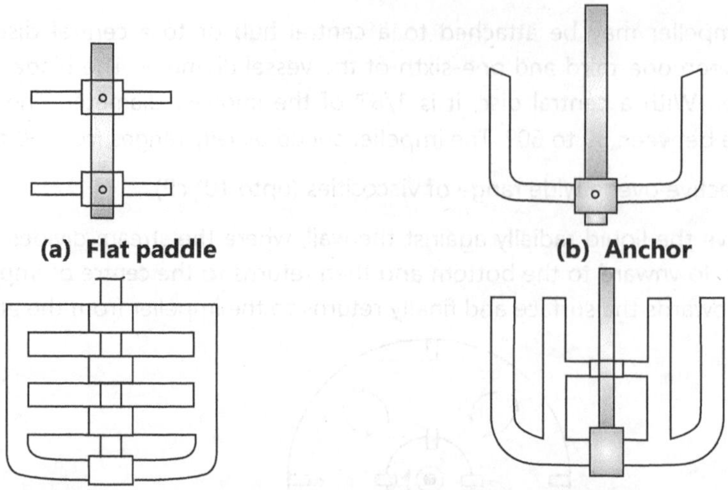

(a) Flat paddle (b) Anchor

(c) Gate (d) Combined anchor and gate

Fig. 6.9 : Paddle agitators

6.5 FLOW PATTERNS IN AGITATED VESSELS

- The factors on which the type of flow pattern in an agitated vessel depends are :
  (i) Type of impeller, (ii) Characteristics of the fluid and
  (iii) Size and proportions of the vessel, baffles and agitator.
- The velocity of the fluid at any point in the agitated vessel has three components, namely radial, longitudinal and tangential. The overall flow pattern depends on the variations in these velocity components from point to point. The radial velocity component acts in a direction perpendicular to the shaft of the impeller. The longitudinal velocity component acts in a direction parallel to the shaft. The tangential or rotational component acts in a direction tangent to the circular path around the shaft. With a vertical shaft in the vessel, the radial and tangential components are in a horizontal plane and the longitudinal component is vertical. Both the radial and longitudinal components are useful and produce the flow necessary for the mixing action.
- If the shaft is vertical and located at the centre in the vessel, the tangential component is generally undesirable since it follows a circular path around the shaft and creates a vortex at the surface of the liquid and tends to continue.
- In circulatory flow, the liquid flows in the direction of motion of the impeller blades. In case of unbaffled vessels, circulatory flow is generated by all types of impellers. When the swirling is strong, the flow pattern in the vessel is virtually the same irrespective of the design of the impeller. At high speeds, the vortex is deep and may reach the impeller. Because of this gas/air from the top of the liquid surface is drawn down into the content of the vessel, which is not desirable.

6.6 CONCEPT OF SWIRLING AND VORTEX (UNBAFFLED TANKS)

- If a low viscosity liquid is stirred in an unbaffled tank by a centrally mounted agitator, there is a tendency for a swirling flow pattern to develop, for the lighter fluid (usually air) to be drawn in to form a vortex at the surface of the liquid. This reduces the degree of agitation and mixing.
- The above said phenomenon takes place in unbaffled tanks regardless of the type of impeller. A typical flow pattern in an unbaffled tank for either axial or radial flow impeller is shown in Fig. 6.10.
- In vortexing, the surface of the liquid takes roughly U-shape and efficient mixing no longer takes place. A vortex is produced owing to the centrifugal force acting on the rotating liquid.
- Thus, there is a limit to the rotational speed that may be used, since once the vortex reaches the impeller, severe air entrainment may occur (air may be sucked in). In addition to this effect, the swirling mass of liquid generates an oscillating surge in the vessel which when coupled with the deep vortex may create a large fluctuating force acting on the agitator shaft.

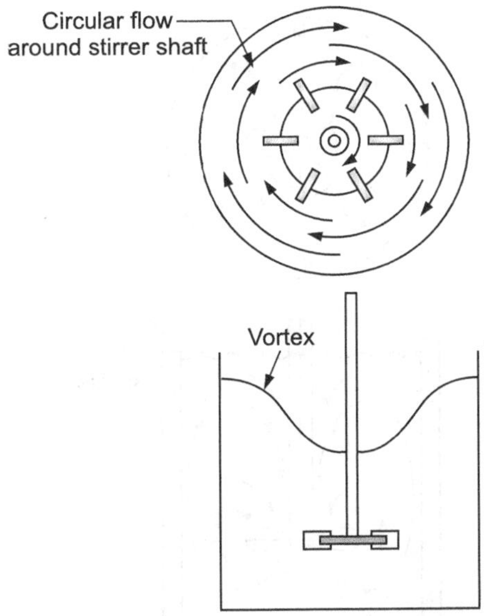

Fig. 6.10 : Vortex formation and circulation pattern in an unbaffled agitated vessel

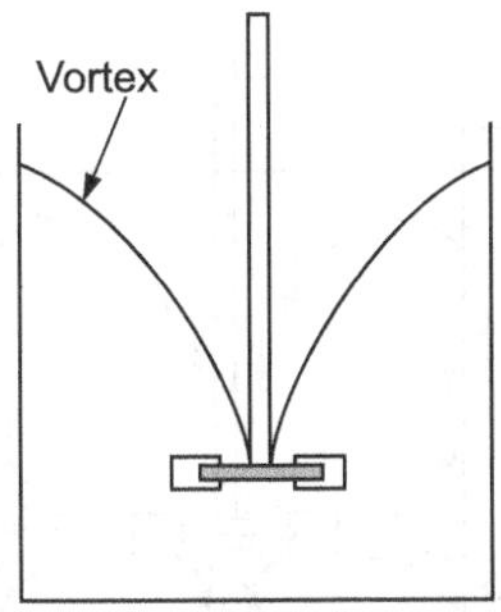

Fig. 6.11 : Vortex at very high impeller speed

6.7 PREVENTION OF SWIRLING AND VORTEX FORMATION

- There are three methods for the prevention of swirling and vortex formation :
  (i) Off-centre mounting of the impeller.
  (ii) Use of baffles.
  (iii) Use of diffuser ring with turbines.
- In small vessels, the impeller can be mounted off-centre as shown in Fig. 6.12. In larger vessels, the agitator may be mounted in the side of the tank with a shaft in horizontal plane but at an angle with radius, as shown in Fig. 6.13.

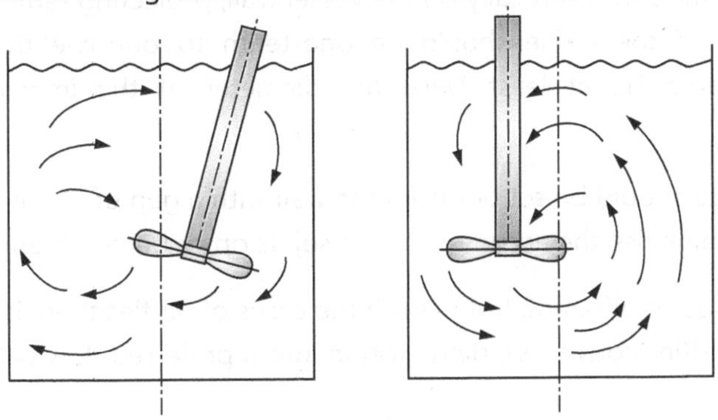

Fig. 6.12 : Propeller off-centre and angular (unbaffled)

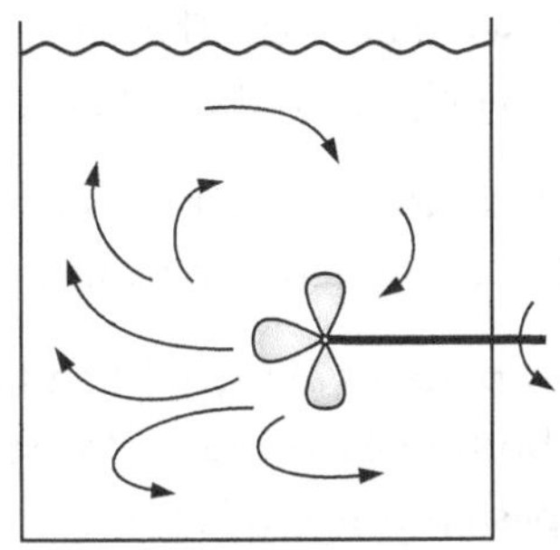

Fig. 6.13 : Side entering propeller

- In large vessels with vertical agitators, the most common method of reducing swirling is to install baffles along the side of the vessel, which hinder rotational flow without disturbing radial or longitudinal flow.
- In an unbaffled vessel, there are strong tangential flow currents and vortex formations at moderate speeds but in the presence of baffles, the vertical flow currents are increased and there is more rapid mixing of the liquid.
- With side entering, inclined and off-centre propellers, baffles are not needed.
- In case of turbines, the principal currents are radial and tangential. The tangential components induce (lead to) swirling and vortexing that must be stopped by the baffles or by the use of diffuser ring for impeller to be most effective.

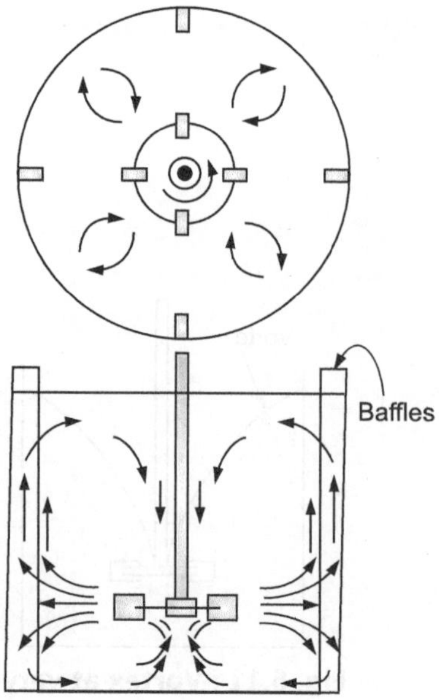

Fig. 6.14 : Flow pattern with turbine impeller in baffled tank

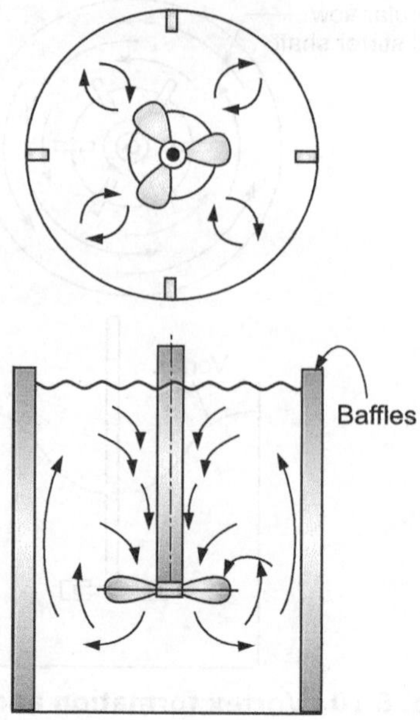

Fig. 6.15 : Flow pattern in a baffled tank with centrally mounted propeller agitator

6.8 BAFFLING

- Use of baffles in a vertical vessel is essential for the efficient mixing action and minimisation of vortex formation.
- Baffles are flat vertical strips that are mounted against the wall of the vessel as shown in Fig. 6.15. It is common practice to use four baffles. They are mounted vertically on the vessel wall, projecting radially from the wall and located 90° apart. The width of the baffle should be one-tenth to one-twelfth of the tank/vessel diameter. The baffle height should be at least twice the diameter of the impeller and approximately centred on the impeller.
- If the solids are to be kept in suspension, baffles should be set out from the wall with a gap of about 1/5th of the baffle width between baffle and vessel to minimise the accumulation of solids on or behind them.
- The flow patterns in Fig. 6.14 and Fig. 6.15 are quite different, but in both the cases of baffles there is a large top to bottom circulation without vortexing. Baffles convert swirling motion into a preferred flow pattern to accomplish process objectives.
- The addition of baffles in a vessel considerably increases the power requirement for mixing.

6.9 POWER CONSUMPTION OF IMPELLERS

- Usually, electrical power is used to drive impellers in agitated vessels.
- An empirical correlation of the power (or power number) with other variables of a system allows us to do fairly accurate prediction of the power requirement of a given impeller to rotate at a given speed. Such correlations can be obtained by using a method of dimensional analysis. The power requirement of the impeller is a function of geometrical details of the impeller and vessel, the viscosity and the density of liquid, and the rotation speed of impeller.

- An empirical correlation that can be obtained for a given system from the dimensional analysis is of the following form :

$$\frac{P}{N^3 D_a^5 \rho} = F\left(\frac{N D_a^2 \rho}{\mu} \frac{N^2.D_a}{\rho}\right) \quad \text{... (6.1)}$$

where $\frac{P}{N^3 D_a^5 \rho}$ is the power number (N_P)

$\frac{N D_a^2 \rho}{\mu}$ is the Impeller Reynolds number (N_{Re})

$\frac{N^2.D_a}{\rho}$ is the Froude number (N_{Fr})

N = rotation speed in revolution per sec.

D_a = diameter of impeller

ρ = density of fluid

and μ = viscosity of fluid

- When $N_{Re} > 10,000$, the flow in the vessel is turbulent and when $N_{Re} < 10$, the flow is laminar. For N_{Re} between 10 and 10,000 the flow is in a transition region in which the flow is turbulent at the impeller and laminar in the remote parts of the vessel.

  Equation (6.1) can also be written as :

$$N_P = f(N_{Re}, N_{Fr}) \quad \text{... (6.2)}$$

- The Froude number, N_{Fr}, represents the influence of gravitation and affects the power consumption only when vortex is present. If the speed of impeller is increased, in unbaffled vessels, the centrifugal force acting in the liquid causes the free surface of the liquid to assume a paraboloid form by raising the liquid level at the wall and lowering the level at the shaft. This is called vortex. The vortex is avoided by use of baffles. For Reynolds number < 300, vortex may not be observed even for the unbaffled vessel. The Reynolds number accounts for the viscous forces and it usually true that in agitated vessels, the viscous forces are significant. Thus, Equation (6.2) reduces to

$$N_P = f(N_{Re}) \quad \text{... (6.3)}$$

- The power consumption is related to the density and the viscosity of the liquid, the rotational speed, and the impeller diameter by plotting power as a function of Reynolds number as shown in Fig. 6.9.

At lower Reynolds number (laminar flow), the relationship between N_P and N_{Re} may be given as

$$N_P = C_o/N_{Re}, \quad \text{... (6.4)}$$

where C_o is a constant for a given impeller and given geometrical details.

Rearranging Equation (6.4), we get

$$N_P.N_{Re} = C_o \quad \text{... (6.5)}$$

Substituting the values of N_P and N_{Re} in Equation (6.5), we get

$$P = C_o \mu D_a^3 N^2 \quad \text{... (6.6)}$$

Equation (6.6) indicates that if the speed is doubled, power consumption will increase by a factor of four.

For higher values of N_{Re}, the Froude number plays important part. In this case, power number is constant i.e.

$$N_P = \text{constant} = C' \quad \text{... (6.7)}$$

$$P = C' \rho D_a^5 N^3 \quad \text{... (6.8)}$$

Equation (6.8) indicates that if the speed is doubled, the power consumption increases by factor of eight in the turbulent flow region.

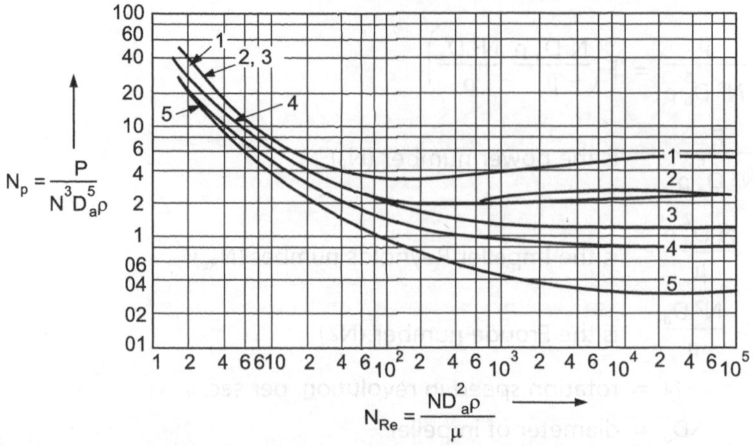

Curve 1 : Curve blade turbine, 4 baffles each width of baffle D_T / 12, D_T - tank diameter.

Curve 2 : Open straight blade (six blade) turbine, 4 baffles each D_T/12.

Curve 3 : Pitched blade turbine, 4 baffles each D_T/12.

Curve 4 : Propeller, 4 baffles each 0.1 D_T . Pitch = 2 D_a.

Curve 5 : Propeller, 4 baffles each 0.1 D_T, Pitch = D_a.

Fig. 6.16 : Power Consumption

6.10 MIXING OF GASES WITH LIQUIDS

- This is usually accomplished by spraying (sparging) a gas under a turbine (flat blade) near the bottom of a cylindrical vessel. Injecting the gas under a propeller is useless because the flow from the propeller is axial and downward. The equipment which can be used for the said purpose consists of a baffled vertical vessel incorporating a flat blade turbine agitator.

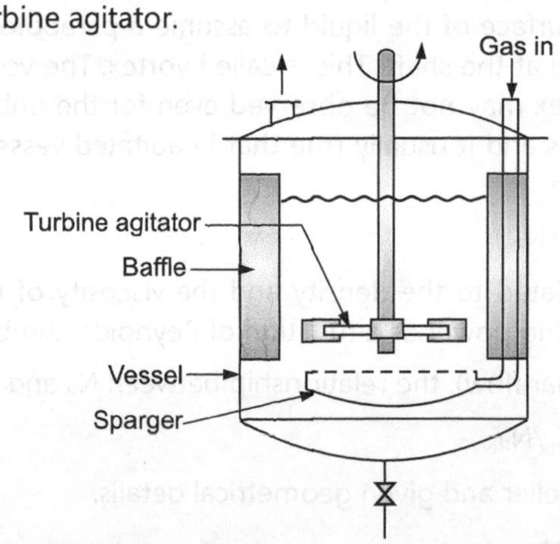

Fig. 6.17 : Mechanically agitated vessel for gas-liquid system

- The diameter of turbine is one-third of the tank diameter. The depth of a pool of liquid is equal to the tank diameter. A sparger (ring shaped) is mounted below the impeller with holes on the top. The diameter of the sparger is equal to or less than the diameter of the impeller. The gas is introduced from the top and injected in a pool of liquid in the form of fine bubbles through the sparger as shown in Fig. 6.17.

6.11 MIXING OF SOLIDS WITH LIQUIDS

- In situations, where the solids are not too coarse, the liquid is not viscous, and the amount of solids per unit volume of liquid is not too great, the solids can be suspended in liquids with the help of a flat blade turbine type of agitator. If any of the above cited conditions do not hold, then for carrying out mixing, one has to look for a kneading machine or some equipment primarily used for mixing solids with solids.

6.12 MIXING OF VISCOUS AND PLASTIC MASSES

- In machines used for viscous and plastic masses, either the material must be brought to the agitator or the agitator must visit all parts of mix. The mixing action in these machines is described as a combination of low-speed shear, smearing, wiping, folding, stretching, and compressing. These machines must be ruggedly built because the forces generated in these mixers are large. The power consumption with these mixers is high. Mixers described in this part are double arm kneaders, banbury mixers and mullers.

6.13 TYPES OF MIXERS

6.13.1 Sigma Mixer/Kneading Machine (Double-Arm Kneader)

- In kneading machines, the mixing action is a combination of bulk movement, smearing, stretching, folding, dividing and recombining as the material is pulled and squeezed against the blades, saddle, and the walls of trough.
- A sigma mixer consists of a short rectangular trough with a saddle shaped bottom [i.e., trough is curved at the bottom to form two longitudinal half cylinders and a saddle section]. Two counter rotating blades (roughly z-shaped outline) are incorporated in the trough. The blades are so placed and so shaped that the material turned up by one blade is immediately turned under the adjacent one. The blades are driven through a gear mechanism provided at either or both ends. The trough may be open or closed and may be jacketed for heating or cooling. The machine is operated in a batchwise fashion.
- The machine can be emptied through a bottom valve where 100% discharge or thorough cleaning, between batches, is not as essential. More commonly, double-arm kneaders are tilted for discharge by power operated jacks. Fig. 6.18 shows a double-arm kneader/sigma mixer employing sigma blades.
- The material to be kneaded is dropped into the trough and mixed for a period of about 5 to 20 minutes or longer. The trough is then unloaded by tilting it.
- It is used for mixing very stiff masses.

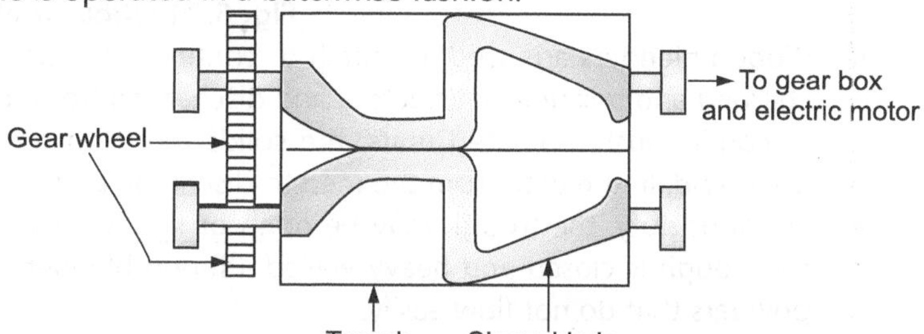

Fig. 6.18 : Kneading Machine/Double arm kneader (sigma mixer) – top view

- Various designs of mixing blades are shown in Fig. 6.19. The sigma blade [Fig. 6.19 (a)] is most widely used. The mixer employing sigma blades is capable of starting and operating with either liquids or solids or a combination of both. The sigma blade has good mixing action and is relatively easy to clean when sticky materials are being handled.
- The dispersion blade (Fig. 6.19) builds up high shear forces required to disperse powder or liquids into plastic or to rubbery masses. The double-naben blade [Fig. 6.19 (c)] is useful for heavy plastic materials.

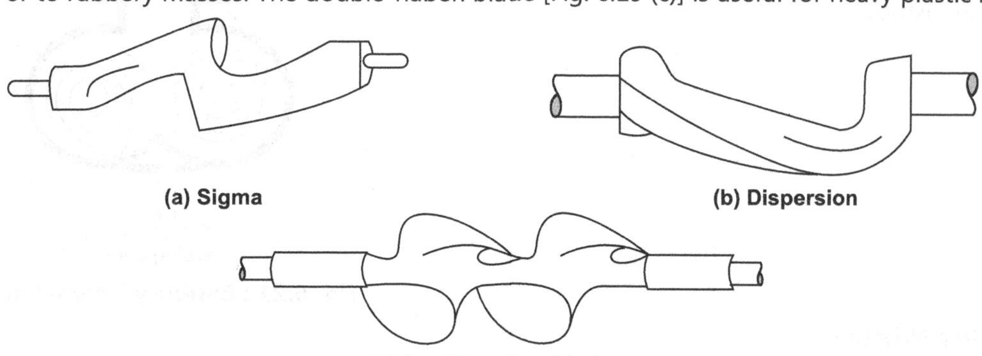

(a) Sigma (b) Dispersion

(c) Double-naben blade

Fig. 6.19 : Blades for double arm kneaders

6.13.2 Ribbon Blenders

- Ribbon blenders mix solids by mechanical shuffling and are used to handle dry powders.
- A ribbon blender consists of a horizontal semicylindrical trough incorporating a central shaft and a helical ribbon agitator. A typical ribbon blender is shown in Fig. 6.20. In this mixer, two counteracting ribbons are mounted on the same shaft. One of the ribbons moves the solids slowly in one direction, while the other one moves the solids quickly in the other direction. The ribbons may be continuous or discontinuous. Mixing takes place due to the turbulence generated by counteracting ribbons and not only by motion of the solids through the trough.

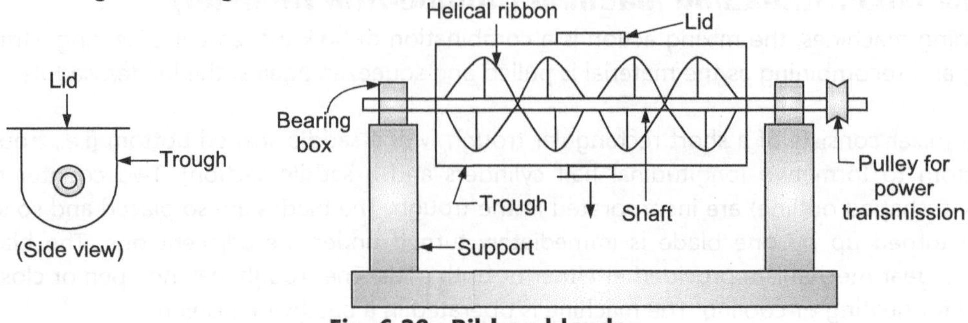

Fig. 6.20 : Ribbon blender

- Ribbon blenders are used for batch or continuous mixing. In batch operated ribbon blenders, the solids are charged and mix until satisfactory and discharged from the bottom.
  In continuously operated units, the solids are fed from one end of the trough and discharged from the other end. In the path from the feed to discharge end, solids are mixed.
- For light duty, the trough may be open or lightly covered, while for operation under pressure or vacuum, the trough is closed and heavy-walled. Ribbon blenders are very effective for handling thin pastes and dry powders that do not flow easily.

6.13.3 Banbury Mixer

- A banbury mixer is the most common internal mixer (See Fig. 6.21). It is a heavy duty machine with two blades each rotating in a cylindrical sheet, but these cylinders partly intersect with each other. In this mixer, the blade is pear shaped, but the projection is spiral along the axis and the two spirals interlock. The machine operates at a speed of 40 r.p.m. or lower. The clearance between the blades and the walls is extremely small, and it is here that the mixing action takes place. The material is fed from above and held in the trough during mixing by an air-operated piston under a pressure of 1 to 10 kgf/cm^2. Mixed material is discharged through a heavy sliding door which is provided at the bottom of the trough. The heat generated is taken out by spraying cooling water on the walls of the mixing chamber and circulating it through the hollow agitator shafts during operation.
- The banbury mixer is used mainly in plastic and rubber industries.

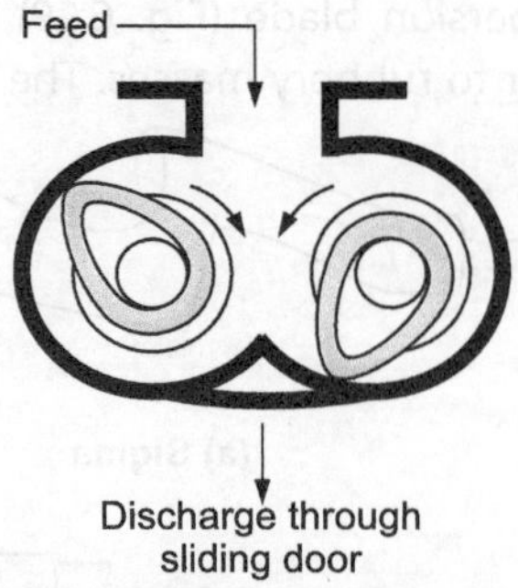

Fig. 6.21 : Banbury internal mixer

6.13.4 Muller Mixers

- Mulling is a smearing or rubbing similar to that in a mortar and pestle.
- A muller mixture consists of a pan incorporating muller wheels.

- In one of the designs of muller mixer, the pan is stationary and wheels rotate (See Fig. 6.22); while in the other design, the pan is rotated and the axis of the wheels is held stationary. In the stationary pan muller mixer, the central vertical shaft is driven, causing the muller wheels to roll in a circular path over a layer of solids on the pan floor. Plows direct the solids under the muller wheels during mixing or to an opening in the pan floor for the discharge of the mixer at the end of the cycle. The muller wheels crush the material, breaking down lumps and agglomerates.

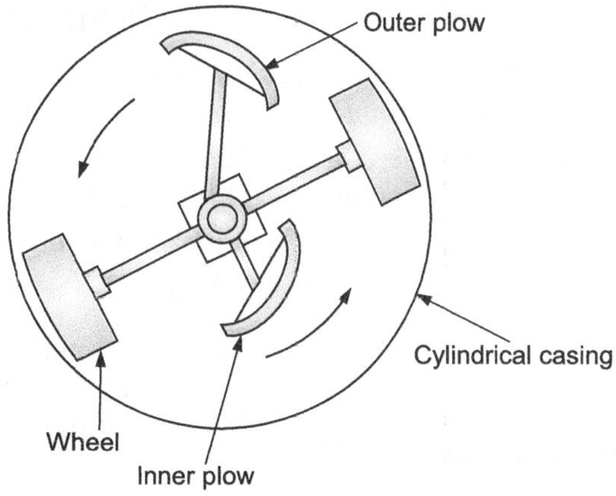

Fig. 6.22 : Muller mixer (Top view)

- Capacity of the muller mixer ranges from a fraction of cubic meter to more than 1.6 m^3 and the corresponding power requirement ranges from 1/3 to 75 hp.
- Mullers are used for handling batches of heavy solids and pastes. These are also effective in uniformly coating the particles of granular solids with a small amount of liquid.

6.13.5 Tumbling Mixers / Tumblers

In tumbling mixers, the mixing results from repeatedly lifting and dropping the material and rolling it over.

A large number of materials are mixed by tumbling them in a partly filled container that rotates about a horizontal axis. Tumbling mixers such as double cone mixer and twin shell blender, shown in Fig. 6.23, are suitable for free flowing dry powders.

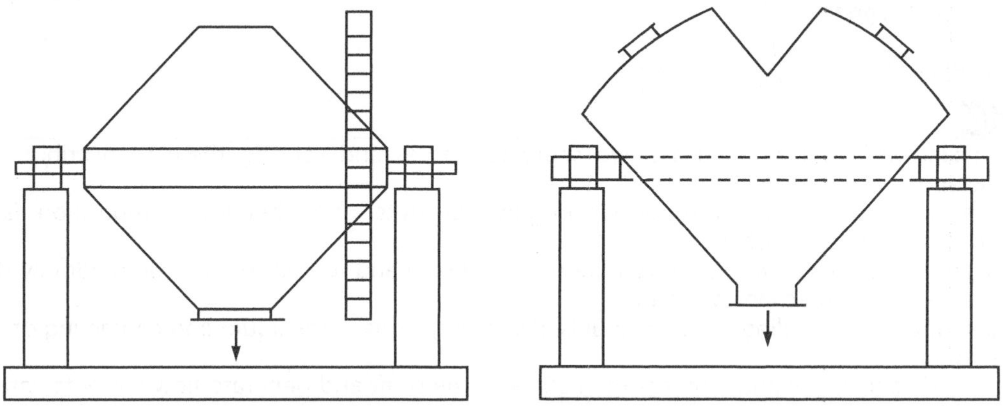

(a) Double cone tumbler **(b) Twin shell tumbler**

Fig. 6.23 : Tumbling Mixers

The double cone mixer [shown in Fig. 6.23 (a)] consists of a container made up of two cones, base to base with or without a cylindrical section in between. The mixer is mounted so that it can be rotated about an axis perpendicular to the line joining the points of the cones. The material to be mixed is charged to the mixer from above until it is 50 to 60% full. The ends of the container are closed and the solids are tumbled for a period of about 5 – 20 min. Finally, mixed material is dropped out from the bottom of the container into a conveyor or bin.

The twin shell blender [shown in Fig. 6.23 (b)] is formed out of two short cylinders. These cylinders are joined to form a V-shaped container (their axes are about 90° to each other) and rotated about a horizontal axis. It may contain internal sprays to introduce small amounts of liquid into the mix or mechanically driven devices to brake agglomerates of solids. Tumbling mixers are capable of handling large volumes, easily cleaned, and require a little less power than ribbon blenders.

SOLVED EXAMPLES

Example 6.1 : *A six-blade turbine agitator of diameter 60 cm is installed centrally in tank with flat bottom of diameter 180 cm, at a height of 60 cm from the bottom. The tank is filled with a solution of viscosity 10 C_p and of 1.45 g/ml density. The speed of agitation is 90 rpm. The tank is baffled. Calculate the power required.*

Data : Power number $= N_p = 1.05$ for $N_{Re} > 300$

Solution :

$$D_a = \text{Impeller diameter} = 60 \text{ cm}, \quad \mu = \text{Viscosity} = 10\ C_p = 0.10 \text{ poise}$$

$$\rho = 1.45 \text{ g/cm}^3$$

$$N = \text{Revolutions per second} = \frac{\text{Speed in rpm}}{60} = \frac{90}{60} = 1.5 \text{ rev. per second}$$

Reynolds number :

$$N_{Re} = \frac{N D_a^2 \rho}{\mu} = \left(\frac{100}{60}\right) \times \frac{(60)^2 \times 1.45}{0.10} = 87000$$

So flow is turbulent.

For turbulent flow : N_p = Power number = C' = constant

and the power is given by $P = C' \rho D_a^5 N^3$

where $N_p = C' = 1.05$

Substituting the values of various terms,

$$P = 1.05 \times 1.45 \times (60)^5 (1.5)^3 = 3.9956 \times 10^9 \approx 4 \times 10^9 \text{ (g.cm}^2\text{)/s}^3$$

$$= 4 \times 10^9 \text{ (g.cm}^2\text{)/s}^3 = 4 \times 10^9 (10^{-3} \text{ kg} \times 10^{-4} \text{ m}^2)/\text{s}^3$$

$$= 4 \times 10^9 \times 10^{-7} \text{ (kg.m}^2\text{)/s}^3 = 400 \text{ (kg.m}^2\text{)/s}^3$$

Let us convert P in the units of power by g_c.

g_c [Newton's law conversion factor] has the units of 1 (kg.m)/(N.s²).

$$\therefore \quad P = \frac{400 \text{ (kg.m}^2\text{)/s}^3}{g_c} = \frac{400 \text{ (kg.m}^2\text{)/s}^3}{1 \text{ (kg.m)/(N.s}^2\text{)}} = 400 \text{ (N.m)/s} = 400 \text{ J/s} = 400 \text{ W}$$

The power required is **400 W**. **... Ans.**

$$\text{Horse power [HP] required} = \frac{400}{746} = 0.536 \text{ HP.}$$

Important Points

- The term mixing implies taking atleast two separate phases and causing them to distribute randomly through one another.
- A substance which is uniform throughout in physical state and chemical composition is called a homogeneous substance or a phase.
- Mixing is a process in which two separate materials are taken and forced them to be randomly distributed through one another by mechanical means.
- Liquids are agitated for blending miscible liquids, dispersing a gas in the liquid and contacting or dispersing immiscible liquids.
- Axial flow impellers make an angle of less than 90° with the shaft and generate flow currents parallel to the axis of the shaft.
- Radial flow impellers have blades parallel to the axis of shaft and generate flow currents in tangential or radial directions.
- The methods used for prevention of swirling and vortex formation in agitated vessels are : off-centre mounting of the impeller, use of baffles and use of diffuser ring with turbines.
- Ribbon blenders are used for handling thin pastes and dry powders that do not easily.

Practice Questions

1. State the various types of impellers and draw a sketch of any one of them.
2. State the methods of avoiding vortex in an agitated vessel.
3. Draw sketches of flow pattern with propeller and turbine impeller.
4. Explain the construction two arm kneaders.
5. Draw sketches of different blades used in the kneading machines.
6. Explain in brief the construction of (i) Ribbon blender, (ii) Muller mixer.

❑❑❑

CHAPTER SEVEN

FLOW OF FLUIDS

Chemical engineers are concerned with transportation of fluids, both liquids and gases, from one location to another through pipes or ducts. This activity requires determination of the pressure drop through the system and consequently of the power required for pumping, selection of a suitable type of pumping device and measurement of the flow rates. In this chapter, we will deal with the types of flow patterns, determination of the pressure drop during fluid flow, methods of flow measurements, etc.

7.1 FLUID

7.1.1 Definitions

- A fluid is a substance which is capable of flowing if allowed to do so.
- A fluid is a substance that has no definite shape of its own, but conforms to the shape of the containing vessel.
- A fluid is a substance which when subjected to a shearing force/shear force, however small, undergoes deformation continuously as long as the force is applied.

It can be verified that, liquids and gases/vapours possess the above cited characteristics and hence, are referred to as *fluids*.

For example, water, ethanol, natural gas, methanol vapours and so on.

7.1.2 Pressure Gradient

- The change in pressure measured across a given distance is called a pressure gradient.
- The pressure gradient is a dimensional quantity expressed in the units of pressure per unit length. The SI unit of pressure gradient is Pa/m.
- It is the pressure difference in pressure between two locations within a given fluid.

$$\text{Pressure gradient} = \frac{\text{Difference in pressure between two locations}}{\text{Distance between two locations}} = \frac{\Delta P}{\Delta L} \quad \text{or} \quad \frac{\Delta P}{L} \qquad \dots (7.1)$$

- The pressure gradient results in a net force called the pressure gradient force. This force is directed from high pressure to low pressure within the fluid. The tendency of fluid to flow is from higher pressure to lower pressure. Refer Fig. 7.1 (B).

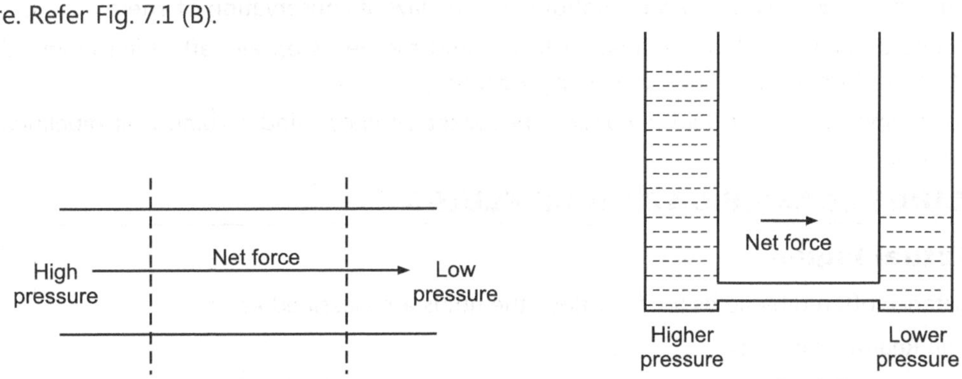

(A) Pressure gradient force **(B) Fluid flow**

Fig. 7.1

7.1.3 Velocity Gradient

The velocity gradient gives an idea of how the velocity of a fluid changes between different points within the fluid.

Velocity gradient is defined *as the difference in velocity between adjacent layers of the fluid.*

The formula for velocity gradient is given as

$$\text{Velocity gradient} = \frac{\Delta u}{\Delta y} \text{ or } \frac{du}{dy} \quad \ldots (7.2)$$

where Δu is the difference in flow velocity between the adjacent layers and Δy is the distance between the adjacent layers.

Therefore, the dimensional formula of velocity gradient is $M^0L^0T^{-1}$ and hence its SI unit is 1/s or s^{-1}.

The velocity gradient is also called as rate of shear or rate of shear deformation.

The study of velocity gradient is useful in analysing path dependent materials and further study of stresses and strains. For example, plastic deformation of metals.

7.2 FLUID STATICS AND DYNAMICS : CONCEPT AND APPLICATIONS

- The branch of engineering science which deals with *the behaviour of fluids at rest or in motion* is called as FLUID MECHANICS. The study of water is referred to as Hydraulics.
- Fluid mechanics is classified as : Fluid Statics and Fluid Dynamics.
- **Fluid statics** (hydrostatics) deals with the study of fluids at rest which involves the study of pressure exerted by a fluid at rest and the variation of fluid pressure throughout the fluid.
- **Fluid dynamics** deals with the study of fluids in motion relative to stationary solid walls or boundaries.
- Fluid statics deals with the behaviour of fluids when they are at rest. In fluid statics, there is no relative motion between adjacent fluid layers and hence there is no shear stress in the fluid that deforms it. The only fluid stress is the normal stress due to pressure, so any force developed is due to normal stress, i.e., pressure. Pressure at any point in a fluid is the same in all directions. Pressure varies within a fluid due to the weight of the fluid, i.e., the pressure varies within the fluid along the depth and is equal to ρgh and is the same at any horizontal level in the fluid.
- **Applications of fluid statics :** Submerged bodies such as ships, parts of ships, submarines, dams and gates and liquid storage tanks (to calculate force). It is also used to determine pressures at different levels of atmosphere and in many pressure measuring devices.
- Fluid dynamics is the study of the flow of fluids (liquids and gases) in and around solid surfaces. The fundamental propositions of fluid dynamics are the law of conservation of mass, law of conservation of linear momentum (known as Newton's law of motion) and the law of conservation of energy.
- Applications of fluid dynamics include calculating forces and moments on aircraft, determining the mass flow rates of petroleum through pipelines, predicting weather patterns.
- Some technological applications of fluid dynamics are rocket engines, wind turbines, oil pipelines and air conditioning systems.

7.3 TYPES OF FLUIDS/CLASSIFICATION OF FLUIDS

7.3.1 Ideal and Actual Fluids

Based upon the resistance offered by fluids for their flow, the fluids are classified as :

(i) Ideal fluids and (ii) Real/actual fluids.

Ideal Fluid

- It is a fluid which does not offer resistance to flow / deformation / change in shape, i.e., it has no viscosity. It is frictionless and incompressible. However, an ideal fluid does not exist in nature and therefore, it is only an imaginary fluid.
- An ideal fluid is the one which offers no resistance to flow/change in shape.

Real Fluid

It is a fluid which offers resistance when it is set in motion. All naturally occurring fluids are real fluids.

7.3.2 Compressible and Incompressible Fluids

Based upon the behaviour of fluids under the action of externally applied pressure and temperature, the fluids are classified as : (i)Compressible fluids (ii) Incompressible fluids.

A fluid possesses a definite density at a given temperature and pressure. Although the density of fluid depends on temperature and pressure, the variation of density with changes in these variables may be large or small.

Compressible Fluid :

- *If the density of a fluid is affected appreciably by changes in temperature and pressure, the fluid* is said to be **compressible**.

If the density of a fluid is sensitive to changes in temperature and pressure, the fluid is said to be **compressible**.

Incompressible Fluid :

- *If the density of a fluid is not appreciably affected by moderate changes in temperature and pressure, the fluid* is said to be **incompressible**.

If the density of a fluid is almost insensitive to moderate changes in temperature and pressure, the fluid is said to be **incompressible.**

Thus, **liquids** are considered to be incompressible fluids, whereas **gases** are considered to be compressible fluids.

7.3.3 Newtonian and Non-Newtonian Fluids

Based upon the behaviour of fluids under the action of shear stress, the fluids are classified as :

(i) Newtonian fluids
(ii) Non-Newtonian fluids.

For most commonly known fluids, a plot of τ versus du/dy results in a straight line passing through the origin and such fluids are called as Newtonian fluids.

Fluids that obey Newton's law of viscosity, i.e., the fluids for which the ratio of the shear stress to the rate of shear or shear rate is constant, are called as **Newtonian fluids**. This is true for all gases and for most pure liquids.

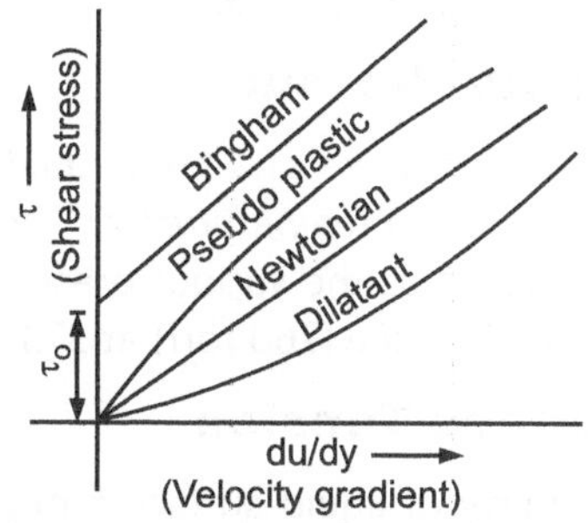

Fig. 7.2 : Shear stress versus velocity gradient for Newtonian and Non-Newtonian fluids

Examples of Newtonian Fluids : All gases, air, liquids, such as kerosene, alcohol, glycerine, benzene, hexane, ether etc., solutions of inorganic salts and of sugar in water.

Fluids for which the ratio of the shear stress to the shear rate is not constant but is considered as a function of rate of shear, i.e., fluids which do not follow Newton's law of viscosity are called as **non-Newtonian fluids**. Generally, liquids particularly those containing a second phase in suspension (solutions of finely divided solids and liquid solutions of large molecular weight materials) are non-Newtonian in behaviour.

Examples of non-Newtonian Fluids : Tooth pastes, paints, gels, jellies, slurries and polymer solutions.

A Newtonian fluid is one that follows Newton's law of viscosity. If viscosity is independent of rate of shear or shear rate, the fluid is said to be Newtonian and if viscosity varies with shear rate, the fluid is said to be non-Newtonian.

7.4 PROPERTIES OF FLUIDS

The properties of fluids are

(i) Mass density (specific mass) or simply density (ρ).
(ii) Weight density (specific weight) (w).
(iii) Vapour pressure.
(iv) Specific gravity.
(v) Viscosity.
(vi) Surface tension and capillarity.
(vii) Compressibility and elasticity.
(viii) Thermal conductivity.
(ix) Specific volume.

7.4.1 Density

Density (ρ) or mass density of a fluid is *the mass of the fluid per unit volume.* In the SI system, it is expressed in kg/m^3. The density of pure water at 277 K (4 °C) is taken as 1000 kg/m^3.

7.4.2 Weight Density

Weight density of a fluid is *the weight of the fluid per unit volume.* In the SI system, it is expressed in N/m^3. Specific weight or weight density of pure water at 277 K (4 °C) is taken as 9810 N/m^3.

The relation between mass density and weight density is

$$w = \rho g$$

where g is the acceleration due to gravity (9.81 m/s^2).

7.4.3 Specific Volume

Specific volume of a fluid is *the volume of the fluid per unit mass.* In the SI system, it is expressed in m^3/kg.

7.4.4 Specific Gravity

The specific gravity of a fluid is *the ratio of the density of the fluid to the density of a standard fluid.* For liquids, water at 277 K (4 °C) is considered/chosen as a standard fluid and for gases, air at NTP (0°C and 760 torr) is considered as a standard fluid.

7.4.5 Vapour Pressure

The vapour pressure of a pure liquid is defined as *the absolute pressure at which the liquid and its vapour are in equilibrium at a given temperature* or The pressure exerted by the vapour (on the surface of a liquid) at equilibrium conditions is called as the vapour pressure of the liquid at a given temperature. Pure air free water exerts a vapour pressure of 101.325 kPa (760 torr) at 373.15 K (100 °C).

7.4.6 Surface Tension

The property of liquid surface film to exert tension is called as the surface tension. It is the force required to maintain a unit length of film in equilibrium. It is denoted by the symbol σ (Greek sigma) and its SI unit is N/m.

7.4.7 Viscosity (Dynamic Viscosity)

- *A fluid undergoes continuous deformation when subjected to a shear stress. The resistance offered by a fluid to its continuous deformation* (when subjected to a shear stress/force) is called viscosity
- The viscosity of a fluid at a given temperature is *a measure of its resistance to flow.*
- The viscosity of a fluid (gas or liquid) is practically independent of the pressure for the range that is normally encountered in practice. However, it varies with temperature. For gases, viscosity increases with an increase in temperature, while for liquids it decreases with an increase in temperature. It is denoted by the symbol μ (Greek mu). In the C.G.S. system, viscosity may be expressed in poise (P) or centipoise (cP).

The unit of viscosity in the SI system is Pa·s (kg/cm.s).

$$1 \text{ poise} = 1 \text{ P} = 100 \text{ cP} = 0.10 \text{ kg/(m.s)} = 0.10 \text{ (N.s)/m}^2 \text{ or Pa.s}$$

7.4.8 Kinematic Viscosity

The kinematic viscosity of a fluid is the ratio of the viscosity to the density of the fluid. It is denoted by the symbol υ (Greek Nu) and has the units of m^2/s.

$$\upsilon = \mu/\rho \quad \dots (7.3)$$

In units, $$m^2/s = \frac{kg/(m.s)}{kg/m^3} \Rightarrow m^2/s$$

7.5 PRESSURE

The basic property of a static fluid is pressure. When a certain mass of fluid is contained in a vessel, it exerts forces at all points on the surfaces of the vessel in contact. The forces so exerted always act in the direction normal to the surface in contact. *The normal force exerted by a fluid per unit area of the surface* is called as the **fluid pressure**. If F is the force acting on the area A, then the pressure or intensity of pressure is given by

$$P = F/A \quad \text{... (7.4)}$$

In a static fluid, the pressure at any given point is the same in all the directions. If the pressure at a given point was not the same in all directions, there would be non-equilibrium and the resultant force should exist. As the fluid is in static equilibrium, there is no net unbalanced force at any point. Hence, the pressure in all directions is the same and thus independent of direction.

In the SI system, the unit N/m^2 is called the pascal, symbol Pa (in honour of the scientist Pascal). As the unit of pressure Pa is small in magnitude, the pressure is usually expressed in SI in kilopascal (kPa). A multiple of pascal is called the bar and it is also used as a unit of pressure.

$$1 \text{ bar} = 10^5 \text{ Pa} = 10^5 \text{ N/m}^2$$

7.5.1 Atmospheric Pressure

The air of the atmosphere exerts pressure on all bodies that are exposed to it. The pressure exerted by the air of the atmosphere is called the atmospheric pressure. At the mean sea level, the pressure exerted by the air is 101325 N/m^2 (101325 Pa = 101.325 kPa). This pressure is known as the standard or normal atmospheric pressure or simply atmospheric pressure. One standard atmosphere (1 atm) balances a column of 760 mm of mercury at 0°C.

$$\begin{aligned} 1 \text{ atm} &= 760 \text{ mm Hg} = 760 \text{ Torr} \\ &= 101325 \text{ Pa or N/m}^2 \\ &= 101.325 \text{ kPa or kN/m}^2 \\ &= 1.103 \text{ bar} = 760 \text{ torr} \\ &= 10.33 \text{ m } H_2O \\ &= 1.033 \text{ kgf/cm}^2 \\ &= 14.7 \text{ psi} \end{aligned}$$

7.5.2 Vacuum (Vacuum Pressure)

Vacuum refers to sub-atmospheric pressure, i.e., pressure below the normal atmospheric pressure. It is usually expressed in the SI system in Torr (in honour of the scientist Torricelli). This unit is used for systems operating under vacuum. The pressure exerted by 1 millimetre column of mercury is named as Torr.

$$1 \text{ Torr} = 1 \text{ mmHg}$$

7.5.3 Absolute Pressure

It is the pressure prevailing in a vessel/pipe line/process equipment. It is the true/actual pressure or pressure above reference zero.

It is the pressure measured from absolute zero.

The absolute pressure at a point in a fluid is equal to the sum of the gauge and the atmospheric pressures.

7.5.4 Gauge Pressure

Normally, the pressure is measured with the help of a pressure gauge. The pressure gauge measures the pressure above the atmospheric pressure. Thus, the atmospheric pressure on the gauge scale is marked as zero (i.e., the atmospheric pressure is taken as datum). The pressure registered by the pressure gauge is called the gauge pressure and therefore the letter 'g' follows the unit of pressure [1.2 (kgf/cm^2) g, 2 atmg, 300 kPag, 30 psig].

The gauge pressure does not indicate the true total pressure (absolute pressure). For obtaining the prevailing true pressure or pressure above reference zero, we have to add the atmospheric or barometric pressure in consistent units to the gauge pressure. If no letter follows the pressure units, it is taken as absolute pressure.

The relationship between absolute and gauge pressure is

Absolute pressure = gauge pressure + atmospheric pressure

Vacuum refers to sub-atmospheric pressure (pressure below the atmospheric pressure).

The relationship between absolute pressure and vacuum is

Absolute pressure = atmospheric pressure – vacuum

The pressure measuring devices are of two types: tube gauges and mechanical gauges.

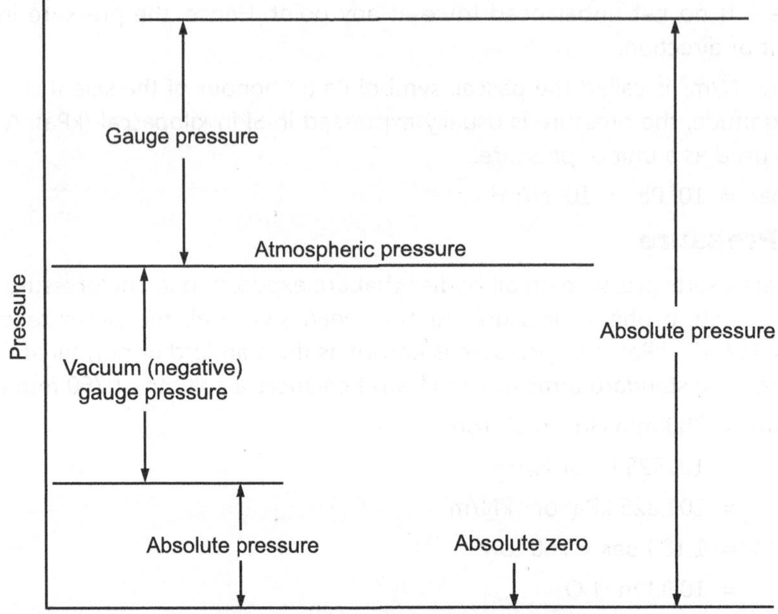

Fig. 7.3 : Pressure relationships

7.6 NEWTON'S LAW OF VISCOSITY

It states that the *shear stress on a layer of a fluid is directly proportional to the rate of shear.* Mathematically,

$$\tau \propto \frac{u}{y}, \quad \text{i.e.,} \quad \tau = \mu \cdot \frac{u}{y}$$

Consider two layers of a fluid 'y' cm apart as shown in Fig. 7.4. Let the area of each of these layers be A cm^2. Assume that the top layer is moving parallel to the bottom layer at a velocity of 'u' cm/s relative to the bottom layer. To maintain this motion, i.e., the velocity 'u' and to overcome the fluid friction between these layers, for any actual fluid, a force of 'F' dyne (dyn) is required.

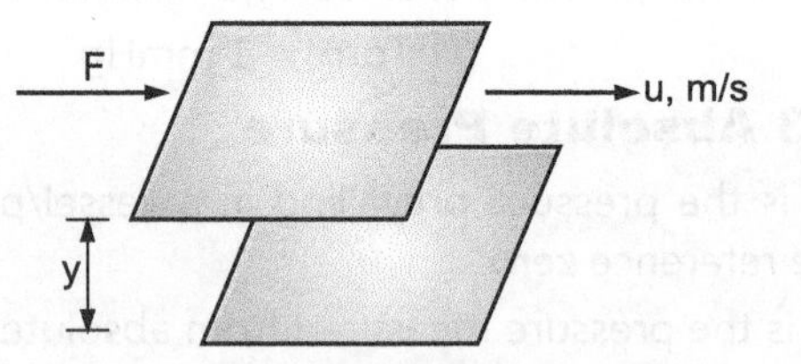

Fig. 7.4 : Definition of viscosity

Experimentally it has been found that the force F is directly proportional to the velocity u and area A, and inversely proportional to the distance y.

Therefore, mathematically it becomes

$$F \propto u.A/y \quad \text{... (7.5)}$$

Introducing a proportionality constant μ (Greek 'mu'), Equation (7.5) becomes

$$F = \mu \, u \, A/y$$

$$F/A = \mu.u/y \quad \text{... (7.6)}$$

Shear stress, τ (Greek 'tau') equal to F/A between any two layers of a fluid may be expressed as

$$\tau = F/A = \mu.u/y \qquad ... (7.7)$$

The above equation in a differential form becomes

$$\tau = \mu \cdot \frac{du}{dy} \qquad ... (7.8)$$

(The ratio u/y can be replaced by the velocity gradient du/dy.)

In the SI system, the shear stress τ is expressed in N/m^2 and the velocity gradient/shear rate or rate of shear deformation is expressed in 1/s or s^{-1}.

Equation (7.8) is called Newton's law of viscosity. In the rearranged form, it serves to define the proportionality constant as

$$\mu = \frac{\tau}{du/dy} \qquad ... (7.9)$$

which is called as the coefficient of viscosity, or the dynamic viscosity (since it involves force), or simply the viscosity of a fluid. Hence, the dynamic viscosity μ, may be defined as the shear *stress required to produce unit rate of shear deformation (or shear rate).*

Viscosity is the property of a fluid and in the SI system it has the units of $(N.s)/m^2$ or Pa.s or kg/(m.s).

As the unit $(N.s)/m^2$ is very large for most of the fluids, it is customary to express viscosity as $(mN.s)/m^2$ or mPa.s, where mN is millinewtons, i.e., 10^{-3} N and mPa is millipascal, i.e., 10^{-3} Pa.

In the C.G.S. system, viscosity may be expressed in poise (P) (the unit poise is named after the French scientist Poiseuille) or centipoise (cP).

$$1 \text{ poise} = 1 \text{ P} = 1 \text{ gm/(cm.s)} = 0.10 \text{ kg/(m.s)} = 0.10 \text{ (N.s)/m}^2 \text{ or Pa.s} = 100 \text{ cP}$$

In many problems involving viscosity, there appears a term kinematic viscosity.

The kinematic viscosity of a fluid is defined as *the ratio of the viscosity of the fluid to its density* and is denoted by the symbol υ (Greek 'nu').

$$\upsilon = \mu/\rho$$

In the SI system, υ has the units of m^2/s. The C.G.S. unit of kinematic viscosity is termed as stoke and is equal to 1 cm^2/s.

7.7 HYDROSTATIC EQUILIBRIUM - CONCEPT AND APPLICATIONS

- The term hydrostatics means the study of pressure exerted by a liquid or a gas at rest. The direction of such a pressure is always perpendicular to the surface on which it acts.
- In fluid mechanics, a fluid is said to be in hydrostatic equilibrium or hydrostatic balance when it is at rest.
- A system is in hydrostatic equilibrium when the force due to pressure exactly balances the force due to gravity.
- The principle of hydrostatic equilibrium is that the pressure at any point within a fluid at rest (when hydrostatic) is only due to the weight of the overlying fluid.
- The pressure at the bottom of fluid in a vertical tank is the weight of a column of the fluid on unit cross-section area of the tank.
- Some applications of hydrostatic equilibrium are :

  (i) Barometric equation.

  (ii) Manometers.

  (iii) Decanters.

  (iv) Centrifuge.

7.8 HYDROSTATIC EQUILIBRIUM-DERIVATION

Consider the vertical column of a single static fluid as shown in Fig. 7.5. In this column of the static fluid, the pressure at any point is the same in all directions. The pressure is also constant at any horizontal plane parallel to the earth's surface, but it varies with the height of the column (it changes along the height of the column). Let the cross-sectional area of the column be A m^2 and the density of the fluid be ρ kg/m^3. Let 'P', N/m^2 be the pressure at a height 'h' (meter) from the base of the column. At a height h + dh from the base of the common (another horizontal plane), let the pressure be P + dP, N/m^2. The forces acting on a small element of the fluid of a thickness dh between these two planes are :

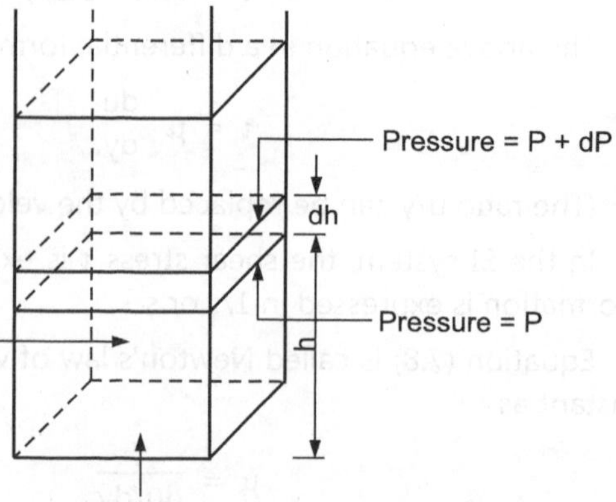

Fig. 7.5 : Hydrostatic equilibrium

(i) Force (P + dP)A is acting downwards. ... taken as +ve.

(ii) Force due to gravity is acting downwards and is equal to mass times acceleration due to gravity : mg = V ρ g = A.dh.ρ.g ... taken as –ve.
where m is the mass of the fluid contained within the two planes.

(iii) Force PA is acting upwards ... taken as –ve.

As the fluid element is in equilibrium, the resultant of these three forces acting on it must be zero. Thus,

$$+ P.A - A.\,dh.\rho.g - (P + dP)\,A = 0$$

$$P.A - A.\,dh.\rho.g - PA - A.dP = 0 \quad \text{... (7.10)}$$

$$-A.dh.\rho.g - A.dP = 0 \quad \text{... (7.11)}$$

$$dP + dh.\rho.g = 0 \quad \text{... (7.12)}$$

Equation (7.12) is the desired basic equation that can be used for obtaining the pressure at any height. Let us apply it to incompressible fluids.

Incompressible Fluids :

For incompressible fluids, density is independent of pressure.

Integrating Equation (7.12), we get

$$dP + g.\rho.dh = 0 \quad \text{... (7.13)}$$

$$\int dP + g.\rho.\int dh = 0 \quad \text{... (7.14)}$$

$$\therefore \quad P + h\rho g = \text{constant} \quad \text{... (7.15)}$$

From Equation (7.15), it is clear that the pressure is maximum at the base of the column or container of the fluid and it decreases as we move up the column or container.

If the pressure at the base of the column is P_1 where h = 0 and the pressure at any height h above the base is P_2 such that $P_1 > P_2$, then

$$\int_{P_1}^{P_2} dP = -g.\rho \int_{0}^{h} dh \quad \text{... (7.16)}$$

Integrating, we get $(P_1 - P_2) = h.\rho.g$ ... (7.17)

where P_1 and P_2 are expressed in N/m^2, ρ in kg/m^3, h in m, 'g' in m/s^2 in SI.

With the help of Equation (7.17), the pressure difference in a fluid between any two points can be obtained by measuring the height of the vertical column of the fluid.

From Equation (7.17), we get $P_1 = P_2 + h\rho g$ ... (7.17 a)

It is used to calculate P_1 knowing P_2 and h.

2. Compressible Fluids :

For compressible fluids, density varies with pressure. For an ideal gas, the density is given by the relation

$$\rho = \frac{PM}{RT} \quad \text{... (7.18)}$$

where
P = absolute pressure
M = molecular weight of gas
R = universal gas constant
T = absolute temperature.

Putting the value of 'ρ' from Equation (7.18) into Equation (7.12),

$$dP + g\,(PM/RT)\,dh = 0 \quad \text{... (7.19)}$$

Rearranging Equation (7.19),

$$\frac{dP}{P} + g \cdot \frac{M}{RT}\,dh = 0 \quad \text{... (7.20)}$$

Integrating Equation (7.20), we get

$$\ln P + g \cdot \frac{M}{RT} \cdot h = \text{constant} \quad \text{... (7.21)}$$

Integrating the above equation between two heights h_1 and h_2 where the pressures acting are P_1 and P_2, we get

$$\ln \frac{P_2}{P_1} = -g \cdot \frac{M\,(h_2 - h_1)}{RT} \quad \text{... (7.22)}$$

$$\frac{P_2}{P_1} = \exp \cdot \left[-g \cdot \frac{M}{RT}\,(h_2 - h_1)\right] \quad \text{... (7.23)}$$

Equation (7.23) is known as the barometric equation and it gives us the idea of pressure distribution within an ideal gas for isothermal conditions.

7.9 MANOMETERS

Manometers are the simplest pressure measuring devices and are used for measuring low pressure or pressure differences.

7.9.1 U-tube Manometer

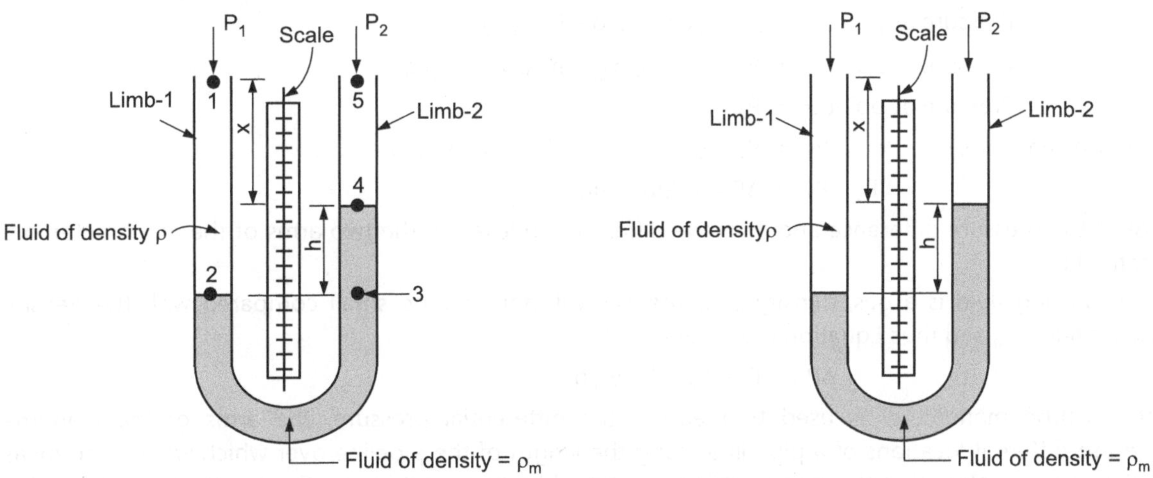

Fig. 7.6 : U-tube Manometer

- A U-tube manometer is the simplest form of manometer. It consists of a small **diameter** U-shaped glass tube. The tube is clamped on a wooden board. A scale is fixed on the same board between the two arms/limbs/legs of the manometer. The U-tube is partially filled with a manometric fluid which is heavier than the process fluid. The manometric fluid is immiscible with the process fluid. The common manometric fluid is **mercury**.

Pressure Measurement by U-tube Manometer

- A manometer works on the principle of hydrostatic equilibrium.
- A pressure measuring device using liquid columns in vertical or inclined tube is called a manometer.
- Manometers are called tube gauges to measure the pressure (P) or differential pressure (ΔP).

U-tube manometers are used to measure the pressure or differential pressure. When U-tube manometer is used to measure the pressure of a fluid flowing through a pipeline, one arm of it is connected to the pipeline (gauge point) and the other arm is open to the atmosphere, in which the manometric fluid can rise freely.

- U-tube manometer is filled with a given manometric fluid (fluid M) upto a certain height. The remaining portion of the U-tube is filled with the process fluid/flowing fluid of density ρ including the tubings. One limb of the manometer is connected to the upstream tap in a pipeline and the other limb is connected to the downstream tap in the pipeline between which the pressure difference $P_1 - P_2$ is required to be measured. Air, if any, is there in the line connecting taps and manometer is removed. At steady state, for a given flow rate, the reading of the manometer, i.e., the difference in the level of the manometric fluid in the two arms is measured and it gives the value of pressure difference in terms of manometric fluid across the taps (stations). It may then be converted in terms of m of flowing fluid.
- Consider a U-tube manometer as shown in Fig. 7.4 connected in a pipeline. Let pressure P_1 be exerted in one limb of the manometer and pressure P_2 be exerted in the another limb of the manometer. If P_1 is greater than P_2, the interface between the two liquids in the limb 1 will be depressed by a distance 'h' (say) below that in the limb 2. To arrive at a relationship between the pressure difference ($P_1 - P_2$) and the difference in the level in the two limbs of the manometer in terms of manometric fluid (h), pressures at points 1, 2, 3, 4 and 5 are considered.

  Pressure at point 1 = P_1

  Pressure at point 2 = $P_1 + (x + h)\ \rho.g$

  Pressure at point 3 = Pressure at point 2 = $P_1 + (x + h)\ \rho.g$

  (as the points 2 and 3 are at the same horizontal plane).

  Pressure at point 4 = $P_1 + (x + h)\ \rho.g - h\ .\ \rho_M.g$

  Pressure at point 5 = $P_1 + (x + h)\ \rho.g - h\ .\ \rho_M.g - x.\rho.g$

  Pressure at point 5 = P_2

Then, we can write,

$$P_2 = P_1 + (x + h)\ .\ \rho.g - h\ \rho_M.g - x\ .\ \rho.g \quad \text{... (7.24)}$$

$$P_1 - P_2 = \Delta P = h\ (\rho_M - \rho)g \quad \text{... (7.25)}$$

where ΔP is the pressure difference and 'h' is the difference in levels in the two arms of the manometer in terms of manometric fluid.

If the flowing fluid is a gas, density ρ of the gas will normally be small compared with the density of the manometric fluid, ρ_M and thus Equation (7.25) reduces to

$$\Delta P = P_1 - P_2 = h\ .\ \rho_M.g \quad \text{... (7.26)}$$

When U-tube manometer is used to measure the differential pressure, the arms of the manometer are connected to different locations of a pipe line along the length of the pipeline over which we have to measure the differential pressure. Note that the manometric fluid should be heavier than the flowing fluid and it should not mix with the flowing fluid.

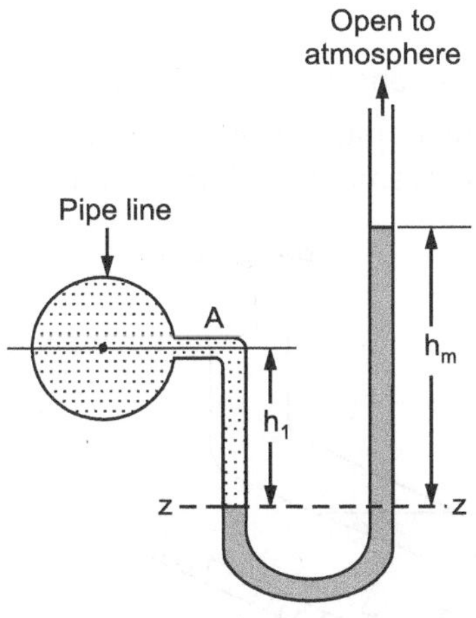

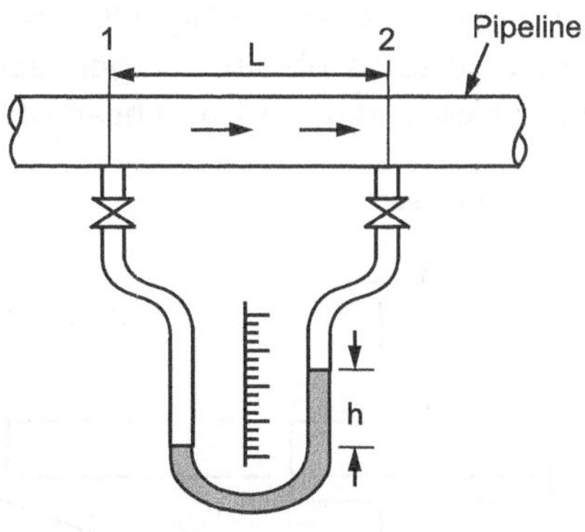

(a) Measurement of pressure (b) Measurement of differential pressure

Fig. 7.7 : Pressure measurement by U-tube manometer

The open U-tube manometer [Fig. 7.7] shows a reading h_m of manometric fluid. The interface between the manometric fluid and the fluid of which the pressure is to be measured is h_1 meters below the point of attachment A.

Let ρ be the density of the flowing fluid and ρ_m be the density of the manometric fluid.

Then, gauge pressure P_A at A is

$$P_A = h_m \rho_m g - h_1 \rho g \quad \text{... (7.27)}$$

Let us obtain the above equation.

Let z-z = the datum line as shown in Fig. 7.7.

h_m = height of the manometric fluid above the datum line.

h_1 = height of the flowing fluid above the datum line.

Pressure in the left arm/limb above the datum line = $P_A + h_1 \rho g$, N/m^2.

Pressure in the right arm/limb above the datum line = $h_m \rho_m g$, N/m^2.

Since the pressure in both the arms above the z-z datum is equal, we can write

$$P_A + h_1 \rho g = h_m \rho_m g$$

$$P_A = h_m \rho_m g - h_1 \rho g$$

Th head in meters of the flowing fluid is given by

$$h_A = \frac{P_A}{\rho g} = \frac{h_m \rho_m g - h_1 \rho g}{\rho g}$$

$$h_A = h_m \left(\frac{\rho_m}{\rho g}\right) - h_1 \text{ ... m of flowing fluid} \quad \text{... (7.28)}$$

The differential pressure or pressure difference between two points within a fluid across which a U-tube manometer is connected is given by the relation :

$$(P_1 - P_2) = \Delta P = h\,(\rho_m - \rho)\,g \quad \text{... (7.29)}$$

where h is the difference in levels in the two arms of the U-tube manometer in terms of the manometric fluid [difference in levels of the manometric fluid in the two arms of the manometer].

Here, $P_1 > P_2$. P_1 and P_2 are expressed in N/m^2, ρ in kg/m^3, h in m and g in m/s^2.

If the flowing fluid is a gas, then the density, ρ, of the gas is normally small compared to the density of the manometric fluid. Thus, it can be neglected and with this Equation (7.29) reduces to :

$$P_1 - P_2 = \Delta P = h\,\rho_m\,g \quad \text{... (7.30)}$$

7.9.2 Inclined Manometer

- Inclined manometers are used for measuring small pressure differences.
- This type of manometer is shown in Fig. 7.8. One arm of the manometer is inclined at an angle of 5 to 10° with the horizontal so as to obtain a larger reading. (e.g., movement of 7 to 10 mm is obtained for a pressure change corresponding to 1 mm head of liquid.)

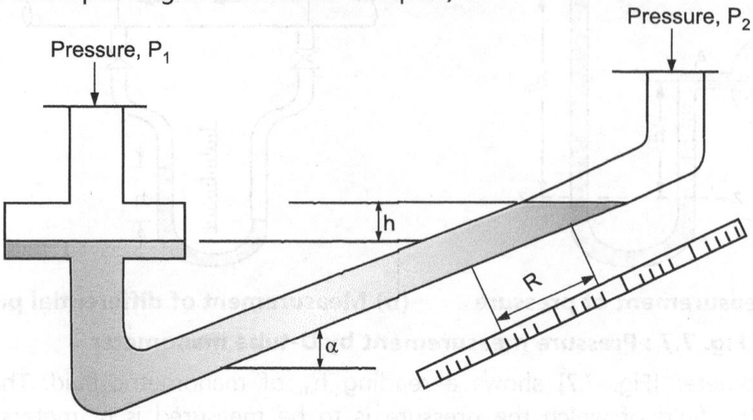

Fig. 7.8 : Inclined Tube Manometer

- In the vertical leg of this manometer an enlargement is provided so that the movement of the meniscus in this enlargement is negligible within the operating range of the manometer. If the reading R(in m) is taken as shown, i.e., distance travelled by the meniscus of the manometric fluid along the tube, then

$$h = R \sin \alpha \qquad \text{... (7.31)}$$

where α = angle of inclination

$$\text{and } (P_1 - P_2) = R \sin \alpha \, (\rho_M - \rho) \, g \qquad \text{... (7.32)}$$

7.9.3 Differential Manometer / Two Liquid Manometer / Multiplying Gauge

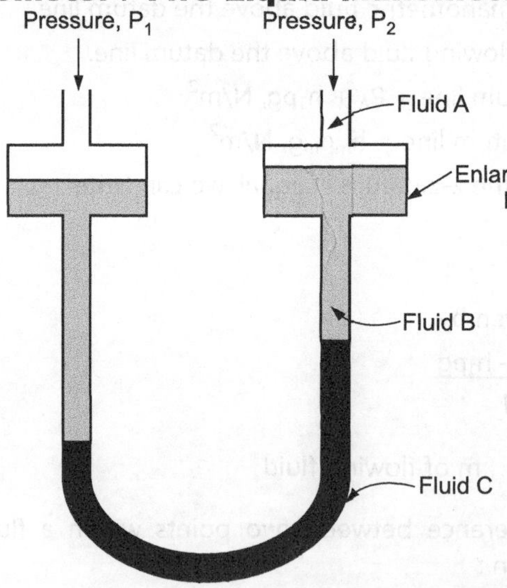

Fig. 7.9 (A) : Differential Manometer

- Differential manometer is used for the measurement of very small pressure differences or for the measurement of pressure differences with a very high precision. It may often be used for gases.
- It consists of a U-tube made of glass. The ends of the tube are connected to two enlarged transparent chambers / reservoirs. The reservoirs at the ends of each arm are of a large cross-section than that of the tube. The manometer contains two manometric liquids of different densities and these are immiscible with each other and with the fluid for which the pressure difference is to be measured. This type of manometer is shown in Fig. 7.9 (A).

- The densities of the manometric fluids are nearly equal to have a high sensitivity of the manometer. Liquids which give sharp interfaces are commonly used, e.g., paraffin oil and industrial alcohol, etc.

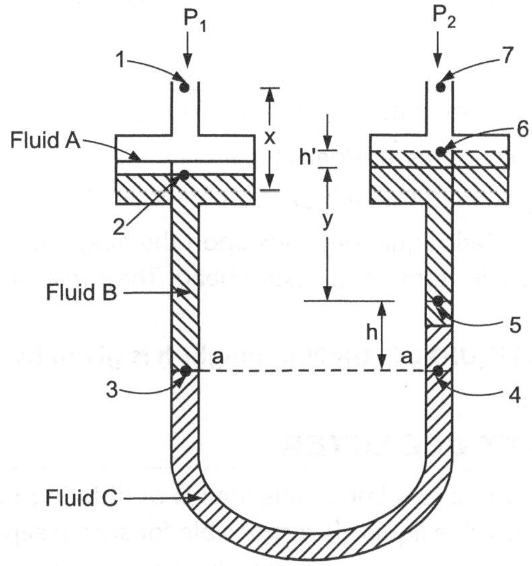

Fig. 7.9 (B) : Differential Manometer (for pressure balance)

Let the flowing fluid be 'A' of density ρ_A and manometric fluids be B and C of densities ρ_B and ρ_C ($\rho_C > \rho_B$), respectively [$\rho_A < \rho_B$ and ρ_C].

The pressure difference between two points (1 and 7) can be obtained by writing down pressures at points 1, 2, 3, 4, 5, 6, and 7 and is given by

$$P_1 - P_2 = h' (\rho_B - \rho_A) g + h (\rho_C - \rho_B) g \quad \text{... (7.33)}$$

If the level of liquid in two reservoirs is approximately same, then $h' \approx 0$ and Equation (7.33) reduces to

$$P_1 - P_2 = h (\rho_C - \rho_B) g \quad \text{... (7.34)}$$

where h is the difference in level in the two arms/limbs of the manometer.

When the densities ρ_B and ρ_C are nearly equal [($\rho_C - \rho_B$) small], then very large values of h can be obtained for small pressure differences.

Alternately, the pressure at the level a – a in Fig. 7.7 must be the same in each of the limbs and therefore,

$$P_1 + [x.\rho_A + h' \rho_A + y.\rho_B + h.\rho_B] g = P_2 + [x.\rho_A + h' \rho_B + y .\rho_B + h.\rho_C] g \quad \text{... (7.35)}$$

$$\therefore \quad (P_1 - P_2) = h' (\rho_B - \rho_A) g + h (\rho_C - \rho_B) g \quad \text{... (7.36)}$$

7.10 PRESSURE EXERTED BY A LIQUID COLUMN

Pressure : The basic property of a static fluid is pressure. When a certain mass of fluid is contained in a vessel, it exerts forces at all points on the surfaces of the vessel in contact. The forces so exerted always act in the direction normal to the surface in contact. *The normal force exerted by a fluid per unit area of the surface* is called as the **fluid pressure**. If F is the force acting on the area A, then the pressure or intensity of pressure is given by

$$P = F/A \quad \text{... (7.37)}$$

In a static fluid, the pressure at any given point is the same in all the directions. If the pressure at a given point was not the same in all directions, there would be non-equilibrium and the resultant force should exist. As the fluid is in static equilibrium, there is no net unbalanced force at any point. Hence, the pressure in all directions is the same and thus independent of direction.

Pressure Head : Frequently, the pressure is expressed in pressure heads.

The vertical height or the free surface above any point in a liquid at rest is called as the pressure head. The pressure head may be expressed as

$$h = \frac{P}{\rho g}, \frac{N/m^2}{kg/m^3 \times m/s^2} = \frac{(kg.m/s^2)/m^2}{(kg/m^3)\ (m/s^2)} = m \quad \ldots (7.38)$$

[pressure head = absolute pressure/density]

where P is in N/m^2, ρ in kg/m^3 and g in m/s^2. The unit of h is m of liquid.

As the pressure at any point in a static liquid depends upon the height of the free surface above the point, it is convenient to express a fluid pressure in terms of pressure head. The pressure head is then expressed in terms of meters of a liquid column.

The PRESSURE EXERTED BY A LIQUID COLUMN of height h is given by

$$P = h\rho g \quad \ldots (7.39)$$

7.11 CONTINUOUS GRAVITY DECANTER

Decantation involves the separation of two immiscible liquids of differing densities from one another. Basically, the difference in densities of two immiscible liquids is responsible for such a separation.

Decanters used for the separation of two immiscible liquids are : (i) gravity decanter and (ii) centrifugal decanter. Decanters utilize either a gravitational force or a centrifugal force to effect the separation.

A gravity decanter is used for the separation of two immiscible liquids when the difference between densities of the two liquids is large. A centrifugal decanter is used for the separation of two immiscible liquids whenever the difference between densities of the two liquids is small. The separating force (centrifugal force) in the centrifugal decanter is much larger than the force of gravity.

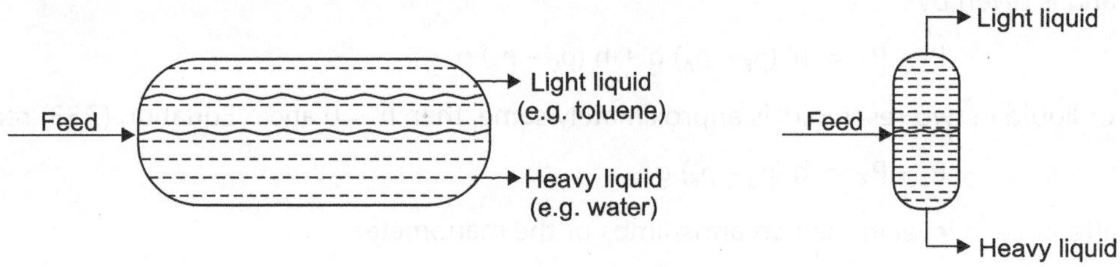

Fig. 7.10 (A) : Continuous Gravity Decanters for immiscible liquids

Separation of two immiscible liquids based on the density difference of the phases involved is commonly encountered in the mass transfer operation such as liquid-liquid extraction.

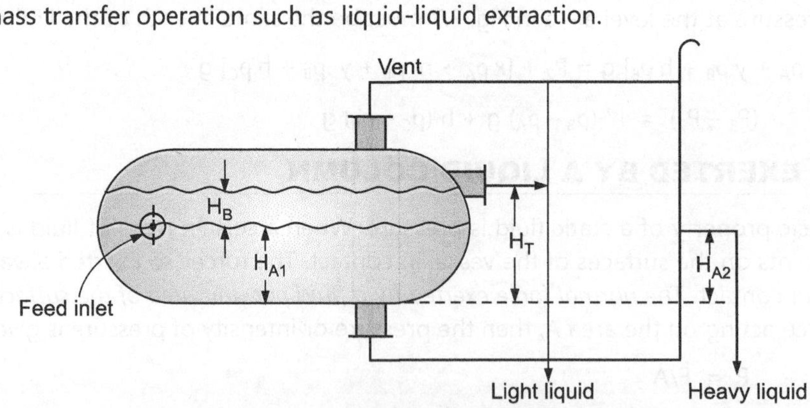

Fig. 7.10 (B) : Continuous Gravity Decanter

Feed enters into the decanter at one end, two immiscible liquids flow slowly, separate into two layers based on the density difference, and then finally the separated layers leave the decanter through the overflow lines at the other end.

Let the densities of the heavy and light liquids be ρ_A and ρ_B respectively. Let the interface between two liquids is at a height H_{A_1} from the bottom of the vessel. The total depth of the liquid in the vessel is H_T and the depth of the layer of the light liquid is H_B. The overflow of the light liquid is at a height H_{A2} and that of the heavy liquid is at a height H_{A_2} from the bottom of the vessel.

Assume that the frictional losses in the overflow discharge lines are negligible, and the overflow lines and the vessel itself are open to the atmosphere through a vent line.

A hydrostatic balance gives

$$H_B.\rho_B + H_{A_1}.\rho_A = H_{A_2}.\rho_A \quad \text{... (7.40)}$$

$$H_{A_1} = H_{A_2} - H_B \cdot \frac{\rho_B}{\rho_A} \quad \text{... (7.41)}$$

$$H_T = H_B + H_{A_1} \quad \text{... (7.42)}$$

$$H_B = H_T - H_{A_1} \quad \text{... (7.43)}$$

Substituting for H_B from Equation (7.43) into Equation (7.41), we get

$$H_{A_1} = H_{A_2} - (H_T - H_{A_1}) \cdot \frac{\rho_B}{\rho_A} \quad \text{... (7.44)}$$

Collecting the terms, we get

$$H_{A_1} (1 - \rho_B/\rho_A) = H_{A_2} - H_T (\rho_A/\rho_B) \quad \text{... (7.45)}$$

$$H_{A_1} = \frac{H_{A_2} - H_T (\rho_A/\rho_B)}{(1 - \rho_B/\rho_A)} \quad \text{... (7.46)}$$

The above equation shows that the position of the interface between the layers in the separator depends on the elevation of the overflow lines and on the ratio of the densities of two liquids.

7.12 CONCEPT OF VOLUMETRIC FLOW RATE

The flow rate of a fluid flowing through a pipeline/vessel may be expressed in terms of volume or mass of the fluid.

The volumetric flow rate of a fluid flowing at a velocity u through a conduit/pipe is given by

$$Q = u \cdot A \quad \text{... (7.47)}$$

It is defined as the *product of the velocity of a fluid flowing a pipe and the cross-sectional area of the pipe.*

The volumetric flow rate of a fluid is the volume of the fluid flowing through a pipe/conduit in a unit time or per unit time.

If V is the volume of the fluid (say liquid) collected in time t, then

$$Q = V/t \quad \text{... (7.48)}$$

The unit of Q in the SI system is m^3/s.

We have : $Q = u \cdot A$

where u is the average velocity of the flowing fluid and A is the cross-sectional area of the conduit or pipe.

If u is in m/s and A is in m^2, then Q will be in m^3/s.

7.13 CONCEPT OF MASS FLOW RATE

The mass flow rate of a fluid flowing through a conduit is the quantity of the fluid flowing through the conduit on weight basis in a unit time or per unit time. It is denoted by the symbol $\dot{m}$ and has the units of kg/s (SI unit).

7.14 INTERCONVERSION OF VOLUMETRIC AND MASS FLOW RATES

The mass flow rate of a fluid flowing through a conduit can be converted to the volumetric flow rate (or vice-versa) by the relation

$$\dot{m} = Q \cdot \rho \qquad \text{... (7.49)}$$

where ρ is the density of the flowing fluid in kg/m^3.

Mass flow rate of fluid = Volumetric flow rate of fluid × Density of fluid

$$\dot{m} = Q \cdot \rho$$

in units, $kg/s = (m^3/s) \cdot (kg/m^3) \Rightarrow kg/s$

If Q is in l/s and ρ in kg/l, then $\dot{m}$ is in kg/s.

While converting the mass flow rate into volumetric flow rate and vice-versa, we have to use consistent units for $\dot{m}$, Q and ρ.

Suppose that Q = 20 l/s and ρ = 1000 kg/m^3 and we have to find $\dot{m}$ in kg/s, then

$$\dot{m} = Q \cdot \rho$$
$$= 20\ l/s \cdot (1000\ kg/m^3)\ (10^{-3}\ m^3/l) = 20\ kg/s$$

(Here the density in kg per m^3 should be converted to kg per l).

7.15 AVERAGE VELOCITY

When the flow is not potential flow, the local velocity of fluid will vary from point to point within a given single cross-section of the conduit (tube/pipe). Hence, for all practical purposes, it is very convenient to express the average velocity.

- The average velocity (u) of the entire fluid stream flowing through the cross-sectional area A is defined by

$$u = \frac{\dot{m}}{\rho . A} \qquad \text{... (7.50)}$$

If $\dot{m}$ is expressed in kg/s, ρ in kg/m^3 and A in m^2, then u will be having the units of m/s.

The average velocity is also equal to the ratio of the volumetric flow rate to the cross-sectional area of the conduit.

$$u = \frac{Q}{A} \qquad \text{... (7.51)}$$

where Q is the volumetric flow rate. If Q is expressed in m^3/s and A in m^2, then u will have the units of m/s.

The flow rate may be expressed in terms of volume or mass of flowing fluid.

The mass flow rate through the conduit is related to the volumetric flow rate by the relation

$$\dot{m} = Q . \rho \qquad \text{... (7.52)}$$

where ρ is the density of the flowing fluid.

7.16 MASS VELOCITY

In the case of flow of compressible fluids, it is a usual practice to use the mass velocity. The mass velocity does not depend upon temperature and pressure when the flow is steady.

- The mass velocity of a fluid is the *ratio of the mass flow rate of the fluid to the cross-sectional area of the conduit.*

$$\text{Mass velocity} = G = \frac{\dot{m}}{A} \qquad \text{... (7.53)}$$

If $\dot{m}$ is expressed in kg/s and A in m^2, then G will have the units of kg/(m.s).

7.17 POINT VELOCITY

It is the velocity of a flowing fluid at a certain location within the stream of the fluid.

7.18 TYPES OF FLOW

7.18.1 Steady Flow

- The flow is said to be steady if it does not vary with time.
- If the mass flow rate and/or the quantities such as temperature and pressure at a point in the flow system do not change/vary with time, then the flow is said to be steady.
- A flow which is invariant with time is steady.

7.18.2 Unsteady Flow

- The flow is said to be unsteady if it varies with time.
- A flow which is time dependent is unsteady.
- If the mass flow rate and/or the quantities such as temperature and pressure at a point in the flow system vary/change with time, then the flow is said to be unsteady.

7.18.3 Potential Flow

The flow of incompressible fluids without the presence of shear is known as potential flow. In potential flow, eddies and cross currents cannot form within the stream and friction cannot develop.

7.18.4 Fully Developed Flow

The flow with unchanging velocity distribution is called fully developed flow.

7.18.5 Compressible Flow

The flow in which the volume of a flowing fluid and thus its density changes during the flow is called compressible flow.

7.18.6 Incompressible Flow

The flow in which the volume of a flowing fluid and thus its density does not change during the flow is called incompressible flow.

7.19 STREAM LINE AND STREAM TUBE

Stream line :

A stream line is an imaginary line in the fluid such that the fluid velocity at any point is tangential to it. As the velocity is tangent to the streamline, no matter can cross it.

Stream tube :

A stream tube is an element of fluid bounded by a number of streamlines. The stream tube may be of any convenient cross-sectional shape and no net flow occurs across it.

7.20 EQUATION OF CONTINUITY OF FLUID FLOW

- When a fluid is in motion, it must move in such a way that its mass is conserved along a flow path.
- If a fluid is continuously flowing through a conduit (pipe or channel) of constant or varying cross-sectional area, the quantity of the fluid passing per second on weight basis is the same at all locations of the conduit.

Mathematically, $\dot{m} = \rho u A = \text{constant}$

This is known as the equation of continuity. It is the first and fundamental equation of fluid flow.

The equation of continuity is a mathematical expression for the law of conservation of mass. According to the law of conservation of mass for a steady flow system, the rate of mass entering the flow system is equal to that leaving as accumulation is either constant or nil in the flow system under steady conditions.

[In a steady state flow system the values of the quantity of a fluid flowing through the system and variables of the system do not change with time.

Consider a flow system (a stream tube of varying cross-section) as shown in Fig. 7.11.

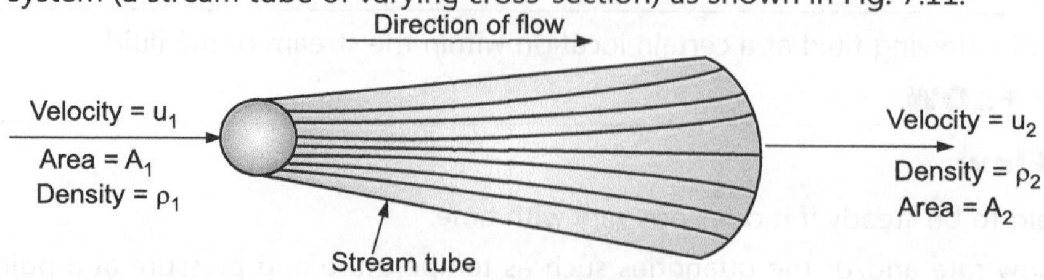

Fig. 7.11 : Continuity

As the flow cannot take place across the walls of the stream tube, the rate of mass entering the stream tube must be equal to that leaving. Let u_1, ρ_1 and A_1 be the average velocity of the fluid, the density of the fluid and cross-section area of the tube at the entrance, and let u_2, ρ_2 and A_2 be the corresponding quantities at the exit of the tube. Assume that the flow to be potential flow and the density to be constant in a single cross-section.

Rate of mass entering the flow system = $\rho_1 u_1 A_1$

Rate of mass leaving the flow system = $\rho_2 u_2 A_2$

Let $\dot{m}$ be the rate of flow in mass per unit time (mass flow rate of the flowing fluid).

Under steady flow conditions, according to the law of conservation of mass, the mass of fluid entering the tube in unit time is the same as that leaving the tube. Therefore,

$$\dot{m} = \rho_1 u_1 A_1 = \rho_2 u_2 A_2 \quad \text{... (7.54)}$$

From Equation (7.54), it follows for a stream tube or a pipe/channel of varying/constant cross-sectional area

$$\dot{m} = \rho uA = \text{constant}$$

OR :

$$\rho_1 u_1 A_1 = \rho_2 u_2 A_2 \quad \text{... (7.55)}$$

Equation (7.55) is known as the *equation of continuity*. It is applicable to compressible as well as to incompressible fluids. In the case of incompressible fluids, $\rho_1 = \rho_2 = \rho$.

The equation of continuity is useful for calculating the velocity of a fluid flowing through pipes of different diameters.

Assume that we know the velocity (u_1) of a fluid through a pipe of diameter D_1 and we have to obtain the velocity of the fluid (u_2) through a pipe of diameter D_2 which is connected to the pipe of diameter D_1. Then, from the equation of continuity,

$$\dot{m} = \rho_1 u_1 A_1 = \rho_2 u_2 A_2 \quad \text{... (7.56)}$$

but $\rho_1 = \rho_2$

$$\therefore \quad u_1 A_1 = u_2 A_2 \quad \text{... (7.57)}$$

where A_1 = cross-section area of the pipe of diameter $D_1 = \pi/4\ D_1^2$

A_2 = cross-section area of the pipe of diameter $D_2 = \pi/4\ D_2^2$

From Equation (7.57), it is clear that $u \propto 1/A$. That is the velocity of a fluid in a steady flow at any point is inversely proportional to the cross-sectional area of the pipe at that point.

Substituting for A_1 and A_2, Equation (7.57) becomes

$$u_1 (\pi/4.D_1^2) = u_2 (\pi/4.D_2^2)$$

$$u_2 = u_1 (D_1^2 / D_2^2) \quad \text{... (7.58)}$$

If $D_2 > D_1$, then $u_2 < u_1$

and if $D_2 < D_1$, then $u_2 > u_1$

When A is expressed in m^2, density in kg/m^3 and velocity in m/s, the unit of mass flow rate is kg/s.

We have : $\dot{m} = \rho u A$

Unit of $\dot{m} \Rightarrow (kg/m^3) \times (m/s) \times (m^2)$

$\Rightarrow kg/s$

7.21 REYNOLDS' EXPERIMENT

This experiment is still used to demonstrate the distinction between the two types of flow, namely laminar flow and turbulent flow. It was first conducted by Sir Osborne Reynolds in 1883.

The experimental set up is shown in Fig. 7.12. It consists of a horizontal glass tube with a flared entrance immersed in a glass walled constant head tank filled with water. The flow of water through the glass tube can be adjusted to any desired value by means of a valve provided at the outlet of the tube. A capillary tube connected to a small reservoir containing water soluble dye is provided at the centre of the flared entrance of the glass tube for injecting a dye solution in the form of a fine or thin filament into the stream of water.

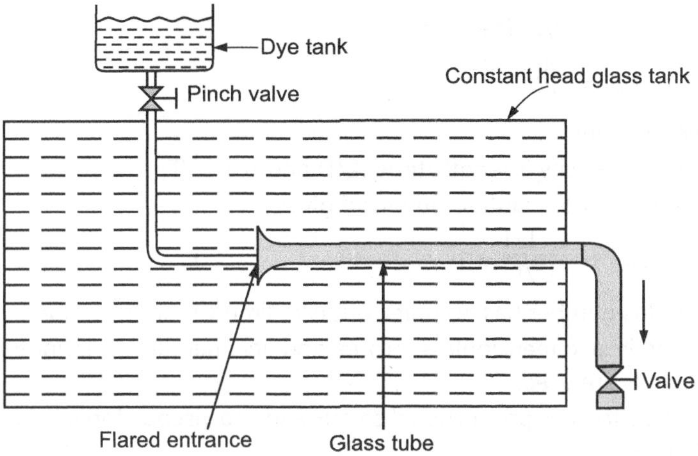

Fig. 7.12 : Reynolds experimental set up for demonstrating type of flow

By introducing a water soluble dye into the flow of water, the nature of flow could be observed. At low flow rates (i.e., at low water velocities), the filament/thread of coloured water flowed along with the stream of water in a thin, straight line without any lateral mixing. This indicated that the water was flowing in the form of parallel streams which did not interfere with each other (i.e., the water was flowing in parallel, straight lines). This type of flow pattern is called *streamline or laminar*. The laminar flow is characterised by the absence of bulk movement at right angles to the main stream direction (lateral movement). As the water flow rate was increased, a definite velocity called the critical velocity was reached, oscillations appeared in the coloured filament/thread and the thread of coloured water became wavy, gradually disappeared and the entire mass of water in the tube became uniformly coloured. In other words, the individual particles instead of flowing in an orderly manner parallel to the axis of the tube, moved erratically in the form of cross-currents and eddies which resulted into complete mixing. This type of flow pattern is known as *turbulent*. The turbulent flow is characterised by the rapid movement of fluid in the form of eddies in random directions across the tube. In between these two types of flow, there exists a transition region wherein the oscillations in the flow were unstable and any disturbance would quickly disappear.

The velocity at which the flow changes from laminar to turbulent is known as the critical velocity.

7.22 REYNOLDS NUMBER

Reynolds found that the critical velocity for the transition from laminar flow to turbulent flow depends on the diameter of the pipe, the average velocity of the flowing fluid, the density and viscosity of the fluid.

He grouped these four variables into a dimensionless group, $Du\rho/\mu$. This dimensionless group is known as the Reynolds number and found that the transition from laminar to turbulent flow occurred at a definite value of this group. The Reynolds number is a basic tool to predict the flow pattern in a conduit and is of a vital importance in the study of fluid flow. When the value of the Reynolds number is less than 2100, the flow is always laminar and when the value of the Reynolds numbers is above 4000, the flow is always turbulent. For Reynolds numbers between 2100 and 4000, a transition region exists and in this region the flow is changing rapidly from laminar to turbulent. The Reynolds number is denoted by the symbol N_{Re}.

$$N_{Re} = \frac{Du\rho}{\mu} \qquad \text{... (7.59)}$$

where D = diameter of the pipe, m

u = average velocity of the flowing fluid, m/s

ρ = density of the fluid, kg/m³

and μ = viscosity of the fluid, kg/(m.s) = (N.s)/m² = Pa.s

The Reynolds number is

$$N_{Re} = \frac{Du\rho}{\mu}$$

But $\upsilon = \mu/\rho$. Therefore,

$$N_{Re} = \frac{Du}{\upsilon}$$

where υ is the kinematic viscosity.

Dimension of D is length (L), velocity is length/time (L/T) and υ is L^2/T.

Substituting the dimensions of the quantities involved gives

$$N_{Re} = \frac{L \cdot L/T}{L^2/T} = 1 \qquad \text{... (7.60)}$$

This shows that the Reynolds number has no units, but is a dimensionless number.

∴ The Reynolds number is a dimensionless group and its magnitude is independent of the units used, provided that the consistent units are used.

For flow in a pipe, the inertia force is proportional to ρu^2 and the viscous force is proportional to $\mu.u/D$.

$$\frac{\text{Inertia force}}{\text{Viscous force}} = \frac{\rho u^2}{\mu\, u/D}$$

$$= \frac{Du\rho}{\mu} = \frac{Du}{\upsilon} = N_{Re} \qquad (\text{since } \upsilon = \mu/\rho)$$

Thus, Reynolds number is the *ratio of the inertia force to the viscous force*. This is an important physical significance of the Reynolds number.

Reynolds number is a useful tool to determine the nature of flow, whether laminar or turbulent, i.e., it gives us information regarding the type of flow whether laminar or turbulent.

7.23 CONCEPT OF LAMINAR AND TURBULENT FLOW

- When a fluid flows steadily through a conduit, either of two different types of flow may occur according to the conditions of flow.
- The flow in which the streamlines remain distinct/separated from one another over their entire length of flow is known as **laminar flow.** This flow is also called as stream line flow or viscous flow. It is characterised by the absence of lateral mixing, cross currents and eddies.
- The flow in which the fluid flows in parallel, straight lines is called laminar flow. This occurs at low fluid velocities.

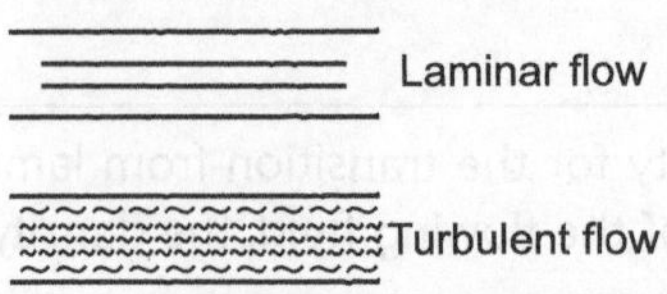

Fig. 7.13 : Type of fluid flow

- The flow in which the fluid instead of flowing in an orderly manner, moves erratically in the form of cross currents and eddies is called **turbulent flow.**
- This occurs at high fluid velocities and there is a lateral mixing in this type of flow.
- In turbulent flow the inertia force is dominating over the viscosity, while in laminar flow the viscosity of fluid is dominating over the inertia force. Turbulent flow is very common in engineering processes.

The flow which lies between laminar flow and turbulent flow is called transition flow.

7.24 CONCEPT OF CRITICAL VELOCITY

The velocity at which the flow changes from laminar to turbulent is known as the critical velocity. The value of critical velocity corresponds to N_{Re} = 2100. It is obtained as follows

$$N_{Rec} = \frac{Du_c\rho}{\mu} = 2100$$

$$\therefore \quad u_c = 2100 \times \mu/D\cdot\rho \qquad \dots (7.61)$$

where u_c is the critical velocity.

7.25 DIFFERENT TYPES OF ENERGIES ASSOCIATED WITH FLOWING FLUIDS

7.25.1 Pressure Energy

It is the work which must be done in order to introduce a fluid into a system without change in the volume. It is the energy of the fluid due to pressure acting on it.

$$\begin{aligned} \text{Pressure energy (flow energy)} &= \text{work done on a fluid} \\ &= \text{force} \times \text{displacement} \\ &= \frac{\text{force}}{\text{area}} \times \text{area} \times \text{displacement} \\ &= \text{pressure} \times \text{volume} \end{aligned}$$

$$\begin{aligned} \text{Pressure energy per unit mass of fluid} &= \frac{\text{pressure} \times \text{volume}}{\text{mass}} = \frac{\text{pressure}}{\text{mass/volume}} \\ &= \frac{P}{\rho}, \text{ J/kg in the SI system.} \end{aligned}$$

7.25.2 Kinetic Energy

It is the energy of a fluid by virtue of its motion with reference to some arbitrarily fixed body.

The kinetic energy of a fluid of mass m moving with velocity u is given by

$$\text{Kinetic energy} = \frac{1}{2} mu^2$$

$$\text{Kinetic energy per unit mass of fluid} = \frac{1}{2} u^2, \text{ J/kg in the S.I. system}$$

7.25.3 Potential Energy

It is the energy of a fluid due to its position in the earth's gravitational field. It is equal to the work that must be done on the fluid in order to raise it to a certain position from some arbitrarily chosen datum level.

At a datum level, the potential energy is taken as zero.

The potential energy of a fluid of mass m situated at a height Z above a datum level is given by

$$\text{Potential energy} = mgZ$$

$$\text{Potential energy per unit mass} = g\,Z, \text{ J/kg in the S.I. system}$$

7.26 BERNOULLI'S EQUATION

7.26.1 Bernoulli's Equation (Theorem) Statement

- Bernoulli's equation states that for an incompressible, steady and inviscid fluid flow, the total mechanical energy of the fluid is constant.
- Bernoulli equation states that for a incompressible fluid flowing steadily through a conduit, the total mechanical energy of the fluid (comprising the energy associated with fluid pressure, the gravitational potential energy and the kinetic energy of fluid motion) remains the same (or remains constant), while the fluid moves from one location to another in the conduit.

The mathematical statement of Bernoulli's equation is

$$\frac{P}{\rho} + gZ + \frac{u^2}{2} = \text{constant} \quad \text{... (7.62)}$$

The unit of each term is J/kg in the SI system, so each term represents energy per unit mass of fluid.

OR :

$$\frac{P}{\rho g} + Z + \frac{u^2}{2g} = \text{constant} \quad \text{... (7.63)}$$

7.26.2 Assumptions in Deriving Bernoulli's Equation

The following assumptions are made in deriving Bernoulli's equation.

(i) The flow is steady [i.e., the properties of a fluid (e.g., density, velocity) at any particular point or location in the flowing fluid do not change with time].

(ii) The flow is incompressible [i.e., eventhough the pressure varies, the density does not vary along a streamline (i.e., it remains constant)].

(iii) The fluid is flowing along the streamline.

(iv) The flow is irrotational (potential flow).

(v) The fluid is inviscid [for a inviscid fluid, the viscosity is zero and hence the flow occurs without friction, i.e., the flow is frictionless. There is no energy dissipation.]

[An inviscid flow is a flow of an ideal fluid which is assumed to have no viscosity.]

7.26.3 Bernoulli Equation Derivation

An important relation, called the Bernoulli equation without friction can be derived on the basis of Newton's second law of motion (force is equal to the rate of change of momentum) for potential flow. It is simply an energy balance. The variation of velocity across a given cross-section, effect of frictional forces are neglected at first and corrections for the same are then made in the equation. Thus, the relation that will be obtained is strictly applicable to an inviscid (frictionless) fluid.

Let us consider an element of length ΔL of a stream tube of constant cross-sectional area as shown in Fig. 7.14.

Let us assume that the cross-sectional area of element be A and the density of the fluid be ρ. Let u and P be the velocity and pressure at the entrance (upstream), and u + Δu, P + ΔP are the corresponding quantities at the exit (downstream).

The forces acting on the element (treating the element as a free body) are

1. The force from the upstream pressure = PA (i.e., the force acting in the direction of flow, taken as positive)
2. The force from the downstream pressure normal to the cross-section of the tube = (P + ΔP) A (i.e., the force opposing the flow, taken as negative)
3. The force from the weight of fluid [i.e., the force of gravity acting downward (taken as negative)] = ρA ΔL.g

The component of this force acting opposite to the direction of flow is ρ AΔL g cos θ.

Of the three forces cited above, the first one helps the flow while the remaining two forces oppose the flow.

Rate of change of momentum of the fluid along the fluid element = $\dot{m}\,[u + \Delta u - u] = \dot{m}\,\Delta u = \rho\, u\, A\, \Delta u$

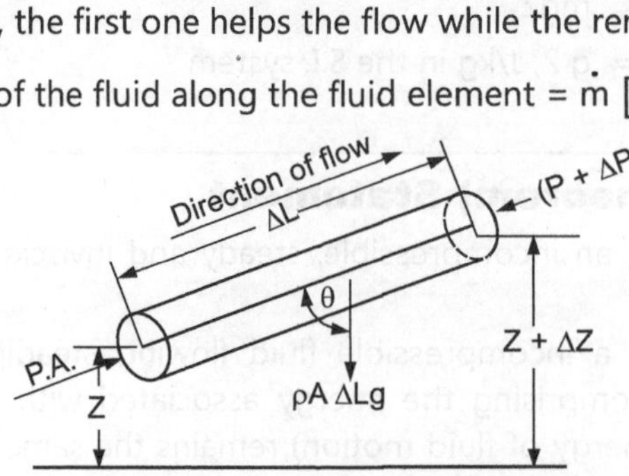

Fig. 7.14 : Force balance for potential flow

According to the Newton's second law of motion,

$$\left\{\begin{matrix}\text{Sum of all forces} \\ \text{acting in the direction of flow}\end{matrix}\right\} = \left\{\begin{matrix}\text{Rate of change of} \\ \text{momentum of a fluid}\end{matrix}\right\} \quad \text{... (7.64)}$$

$$PA - (P + \Delta P)\,A - \rho\,A\,\Delta L\,g\cos\theta = \rho\,.\,u.A\,\Delta u$$

$$-\Delta P.A - \rho\,A\,\Delta L\,g\cos\theta = \rho\,.\,u.A\,\Delta u$$

$$\therefore \quad \Delta P.A + \rho A\,\Delta L\,g\cos\theta + \rho.u.A\,\Delta u = 0 \quad \text{... (7.65)}$$

Dividing each term of Equation (7.65) by A.ΔL.ρ, we get

$$\frac{\Delta P}{\rho \Delta L} + g\cos\theta + \frac{u.\Delta u}{\Delta L} = 0 \quad \text{... (7.66)}$$

But $\cos\theta = \Delta Z/\Delta L$

Substituting the value of cos θ in Equation (7.66) gives

$$\frac{1}{\rho}\frac{\Delta P}{\Delta L} + g\frac{\Delta Z}{\Delta L} + u\frac{\Delta u}{\Delta L} = 0 \quad \text{... (7.67)}$$

If we express the changes in the pressure, velocity, height, etc. in differential form, then Equation (7.67) becomes

$$\frac{1}{\rho}\frac{dP}{dL} + g\frac{dZ}{dL} + \frac{d\,(u^2/2)}{dL}$$

which can be rewritten as :

$$\frac{dP}{\rho} + g\,dZ + d\,(u^2/2) = 0 \quad \text{... (7.68)}$$

Equation (7.68) is known as the *Bernoulli equation*. It is the differential form of the Bernoulli equation. For incompressible fluids, density is independent of pressure and hence, the integrated form of Equation (7.68) is

$$\frac{P}{\rho} + gZ + \frac{u^2}{2} = \text{constant} \quad \text{... (7.69)}$$

Thus, the Bernoulli equation, Equation (7.69), relates the pressure at a point in the fluid to its position and velocity, i.e., pressure energy at a point within a fluid to its potential energy and kinetic energy.

$$\frac{P}{\rho} = \text{pressure energy}$$

$$gZ = \text{potential energy}$$

and $$\frac{u^2}{2} = \text{kinetic energy}$$

Each term in the Bernoulli equation [Equation (7.69)] represents energy per unit mass of the fluid and has the units of J/kg in the SI system.

Let us check the unit of each term.

The unit of $\frac{P}{\rho}$ is : $$\left(\frac{N}{m^2}\right) \times \frac{1}{(kg/m^3)} = \frac{N.m}{kg} = J/kg$$

The unit of g Z is : $$\left(\frac{m}{s^2}\right)(m) = \left(\frac{kg}{kg}\right)\left(\frac{m}{s^2}\right)(m) = \left(\frac{kg.m}{s^2}\right)\left(\frac{m}{kg}\right)$$

$$= \frac{N.m}{kg} = J/kg$$

The unit of $u^2/2$ is : $$\frac{m^2}{s^2} = \left(\frac{kg}{kg}\right)\left(\frac{m^2}{s^2}\right) = \left(\frac{kg.m}{s^2}\right)\left(\frac{m}{kg}\right) = \frac{N.m}{kg} = J/kg$$

[As 1 N = 1 (kg.m)/s^2]

Equation (7.69) can be written in an alternate form as :

$$\frac{P}{\rho g} + Z + \frac{u^2}{2\,g} = \text{constant} \qquad \dots (7.70)$$

Equation (7.70) is obtained by dividing each term in Equation (7.69) by g.

Equation (7.70) is the alternate form of the Bernoulli equation and each term in this equation represents energy per unit weight of the fluid and has the dimensions of length. Hence, each term of Equation (7.70) is regarded as the head that is contributing to the *total fluid head.*

$$\text{The unit of } \frac{P}{\rho g} \text{ is } \left(\frac{N}{m^2}\right)\left(\frac{1}{\frac{kg}{m^3}}\right)\left(\frac{1}{m/s^2}\right) \Rightarrow \frac{N.s^2}{kg} \Rightarrow \frac{kg.m}{s^2} \times \frac{s^2}{kg} = m$$

$$\text{The unit of } u^2/2g \text{ is } (m/s)^2 \times \frac{1}{(m/s^2)} \Rightarrow m$$

Therefore, $\frac{P}{\rho g}$ is the pressure head / static head, Z is the potential head and $u^2/2g$ is the velocity head/kinetic head.

Weight density (w) = mass density (ρ) × g

Kinetic energy per unit weight = kinetic head or velocity head

$$= \frac{1/2\,.\,mu^2}{mg} = u^2/2g\text{ , m}$$

Potential energy per unit weight = Potential head

$$= \frac{mg\,Z}{mg} = Z\text{ , m}$$

Pressure energy per unit weight = Pressure head

$$= \frac{\text{pressure} \times \text{volume}}{\text{weight}} = \frac{\text{pressure}}{\text{weight/volume}} = \frac{P}{\text{weight density}}$$

$$= \frac{P}{\rho\,g}\text{ , m}$$

The sum of the pressure head, velocity head, and potential head is known as the total head or the total energy per unit weight of the fluid. The Bernoulli equation states that *in a steady irrotational flow of an incompressible fluid, the total energy at any point is constant.*

The sum of the pressure head and potential head, i.e., $P/\rho g + Z$ is termed as the piezometric head.

Equation (7.69) when applied between stations 1 and 2 in the direction of flow becomes

$$\frac{P_1}{\rho} + g\,Z_1 + \frac{u_1^2}{2} = \frac{P_2}{\rho} + g\,Z_2 + \frac{u_2^2}{2} \qquad \dots (7.71)$$

7.27 CORRECTIONS FOR BERNOULLI'S EQUATION

7.27.1 Kinetic Energy Correction

In the previous discussion, it is assumed that the velocity u to be constant over the area A. But in actual practice, the velocity varies over a single cross section and we have a velocity profile over the cross section. The velocity of the fluid is zero at the wall surface and maximum at the centre of the pipe. Hence, allowance must be made for the velocity profile in the kinetic energy term. This can be done by introducing a correction factor α into the kinetic energy term. The kinetic energy term would be written as $\frac{\alpha u^2}{2}$. For the flow of a fluid through a circular cross-section, α = 2 for laminar flow and α = 1 for turbulent flow.

7.27.2 Correction for Fluid Friction

The Bernoulli equation is derived for the frictionless fluid. Therefore, it must be corrected for the existence of the fluid friction whenever boundary layer forms. Fluid friction is an irreversible conversion of mechanical energy into heat. Thus, the quantity $P/\rho + u^2/2 + gZ$ is not constant but always decreases in the direction of flow.

The Bernoulli equation for incompressible fluids is corrected for friction by adding a friction term on the R.H.S. of Equation (7.71).

The Bernoulli equation between stations 1 and 2, after making necessary corrections, in terms of energy per unit mass (J/kg) is

$$\frac{P_1}{\rho} + gZ_1 + \frac{\alpha_1 u_1^2}{2} = \frac{P_2}{\rho} + gZ_2 + \frac{\alpha_2 u_2^2}{2} + h_f \quad \text{... (7.72)}$$

The Bernoulli equation is a special form of a mechanical energy balance.

where P_1, P_2 are the pressures at stations 1 and 2, respectively in N/m^2,

α_1 and α_2 are the kinetic energy conversion factors,

Z_1 and Z_2 are the heights of stations 1 and 2 from some arbitrarily chosen datum level respectively in m,

u_1 and u_2 are the average velocities at stations 1 and 2, respectively in m/s,

h_f is the total frictional loss of energy due to friction between stations 1 and 2 in J/kg.

The term h_f indicates the friction generated per unit mass of fluid that occurs in the fluid between stations 1 and 2.

Each term involved in the Bernoulli equation has the units of J/kg.

7.28 PUMP WORK IN BERNOULLI EQUATION

A pump is installed in a flow system for increasing the mechanical energy of the fluid to maintain its flow.

Assume that a pump is installed in the flow system between the stations 1 and 2 as shown in Fig. 7.15.

Let W_p be the work done by the pump per unit mass of fluid.

Let h_{fp} be the total friction in the pump per unit mass of fluid (friction in bearings, seals or stuffing box).

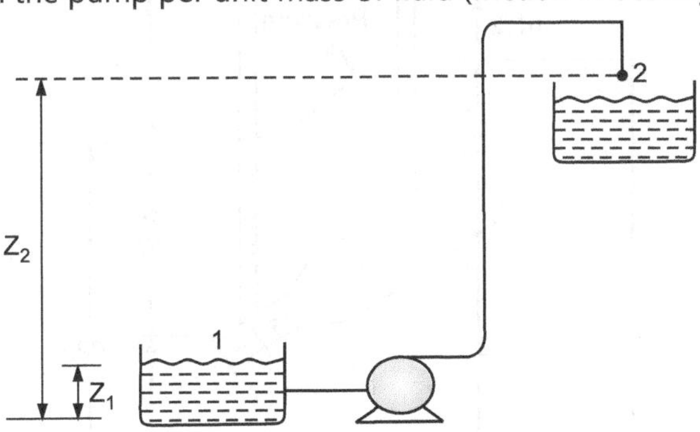

Fig. 7.15 : Pump work in Bernoulli equation

The net mechanical energy delivered to the flowing fluid is the difference between the mechanical energy supplied to the pump and frictional losses within the pump. i.e., $W_p - h_{fp}$. But to obtain the net mechanical energy (net work) delivered to the fluid, instead of using h_{fp}, a pump efficiency designated by the symbol η is used. It is defined as

$$W_p - h_{fp} = \eta W_p \quad \text{... (7.73)}$$

$$\eta = \frac{W_p - h_{fp}}{W_p} \quad \text{... (7.74)}$$

Since η is always less than one, the mechanical energy delivered to the fluid (ηW_p) is less than the work done by the pump. The Bernoulli equation corrected for the pump work between stations 1 and 2 is thus given by

$$\frac{P_1}{\rho} + gZ_1 + \frac{\alpha_1 u_1^2}{2} + \eta W_p = \frac{P_2}{\rho} + g Z_2 + \frac{\alpha_2 u_2^2}{2} + h_f \qquad \text{... (7.75)}$$

7.29 BERNOULLI'S EQUATION : SIGNIFICANCE AND APPLICATIONS

Whenever the fluid is inviscid (non-viscous) and incompressible and the flow of fluid is steady, the flow can be taken into consideration from an energy perspective.

In fluid dynamics, Bernoulli's equation is a relation among the pressure, velocity and elevation in a flowing fluid, the compressibility and viscosity of the fluid are negligible and the flow is steady.

Therefore, Bernoulli's equation (or theorem) implies that if the fluid is flowing (e.g., through a horizontal pipe) so that no change in gravitational potential energy occurs, then a decrease in fluid pressure is associated with an increase in fluid velocity.

Applications :

(i) Pumps.
(ii) Ejectors.
(iii) Siphon
(iv) Engine carburetors.
(v) Flow meters, e.g., venturi meter.
(vi) Automisers.
(vii) Bunsen burners.
(viii) Air foils.
(ix) Draft.
(x) Wind mills.
(xi) Aeroplane wings.

The working of all these is based on Bernoulli's principle/theorem.

7.30 GRAPHICAL REPRESENTATION OF ENERGY VARIATION

A graphical representation of terms P/ρ, Z and $u^2/2$ for the flow constriction in a pipe is shown in Fig. 7.16.

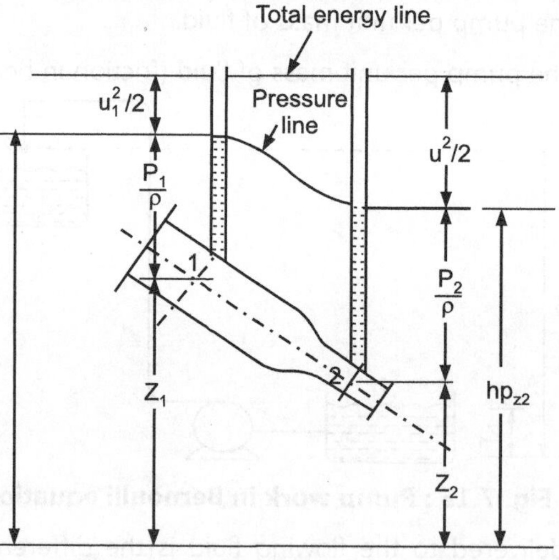

Fig. 7.16 : Energy variation in flow

The total energy line of the system shown in Fig. 7.16 is drawn horizontal since the flow is assumed to be frictionless (ideal). The flow through the system follows the continuity principle. Based on the flow rate, kinetic energy, $u^2/2$ is determined at each section. The piezometric line is drawn at a distance of $u^2/2$ below the total energy line. The distance between the piezometric line is the pressure energy P/ρ at any section.

The difference in kinetic energy $\Delta(u^2/2)$ at two locations across the constriction results in a change in potential energy, $\Delta hpz = \Delta(P/\rho) + \Delta z$.

7.31 CONCEPT OF FRICTION IN FLUID FLOW

When a fluid is flowing in a pipe, the fluid experiences some resistance to its motion due to friction of the pipe, change in cross-section and obstruction in the flow and this results in a loss of energy or head of the flowing fluid. Frictional loss is the loss of energy or head which depend on the conditions of flow and the physical properties of the system.

In a straight pipe, there is a loss of head due to frictional resistance of the pipe only.

7.31.1 Types of Friction

In fluid flow through pipes, the losses due to friction are of two kinds : skin friction and form friction, i.e., there are head losses due to skin friction and form friction. The former is due to the roughness of the inner part of the pipe where the fluid comes in contact with the pipe material, while the latter is due to obstructions present in the line of flow-may be a bend, control valve or any thing that changes the course of motion of the flowing fluid.

Skin friction : Friction between a moving fluid and the wall of a pipe. It arises due to roughness of a pipe.

Form friction : Friction caused by the eddies set-up due to an obstruction present in the line of flow.

7.31.2 Shear-stress Distribution in a Cylindrical Tube/Pipe (Flow of Incompressible Fluids)

Consider the steady flow of an incompressible fluid through a horizontal tube. Imagine a disk-shaped element of fluid, of radius r and length dL, concentric with the axis of the tube as shown in Fig. 7.17. Let P and P + dP be the fluid pressures on the upstream and downstream faces of the disk respectively. Assume fully developed flow and the density of fluid to be constant.

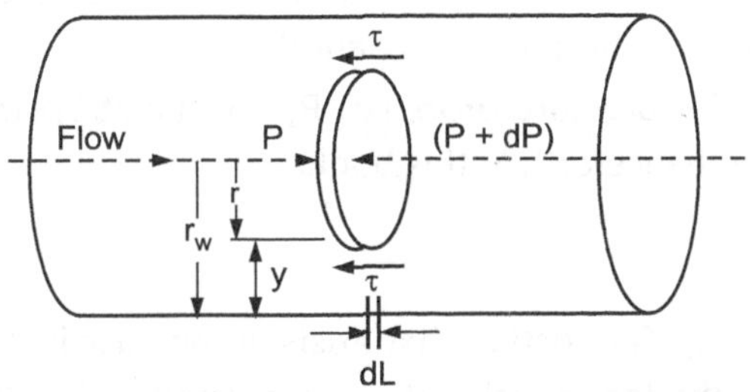

Fig. 7.17 : Fluid element (of radius r and length dL) in steady flow through a pipe

The sum of all the forces acting on this disk shaped element of fluid is equal to zero as the flow is fully developed.

$$\left[\begin{matrix}\text{Pressure force on the upstream} \\ \text{face of the disk}\end{matrix}\right] = P\pi r^2 \quad \text{... in the direction of flow (+ve)}$$

$$\left[\begin{matrix}\text{Pressure force on the downstream} \\ \text{face of the disk}\end{matrix}\right] = (P + dP)\pi r^2 \quad \text{... opposing the flow (−ve)}$$

$$\left[\begin{matrix}\text{Shear force opposing flow} \\ \text{acting at the outer surface of the disk} \\ \text{due to the viscosity of fluid}\end{matrix}\right] = \text{shear stress} \times \text{cylindrical area}$$

$$= \tau\,.\,(2\pi r\, dL) \quad \text{... opposing the flow (−ve)}$$

We have :

$$\sum F = 0 \quad \text{... (7.76)}$$

$$\therefore \quad + \pi r^2 P - \pi r^2 (P + dP) - 2\pi r\, dL\, \tau = 0$$

$$\therefore \quad - \pi r^2\, dP - 2\pi r\, dL\, \tau = 0$$

$$\pi r^2\, dP + 2\pi r\, dL\, \tau = 0 \quad \text{... (7.77)}$$

Dividing by $\pi r^2\, dL$, we get

$$\frac{dP}{dL} + \frac{2\tau}{r} = 0 \quad \text{... (7.78)}$$

In steady-state laminar or turbulent flow, the pressure at any given cross-section of a stream tube is constant, therefore dP/dL is independent of r. For the entire cross-section of the tube, Equation (7.78) can be written by taking $\tau = \tau_w$ at $r = r_w$, where τ_w is the shear stress at the wall of the tube and r_w is the radius of the tube.

$$\therefore \quad \frac{dP}{dL} + \frac{2\tau_w}{r_w} = 0 \qquad \dots (7.79)$$

Subtracting Equation (7.78) from Equation (7.79), we get

$$\frac{\tau_w}{r_w} = \frac{\tau}{r} \qquad \dots (7.80)$$

For r = 0, τ = 0 (i.e., shear stress is zero at the centreline of the tube). The simple linear relation between τ and r throughout the tube cross-section is shown graphically in Fig. 7.18.

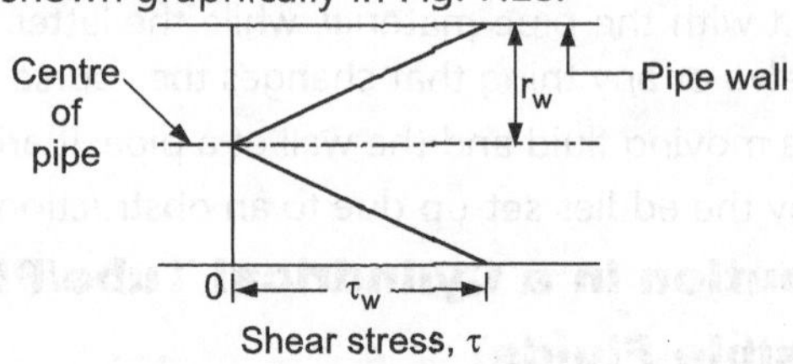

Fig. 7.18 : Shear stress variation in pipe

7.31.3 Relation between Skin Friction and Wall Shear Stress

The Bernoullis equation can be written over a definite length, ΔL for the complete stream.

Let P_1 be the pressure at the upstream of length ΔL and P_2 be the pressure at the downstream of length ΔL such that $P_1 > P_2$, so that ΔP is the pressure drop over the length ΔL.

The term ΔP is commonly used for pressure drop, i.e., $P_1 - P_2$ (inlet/upstream pressure – outlet/downstream pressure) and this terminology is used henceforth in this book.

If $P_1 = P$, then $P_2 = P - \Delta P$

since $P_1 - P_2 = \Delta P$

In this case, $Z_1 = Z_2 = 0$ and $u_1 = u_2$. The friction that exists in this case is the skin friction between the wall and the fluid stream (h_{fs}). Skin friction is the tangential friction associated with a fluid flowing over a smooth surface. When a fluid is flowing through a pipe, it is only the skin friction that exists.

Then, the Bernoulli equation over the length ΔL becomes

$$\frac{P}{\rho} = \frac{P - \Delta P}{\rho} + h_{fs} \qquad \dots (7.81)$$

$$\therefore \quad h_{fs} = \frac{\Delta P}{\rho} \qquad \dots (7.82)$$

In the above equation, each term has the units of J/kg.

If ΔP is expressed in N/m^2 and ρ in kg/m^2, then the frictional loss h_{fs} has the units of J/kg.

The Bernoulli equation, over the length ΔL, when each term in it is expressed as the head (in m of flowing fluid) is :

$$\frac{P}{\rho g} = \frac{P - \Delta P}{\rho g} + \frac{h_{fs}}{g} \qquad \dots (7.83)$$

[Equation (7.83) is obtained by dividing each term of Equation (7.81) by g]

$$\therefore \quad \frac{h_{fs}}{g} = \frac{\Delta P}{\rho g}$$

h_{fs}/g represents the head loss due to friction.

$$\begin{bmatrix} \text{Head loss due to friction} \\ \text{(frictional head loss)} \end{bmatrix} = h'_{fs} = \frac{h_{fs}}{g} = \frac{\Delta P}{\rho g} \qquad \dots (7.84)$$

Combining Equations (7.79) and (7.82) for eliminating ΔP from these equations, we get

$$h_{fs} = \frac{2\,\tau_w}{\rho_{r_w}}\Delta L = \frac{4\,\tau_w}{\rho\,D}\Delta L \quad \text{... (7.85)}$$

where D is the diameter of the pipe ($D = 2r_w$).

Combining Equations (7.84) and (7.85), we get

$$\frac{\Delta P}{\rho} = \frac{4\,\tau_w}{\rho D}\Delta L$$

$$\therefore \quad \tau_w = \frac{\Delta P D}{4\Delta L} = \frac{\Delta P\, r_w}{2\Delta L} \quad \text{... (7.86)}$$

7.31.4 Fanning Friction Factor (f)

It is especially useful in the study of turbulent flow. It is defined as *the ratio of the wall shear stress to the product of the kinetic energy of fluid and the density.*

$$f = \frac{\tau_w}{\rho.u^2/2} = \frac{2\,\tau_w}{\rho u^2} \quad \text{... (7.87)}$$

The unit of f is ⇒

$$\frac{N/m^2}{(kg/m^3)\,(m^2/s^2)} = \frac{kg.m}{s^2.m^2} \times \frac{1}{\frac{kg}{m^3} \times m^2/s^2} = \frac{kg.m^4.s^2}{kg.m^4.\,s^2}$$

Since the term $2\tau_w/\rho u^2$ (which defines f) is not having units, f is a dimensionless quantity.

From Equation (7.87), we have

$$\tau_w = f\,\rho\,u^2/2 \quad \text{... (7.88)}$$

Substituting the value of τ_w from Equation (7.88) into Equation (7.85), we get

$$h_{fs} = \frac{4\,f\,\rho\,u^2\,\Delta L}{2\,\rho\,D}$$

$$h_{fs} = \frac{4\,f\,u^2\,\Delta L}{2\,D} \quad \text{... (7.89)}$$

ΔL can be replaced by L (the length of the pipe).

$$\therefore \quad h_{fs} = \frac{4\,f\,L\,u^2}{2\,D} \quad \text{... (7.90)}$$

7.31.5 Fanning Friction Factor for Laminar Flow

The head loss due to friction in laminar flow is given by

$$h'_{fs} = h_{fs}/g = \frac{\Delta P}{\rho\,g} = \frac{32\,\mu\,L\,u}{\rho\,g\,D^2} \quad \text{... (7.91)}$$

We know that, $$u = \frac{\tau_w\,.\,r_w}{4\,\mu}$$

$$u = \frac{\tau_w.D}{8\,\mu} \quad [\text{as } D = 2\,r_w] \quad \text{... (7.92)}$$

Rearranging Equation (7.92), we get

$$\tau_w = \frac{8\,u.\mu}{D} \quad \text{... (7.93)}$$

Substituting the value of τ_w from Equation (7.93) into Equation (7.87), we get

$$f = \frac{2}{\rho u^2}\left(\frac{8\,u\,\mu}{D}\right) \quad \text{... (7.94)}$$

$$f = 16\left(\frac{\mu}{D\,\rho\,u}\right)$$

$$f = \frac{16}{\left(\frac{Du\rho}{\mu}\right)}$$

$$f = \frac{16}{N_{Re}} \quad \text{... (7.95)}$$

Thus, for laminar flow through pipes, the Fanning friction factor is calculated with the help of Equation (7.95) knowing N_{Re} (Reynolds number).

7.31.6 Fanning Friction Factor for Turbulent Flow

The turbulent flow is characterised by the presence of eddies (eddy - a circular movement of, say, water causing a small whirlpool). The eddies are of various sizes and they coexist. The fluctuations/oscillations are present in all the three directions. The fluctuations cause the mixing of different fluid portions by lateral movement. The turbulent flow is of a great importance because it brings about mixing of the fluid elements which results in higher rates of heat and mass transfer.

Fanning friction factor as a function of the Reynolds number for turbulent flow through a smooth pipe is given by

$$\frac{1}{\sqrt{f}} = 4\log\left(N_{Re}\sqrt{f}\right) - 0.40 \quad \text{... (7.96)}$$

Equation (7.96) is known as the Nikuradse equation. This equation requires a trial and error procedure for estimating the friction factor.

The other empirical relation that can be used for estimating the friction factor from the Reynolds number for turbulent flow is

$$f = \frac{0.078}{(N_{Re})^{0.25}} \quad \text{... (7.97)}$$

7.32 EFFECT OF ROUGHNESS

Our previous discussion was restricted to smooth conduits. In turbulent flow, a rough pipe results in a larger friction factor than that with a smooth pipe, for a given Reynolds number. If a rough pipe is made smooth, the friction factor is reduced and ultimately a stage will come when further smoothening of the pipe does not reduce the friction factor for a given Reynolds number. The pipe is then said to be *hydraulically smooth*.

A roughness parameter (k) is a length representing the magnitude of surface roughness. Relative roughness is the *ratio of the roughness parameter to the diameter of pipe*, i.e., k/D. The friction factor dependency on the surface roughness is given through the relative roughness. For low values of Reynolds number (N_{Re} < 2100), the friction factor is independent of the surface roughness. For large values of Reynolds number (N_{Re} > 2500), the friction factor is also a function of the surface roughness. At very large values of Reynolds number, the friction factor is independent of N_{Re} but is a function of the surface roughness.

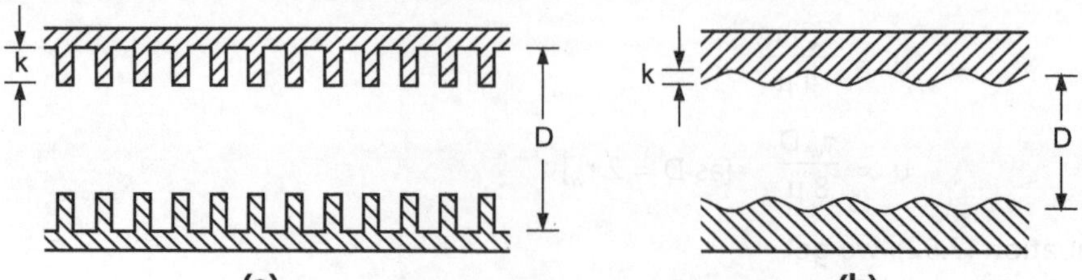

Fig. 7.19 : Types of roughness

7.33 FRICTION FACTOR CHART

It is a logarithmic plot of friction factor v/s Reynolds number over a wide range of the Reynolds number for flow of fluids in smooth as well as rough pipes. The data for the plot are taken over a wide range of each variable such as velocity, density and pipe diameter, etc. using liquids and gases. This chart is also termed as the friction factor – Reynolds number correlation chart. Such a plot is shown in Fig. 7.20. This chart is a graphical representation of the relationship : f = (N_{Re}, k/D), i.e., the relationship between the friction factor and the Reynolds number with k/D as a parameter on log-log paper.

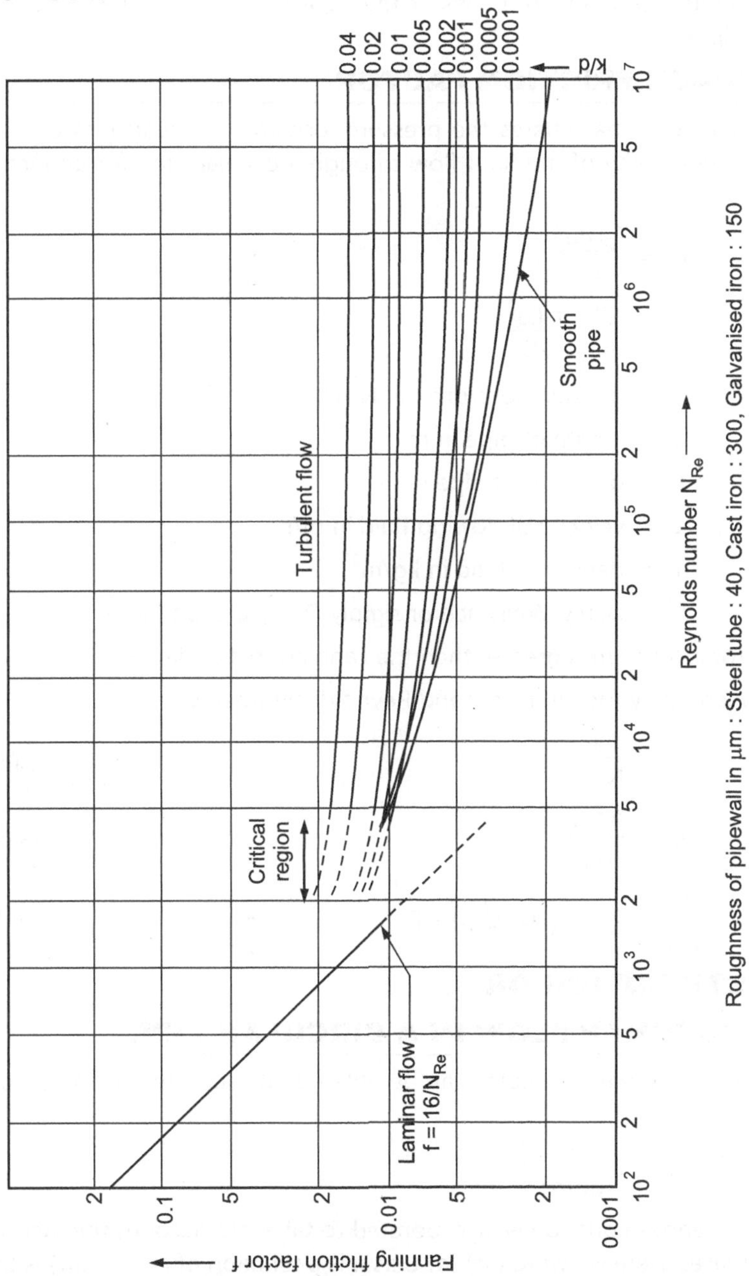

Fig. 7.20 : Friction factors for fluids flowing through pipes / Friction factor chart

The friction factor is a function of both the Reynolds number and relative roughness. For laminar flow region, the friction factor is not affected by the relative roughness of a pipe, therefore, only one line is shown for Reynolds number upto about 2100.

For the laminar flow region, the relation $f = 16/N_{Re}$ expresses the dependency of f on N_{Re}. A log-log plot of the above relation results in a straight line with a slope equal to minus one.

For turbulent flow, the relationship between 'f' and N_{Re} is given by the Nukuradse equation. Since 'f' is also a function of the relative roughness for turbulent flow, curves with different values of the dimensionless ratio k/D are presented in the chart for $N_{Re} > 2100$ assuming that turbulent flow exist at all Reynolds number greater than 2100. It is clear from Fig. 7.20 that the friction factor is low for a smooth pipe as compared to a rough pipe and increases as the relative roughness increases for a given Reynolds number. For Reynolds number above 4000, the flow is turbulent and values of f should be read from lines at the right of Fig. 7.20, where the lowest line represents the friction factor for smooth pipes.

7.34 DARCY - WEISBACH FRICTION FACTOR

The Darcy-Weisbach equation that relates the pressure loss/drop or head loss due to friction along a given length of pipe to the average velocity of the fluid flow through a circular pipe for an incompressible fluid is given as :

$$\Delta P = \frac{f_D L \rho u^2}{2D} \quad \text{...(7.98)}$$

$$f_D = \frac{\Delta P}{\rho} = \frac{f_D L u^2}{2D} \quad \text{... (7.99)}$$

where,

ΔP = pressure loss/drop in N/m^2

L = length of pipe in m

D = diameter of pipe in m

u = velocity of flowing fluid in m/s

ρ = density of fluid in kg/m^3

f_D = Darcy-Weisbach or simply Darcy friction factor

The Darcy friction factor is four times greater than the Fanning friction factor.

The relationship between Darcy friction factor and Reynolds number is

$$f_D = \frac{64}{N_{Re}} \quad \text{... for laminar flow ... (7.100)}$$

$$f_D = \frac{0.316}{N_{Re}} \quad \text{... for turbulent flow ... (7.101)}$$

OR :

$$\frac{1}{\sqrt{f_D}} = 2 \log \left(N_{Re} \sqrt{f_D}\right) - 8 \quad \text{... for turbulent flow ... (7.102)}$$

7.35 VELOCITY DISTRIBUTION OR PROFILE FOR LAMINAR FLOW IN A CIRCULAR PIPE

The velocity distribution for Newtonian fluids can be obtained through the definition of viscosity, which can be written as

$$\mu = -\frac{\tau}{du/dr} \quad \text{... (7.103)}$$

The negative sign in the above equation is incorporated to take into account the fact that in a pipe, u (velocity) decreases as r (radius) increases. Here we write u(r) for u as it is a function of radius, u(r) is the local velocity.

Rearranging Equation (7.103), we get

$$\frac{u(r)}{dr} = -\frac{\tau}{\mu} \quad \text{... (7.104)}$$

Substituting the value of τ from Equation (7.80) into Equation (7.104) gives

$$\frac{du(r)}{dr} = -\frac{\tau_w}{r_w \mu} \cdot r$$

$$du(r) = -\frac{\tau_w}{r_w \cdot \mu} \cdot r\, dr \quad \text{... (7.105)}$$

Integrating Equation (7.105) with the boundary condition : At $r = r_w : u(r) = 0$, we get

$$\int_0^u du(r) = -\frac{\tau_w}{r_w \mu} \cdot \int_{r_w}^{r} r.dr$$

$$u(r) = \frac{\tau_w}{2\, r_w \mu}\left[r_w^2 - r^2\right] \quad \text{... (7.106)}$$

The maximum value of the local velocity (u_{max}) is located at the centre of the pipe.

At the centre of the pipe, r = 0 and $u(r) = u_{max}$. Thus, from Equation (7.106), we get

$$u_{max} = \frac{\tau_w \cdot r_w}{2\, \mu} \quad \text{... (7.107)}$$

Substituting for τ_w as $\tau_w = \dfrac{\Delta P . r_w}{2\, \Delta L}$ into Equation (7.107) yields

$$u_{max} = \frac{\tau_w\, r_w}{2\, \mu} \quad \text{... (7.108)}$$

$$u_{max} = \frac{\Delta P \,.\, r_w^2}{4\mu\, \Delta L} \quad \text{... (7.109)}$$

$$u_{max} = \frac{\Delta P \,.\, D^2}{16\, \mu . \Delta L} \quad [\text{as } D = 2\, r_w] \quad \text{... (7.110)}$$

Dividing Equation (7.106) by (7.108), we get

$$\frac{u(r)}{u_{max}} = \left[1 - \left(\frac{r}{r_w}\right)^2\right]$$

or

$$\boxed{u(r) = u_{max}\left[1 - \left(\frac{r}{r_w}\right)^2\right]} \quad \text{... (7.111)}$$

It is clear from the form of Equation (7.111) that for laminar flow of Newtonian fluids through a circular pipe, the velocity distribution/profile with respect to radius is a parabola with the apex at the centreline of the pipe. Equation (7.111) relates the local velocity to the maximum velocity.

The velocity profile of a fully developed laminar flow in a circular pipe is shown in Fig. 7.21.

The nature of velocity profile in a laminar flow through a circular pipe is paraboloid with velocity zero at the wall and maximum at the centre-line. It is given by the equation :

$$u(r) = u_{max}\left[1 - \left(\frac{r}{r_w}\right)^2\right] \quad \text{... (7.112)}$$

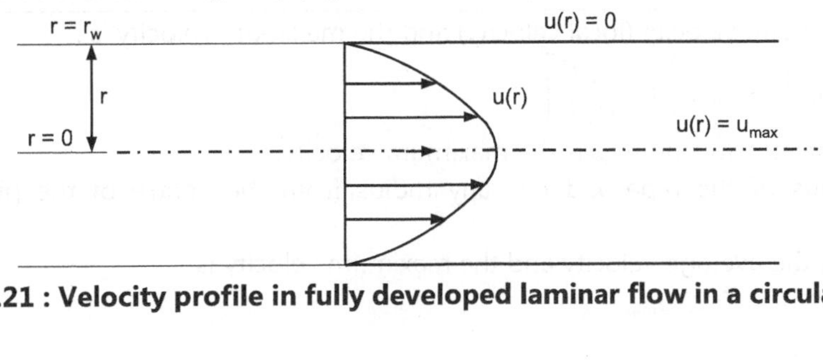

Fig. 7.21 : Velocity profile in fully developed laminar flow in a circular pipe

The average velocity u of the entire stream flowing through any given cross-sectional area (A) is defined by

$$u = \frac{1}{A}\int_A u.dA \qquad \text{... (7.113)}$$

Substituting the values of A, u and dA into Equation (7.113), we get

$$u = \frac{1}{2\pi r_w}\int_0^{r_w} u_{max}\,[1-(r/r_w)^2]\cdot 2\pi r\,dr$$

Substituting the value of u_{max} from Equation (7.113) gives

$$u = \frac{1}{2\pi r_w}\int_0^{r_w} \frac{\tau_w r_w}{2\mu}\left(\frac{r_w^2 - r^2}{r_w^2}\right)\cdot 2\pi r dr$$

$$u = \frac{\tau_w}{r_w^3\,\mu}\int_0^{r_w}\left(r_w^2 - r^2\right) r.dr \qquad \text{... (7.114)}$$

$$u = \frac{\tau_w\, r_w}{4\,\mu} \qquad \text{... (7.115)}$$

[As $A = \pi r_w^2$ and $dA = 2\,\pi\,r\,dr$ = area of elementary ring of radius r and width dr and $u_{max} = \tau_w \cdot r_w/2\mu$]

From Equations (7.108) and (7.115), we get

$$\frac{\mathbf{u}}{\mathbf{u_{max}}} = \mathbf{0.5} \qquad \text{... (7.116)}$$

i.e., *the average velocity is one half the maximum velocity.*

Eliminating τ_w by replacing it by ΔP, using $\tau_w = \frac{\Delta P\, r_w}{2\,\Delta L}$, from Equation (7.115) and replacing r_w by D/2, we get

$$u = \frac{\Delta P\, D^2}{32\,\Delta L\,\mu} \qquad \text{... (7.117)}$$

Rearranging Equation (7.117), we get

$$\Delta P = \frac{32\,\Delta L\,\mu\,u}{D^2} \qquad \text{... (7.118)}$$

ΔL can be replaced by L.

$$\therefore \quad \boxed{\Delta P = \frac{32\,L\,\mu\,u}{D^2}} \qquad \text{... (7.119)}$$

where ΔP is the pressure drop over the length L of the pipe.

Equation (7.119) is the **Hagen-Poiseuille equation for laminar flow in a circular pipe**. The Hagen-Poiseuille equation is useful to determine experimentally the viscosity of a fluid by measuring the pressure drop and the volumetric flow rate through a tube of a given length and diameter. This equation also useful for the calculation of the pressure drop due to friction in laminar/viscous flow if the viscosity is known.

The Hagen-Foiseullie equation gives us the values of loss of head due to viscosity in a laminar flow.

7.36 POINT VELOCITY, AVERAGE VELOCITY & MAXIMUM VELOCITY RELATIONS

- The relation between point velocity (local velocity) and the maximum velocity is :

$$u(r) = u_{max}\left[1-\left(\frac{r}{r_w}\right)^2\right] \qquad \text{... (7.120)}$$

where u(r) is the point velocity and u_{max} is the maximum velocity.

r_w is the inside radius of the pipe and r is any radius from the centre of the pipe that varies from r = 0 to r = r_w.

- The relation between the average velocity and the maximum velocity is

$$u = 0.50\, u_{max} \qquad \text{... (7.121)}$$

7.37 FRICTION LOSSES

When a fluid is flowing through a straight pipe, only skin friction exists and so far we have considered only this kind of friction. Whenever there are disturbances in a fluid flow path due to a change in the direction of flow or a change in the size of the pipe or due to the presence of fittings and valves, friction is generated in addition to the skin friction, i.e., additional friction losses occur. When the cross-section of the pipe changes gradually to a new cross-section or when there is a gradual change in the direction of flow, disturbance to the normal flow pattern can be small and the amount of mechanical energy loss as friction is negligible. If the change is sudden, an appreciable amount of mechanical energy is lost as friction (heat). Similarly, the presence of fittings and valves disturbs the normal flow pattern and this can cause friction losses. All these losses should be included in the friction term, h_f, of the Bernoulli equation (Equation 7.97).

Now, we will deal with the friction losses occurring as a result of a sudden enlargement or contraction of the cross-section of the pipe and the losses due to fittings.

7.37.1 Friction Loss from Sudden Enlargement

If the cross-section of the pipe is suddenly enlarged (increased), as shown in Fig. 7.22, there is a friction loss (mechanical energy loss) due to eddies which are greater at the point of sudden enlargement than in the straight pipe. Here the fluid stream separates from the wall of the pipe and issues in the enlarged section as a jet, which then expands and consequently fills the entire cross-section of the larger pipe. The space between the expanding jet and the pipe wall is filled with eddies (fluid vortices) and a large amount of friction is generated there.

The friction head loss from a sudden enlargement (expansion) of the cross-section of the pipe for turbulent flow is proportional to the velocity head of the fluid in the smaller pipe and is given by

$$h_{fe} = K_e \cdot \frac{u_1^2}{2g} \qquad \text{... (7.122)}$$

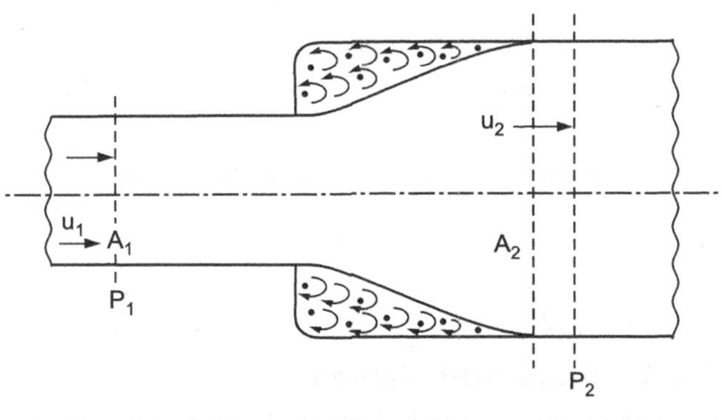

Fig. 7.22 : Flow through sudden enlargement / expansion in a pipeline

In Equation (7.122), the friction loss is expressed in m of flowing fluid.

In the energy units (J/kg), $h_{fe} = K_e' \frac{u_1^2}{2}$ . ... (7.123)

where K_e is a proportionality constant and known as the expansion-loss coefficient and u_1 is the average velocity of fluid in the smaller pipe/upstream pipe.

The expansion-loss coefficient is calculated by the following relation :

$$K_e = \left(1 - \frac{A_1}{A_2}\right)^2 \qquad (7.124)$$

where A_1 is the cross-sectional area of the smaller pipe and A_2 is the cross-sectional area of the larger pipe.

7.37.2 Friction Loss from Sudden Contraction

When the cross-section of the pipe is suddenly reduced, as shown in Fig. 7.23, the fluid stream gets separated from the wall of the pipe and a fluid jet is formed. This jet first contracts, at a short distance from the sudden contraction, in the smaller pipe and then expands to fill the entire cross-section of the smaller pipe. *The cross-section of minimum effective flow area at which the fluid jet changes from a contraction to an expansion is known as the **vena contracta**.* Upto the vena contracta, the fluid is accelerated and losses are very small. But beyond it, the velocity decreases as the flow area increases and conditions are equivalent to those for a sudden expansion. The space between the wall and the jet is filled with eddies (as shown in Fig. 7.23).

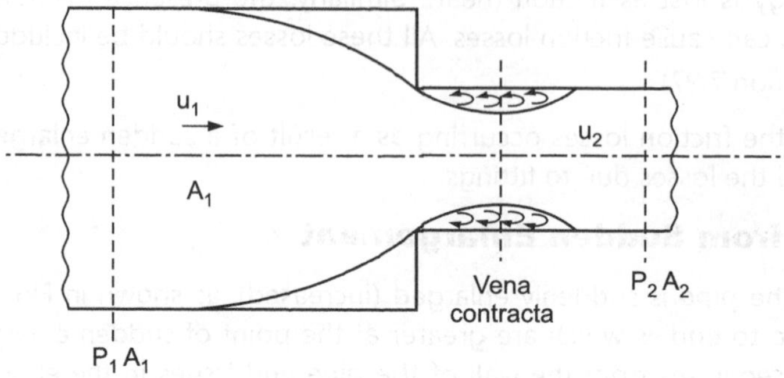

Fig. 7.23 : Flow through sudden contraction of cross-section/contraction in a pipeline

The friction loss or frictional head loss from a sudden contraction of the cross-section of the pipe is proportional to the velocity head of the fluid in the smaller diameter pipe. It is given by the relation

$$h_{fc} = K_c \frac{u_2^2}{2g}, \text{ in m of flowing fluid} \quad \text{... (7.125)}$$

where K_c is a proportionality constant and is known as the contraction-loss coefficient, u_2 is the average velocity in the smaller or downstream pipe.

For turbulent flow, K_c is given by the following relation :

$$K_c = 0.4\left[1 - \frac{A_2}{A_1}\right] \quad \text{... (7.126)}$$

where A_1 is the cross-sectional area of the larger pipe/upstream pipe and A_2 is the cross-sectional area of the smaller pipe/downstream pipe.

The frictional loss in J/kg is given by

$$h_{fc} = K_c' u_2^2/2 \quad \text{... (7.127)}$$

7.37.3 Friction Losses in Fittings and Valves

Various types of fittings and valves are used in industrial piping systems to change the direction of flow, for connecting pipes of different diameters, etc. and valves are used to control the flow or to stop the flow of a fluid. Fittings and valves disturb the normal flow-line and cause friction and may lead to greater frictional loss than that caused by the straight pipe. The frictional loss due to fittings and valves may be given either in terms of the velocity head or in terms of the equivalent length.

The frictional head loss in terms of the velocity head is given by the following equation :

$$h_{ff} = K_f \frac{u_1^2}{2\,g} \quad \text{... (7.128)}$$

The frictional loss in J/kg is given by

$$h_{ff} = K_f' \frac{u_1^2}{2} \quad \text{... (7.129)}$$

where K_f is the loss factor for the fitting and u_1 is the average velocity of the fluid in the piping leading to the fitting.

7.38 EQUIVALENT LENGTH

Another way to express the frictional loss in fittings and valves is through the equivalent length of fittings. The **equivalent length** of a fitting is *that length of straight pipe of the same nominal size as that of the fitting, which would cause the same friction loss as that caused by the fitting.*

The equivalent length of a fitting/valve is usually expressed as a certain number of pipe diameters. Thus, if a bend has an equivalent length of X diameters and if the nominal pipe diameter is 5 cm, then the bend will cause the frictional loss equivalent to that caused by the straight pipe of length 5 X cm.

The equivalent length of a fitting is to be added to the length of a straight pipe to get the total equivalent length of a flow system composed of the straight pipe and the fitting.

Equivalent lengths of some pipe fittings are :

Fitting	Equivalent length, pipe diameters
45° Elbow	15
90° Elbow	32 (upto D = 40 mm)
Tee	60 to 90
Coupler and union joint	Negligible
Gate valve (fully open)	10
Globe valve (fully open)	300

7.39 FORM-FRICTION LOSSES IN THE BERNOULLI EQUATION

Consider a flow system for the flow of an incompressible fluid as shown in Fig. 7.24, which is composed of two enlarged headers, the connecting pipes and a globe valve.

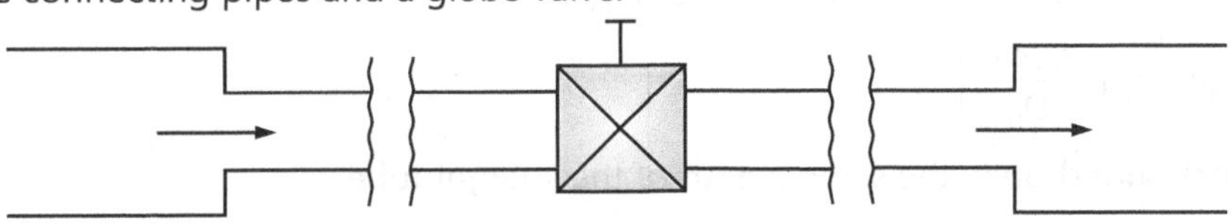

Fig. 7.24 : Typical flow system for incompressible fluid

For obtaining the total frictional loss, the losses in the expansion and contraction headers and the valve must be combined with the skin friction loss of the straight pipe.

$$\text{Skin friction loss in J/kg} = \frac{4\,fL\,u^2}{2\,D} \quad \text{... (7.130)}$$

where D is the diameter of the straight pipe, u is the average velocity of the fluid through the pipe and L is the length of the straight pipe.

$$h_f = \text{total friction loss in J/kg} = \left[\frac{4fL}{D} + K_c' + K_e' + K_f'\right]\frac{u^2}{2} \quad \text{... (7.131)}$$

Assume that the piping system under consideration involves a straight pipe of length L, two globe valves and one elbow. Further assume that the equivalent length of the globe valve is given as X pipe diameters (e.g., 190 D, so X = 190) and that of the elbow is given as Y pipe diameters.

In this situation, the total friction loss is obtained as follows :

Equivalent length of elbow = YD

Equivalent length of globe valve = XD

Total equivalent length of the piping system = L' = L + YD + 2 XD

$$h_f = \text{total friction loss in J/kg} = \frac{4\,fL'\,u^2}{2\,D} \quad \text{... (7.132)}$$

Total head loss due to friction in m of flowing fluid for the system of Fig. 7.24 is given by

$$h_f = \left[\frac{4\,fL}{D} + K_c + K_e + K_f\right]\frac{u^2}{2g} \quad \text{... (7.133)}$$

where u is the average velocity of the fluid through the pipe of diameter D and f is the Fanning friction factor.

7.40 FRICTION LOSS AT THE EXIT FROM A PIPE

The outlet end of a pipe carrying fluid may be either left free so that the fluid is discharged freely in the atmosphere or it may be connected to a large reservoir where the pipe outlet is submerged and fluid is discharged into a large body of static fluid.

In such a case, the loss of head at the exit of the pipe is equal to $u^2/2g$, where u is the average velocity of flow in the pipe.

When the end of a pipe is connected to a large reservoir, tank, etc. where the pipe end is not submerged, then the flow pattern is similar to that for a sudden enlargement. Hence, the loss of head in this case can be determined by the equation given for the head loss for a sudden enlargement.

7.41 FRICTION LOSS AT THE ENTRANCE OF PIPE FROM A LARGE VESSEL

When a fluid enters a pipe from a large vessel (or a tank or a reservoir), some loss of energy occurs at the entrance of the pipe which sometimes known as the inlet loss of energy. The flow pattern at the entrance to the pipe is similar to that in the case of a sudden contraction. So the equations given for the friction loss or head loss due to friction for a sudden contraction are applicable for this situation.

7.42 FRICTIONAL LOSSES IN COILS

In some situations a fluid flows through a circular tube arranged/set out in the form of a coil (e.g., a helical cooling coil incorporated in a batch reactor in which an exothermic reaction is occurring). The coil accommodates a long tube in a small space. The frictional pressure drop in a coil (ΔP_{coil}) is greater than that in a straight tube of the same length (ΔP_{st}). The relationship between ΔP_{coil} and ΔP_{st} is

$$\Delta P_{coil} = \Delta P_{st}\left[1 + \frac{3.54\ D}{D_H}\right] \quad \text{... (7.134)}$$

where ΔP_{st} = pressure drop in the same length of the straight tube

D = diameter of the tube

D_H = diameter of a turn of the coil or a helix of the coil (helix diameter)

7.43 PRESSURE DROP IN TURBULENT FLOW OF FLUIDS THROUGH NON-CIRCULAR CONDUITS/DUCTS

For turbulent flow in ducts of constant non-circular cross-sections, the frictional losses can be estimated by using the equations that are applicable for circular pipes if the diameter in the Reynolds number and in the definition of the friction factor is replaced by an *equivalent diameter*, which is defined as

$$\text{Equivalent diameter} = 4\ (\text{Hydraulic radius}) \quad \text{... (7.135)}$$

$$D_e = 4\ r_H \quad \text{... (7.136)}$$

$$= 4\ \frac{\text{Wetted cross-sectional area of flow}}{\text{Wetted perimeter (for fluid flow)}} \quad \text{... (7.137)}$$

The equivalent diameter is defined as four times the hydraulic radius or defined as four times the cross-sectional area of the duct divided by the wetted perimeter.

(i) For an annular space between two concentric pipes, the equivalent diameter is given by

$$D_e = 4\left(\frac{\pi/4\ D_2^2 - \pi/4\ D_1^2}{\pi\ D_1 + \pi\ D_2}\right)$$

$$\therefore \quad D_e = (D_2 - D_1) \quad \text{... (7.138)}$$

where D_1 be the outside diameter of the inside pipe (inside diameter of the annulus), and D_2 be the inside diameter of the outside pipe (outside diameter of the annulus).

(ii) For a square duct of the side x,

$$D_e = 4\left[\frac{x^2}{4\,x}\right] = x \qquad \text{... (7.139)}$$

(iii) For a rectangular duct of the size x by y :

$$D_e = \frac{4\,[x.y]}{2\,[x+y]} = \frac{2\,xy}{x+y} \qquad \text{... (7.140)}$$

(iv) For a circular pipe :

$$D_e = 4\,\frac{\pi/4 \cdot D^2}{\pi D}$$

$$D_e = D$$

Therefore, the equivalent diameter is the same as the pipe diameter for any circular pipe.

7.44 PRESSURE DROP IN FLOW THROUGH POROUS MEDIA

Porous medium : It is a continuous solid phase with many pores or void spaces in it. Typical examples of porous media include sponges, filters and packed beds used for absorption and distillation, etc.

Porous media which are not having interconnected pores are said to be impermeable to fluid flow. In such cases, there is no possibility of a fluid to flow through them, e.g., foamed polystyrene drinking cups.

Porous media which are having interconnected pores are said to be permeable to fluid flow. In such cases, there is a flow of fluid through them, e.g., a bed of granular solids.

The flow of fluids through permeable porous media is of great practical importance in filters, packed columns used for absorption and distillation and fixed-bed catalytic reactors.

Consider the flow of a homogeneous fluid through a bed of spherical particles of uniform size. If the average velocity at any cross-section perpendicular to the flow is based on the entire cross-sectional area of the pipe/tower/bed (i.e., empty tower/bed cross-section), then it is called as the superficial velocity and is given by

$$u_o = Q/A \qquad \text{... (7.141)}$$

On the other hand, the average velocity may be based on the area actually available for the flowing fluid, in which case it is called as the interstitial velocity. It is given by

$$u = \frac{u_o}{e} = \frac{Q}{e.A} \qquad \text{... (7.142)}$$

where e is the porosity, or void fraction or voidage of the bed.

$$e = \frac{\text{total volume of bed} - \text{volume of solids in bed}}{\text{total volume of bed}} = \frac{\text{free volume available for flow}}{\text{total volume of fixed bed}} \qquad \text{... (7.143)}$$

If D is the diameter of a pipe/bed/tower, h is the height of a bed in the pipe/tower, n is the number of solids in the bed and v_p is the volume of one solid particle then

$$e = \frac{\pi/4\,D^2 h - n\,v_p}{\pi/4\,D^2\,h} \qquad \text{... (7.144)}$$

From the theoretical point of view, the interstitial velocity is more important since it determines the kinetic energy, the fluid forces and nature of the flow (i.e., whether the flow is laminar or turbulent). From the practical point of view, the superficial velocity is usually more useful.

For a packed bed, the hydraulic radius varies along the bed height and is given by

$$r_H = \frac{\text{Wetted cross-sectional area normal/perpendicular to fluid flow}}{\text{Wetted perimeter}}$$

Multiplying both the numerator and the denominator by the length of the bed / the height of the bed, we get

$$r_H = \frac{\text{Volume open to flow/available for flow}}{\text{Total wetted surface area of packing}}$$

For a porous medium made of spherical particles of uniform size, the hydraulic radius is given by

$$r_H = \frac{\text{Total volume of bed} \times e}{\text{Number of particles} \times \text{Surface area of one particle}} \quad \text{... (7.145)}$$

The number of solid particles in the bed are given by

$$\text{Number of particles} = \frac{\text{Volume of bed} \times (1-e)}{\text{Volume of one particle}} \quad \text{... (7.146)}$$

Therefore, the hydraulic radius becomes

$$r_H = \frac{e}{(1-e) \times \left(\frac{\text{surface area}}{\text{volume}}\right) \text{of a spherical particle}} = \frac{e}{(1-e)\,(\pi\, d_p^2 / \pi/6\, d_p^3)}$$

$$r_H = \frac{d_p}{6}\left(\frac{e}{1-e}\right) \quad \text{... (7.147)}$$

where d_p is the diameter of the spherical particle.

For flow through packed beds, the Reynolds number and the friction factor become

$$N_{Re} = D_e\, u\rho/\mu = (4\, r_H)\,(u_o/e)\,\rho/\mu$$

Substituting for r_H gives

$$N_{Re} = 4\frac{d_p}{6} \times \frac{e}{1-e} \times \frac{u_o}{e} \times \frac{\rho}{\mu}$$

$$N_{Re} = \frac{2\, d_p\, u_o \rho}{3\,\mu\,(1-e)} \quad \text{... (7.148)}$$

We have, $\frac{\Delta P}{\rho} = \frac{4\, f\, L\, u^2}{2\, D}$

Replacing D by D_e and rearranging, we get

$$f = \left(\frac{\Delta P}{\rho\, L}\right)\left(\frac{D_e}{2\, u^2}\right)$$

Replacing D_e by 4 r_H gives

$$f = \left(\frac{\Delta P}{\rho\, L}\right)\frac{(4\, r_H)}{2\,(u_o/e)^2}$$

Substituting for r_H gives

$$f = \left(\frac{\Delta P}{\rho\, L}\right) 4\,\frac{d_p}{6} \times \frac{e}{1-e} \times \frac{e^2}{2\, u_0^2}$$

$$f = \frac{1}{3}\left(\frac{\Delta P}{\rho L}\right)\frac{d_p}{u_0^2} \times \frac{e^3}{1-e} \quad \text{... (7.149)}$$

Hence, the modified friction factor and the Reynolds number for flow through packed beds are defined as

$$f_m = \left(\frac{\Delta P}{\rho L}\right)\frac{d_p}{u_o^2} \times \frac{e^3}{1-e} \quad \text{... (7.150)}$$

and

$$N_{Re,m} = \frac{d_p\, u_o\, \rho}{\mu\,(1-e)} \quad \text{... (7.151)}$$

The Ergun equation for the flow of homogeneous fluids through packed beds relating f_m and $N_{Re,m}$ is

$$f_m = \frac{150}{N_{Re,m}} + 1.75 \quad \text{... (7.152)}$$

Equation (7.152) fits experimental data well for $N_{Re,\,m}$ between 1 and 2000.

For $N_{Re, m} < 10$ (i.e., when the flow is laminar/streamline through packed beds), the viscous forces control and Equation (7.152) reduces to

$$f_m = \frac{150}{N_{Re,m}} \quad \text{... (7.153)}$$

Substituting the values of f_m and $N_{Re,m}$, we get

$$\left(\frac{\Delta P}{\rho L}\right)\frac{d_p}{u_0^2} \times \frac{e^3}{1-e} = \frac{150}{d_p\, u_0\, \rho / \mu\, (1-e)}$$

$$\therefore \quad \left(\frac{\Delta P}{\rho}\right) = \frac{150\, \mu\, u_0\, L}{d_p^2\, \rho} \times \frac{(1-e)^2}{e^3} \quad \text{... (7.154)}$$

Equation (7.154) is known as the Blake-Kozeny equation, or Kozeny-Carman equation.

The flow of fluids through industrial filters is usually streamline/laminar and thus Equation (7.154) applies.

For $N_{Re, m} > 1000$ (i.e., when the flow is turbulent), the effect of viscous forces is negligible and inertial forces control and Equation (7.152) reduces to

$$f_m = 1.75 \quad \text{... (7.155)}$$

or

$$\frac{\Delta P}{\rho} = 1.75 \frac{u_0^2\, L}{d_p} \times \frac{(1-e)}{e^3} \quad \text{... (7.156)}$$

Equation (7.156) is known as the Burke-Plummer equation.

For non-spherical particles, the diameter d_v of a non-spherical particle may be defined as the *diameter of a sphere having the same volume as that of the particle.*

For non-spherical particles, the shape factor or sphericity ϕ_s is defined as below :

$$\phi_s = \frac{\text{Surface area of a sphere having the same volume as that of the particle}}{\text{Surface area of a particle}} \quad \text{... (7.157)}$$

For spheres : $\phi_s = 1$ and for all other particle shapes : $0 < \phi_s < 1$

For beds of non-spherical granular solids, d_p in Equations (7.147) and (7.156) is to be replaced by $\phi_s\, d_v$ so that the Ergun Equation (7.152) becomes

$$\frac{\Delta P}{\rho L} = 150 \frac{\mu\, u_0}{\rho\, (\phi_s\, d_v)^2} \frac{(1-e)^2}{e^3} + 1.75 \frac{u_0^2}{\phi_s\, d_v} \times \frac{1-e}{e^3} \quad \text{... (7.158)}$$

7.45 PRESSURE DROP IN FLUIDISATION

When a fluid (gas or liquid) is passed up through a bed of solid particles at very low velocity, the particles do not move and the pressure drop is given by the Ergun equation. If the fluid velocity is steadily increased, the pressure drop and drag on the individual particles increase, and eventually the particles start to move and become suspended in the fluid stream.

An operation by which fine granular solids are transformed into a fluid-like state through contact with a fluid is known as fluidisation.

The process occurs when a fluid (gas or liquid) passes up through the granular material. The term fluidised bed is used to describe the condition of fully suspended particles in a fluid stream.

Fluidisation is an operation in which the solid material is kept in suspension by the fluid flowing upward through a bed of the solid material.

Advantages of fluidisation : (i) good contact of the fluid with all parts of the solid particles, (ii) prevents segregation of the solid particles by thoroughly agitating the bed, (iii) temperature is uniform even in a large reactor due to complete mixing and (iv) high heat and mass transfer rates.

Disadvantages of fluidisation : (i) greater power requirement, (ii) higher breakage of solid particles, (iii) serious erosion of pipelines and vessels and (iv) necessity of recovery systems.

Minimum Fluidisation Velocity

- The smallest velocity at which fluidisation occurs is the minimum fluidisation velocity.
- The minimum velocity at which a bed of solid particles fluidises.
- The minimum fluidisation velocity is the velocity of the fluid in a fluidised bed at which the pressure drop across the bed is equal to the weight of the solid particles (i.e., the weight of the bed).

Applications of Fluidisation

Industrial applications of fluidisation include :

- Catalytic cracking of heavy petroleum fractions, calcination and roasting of minerals, combustion of coal for power generation, drying, transportation of solids, chemical reactors (e.g., commercial production of phthalic anhydride) and incineration of municipal and hazardous wastes.

Consider a vertical tube partially filled with a fine granular particles. The tube is open at the top and has a porous plate to support the bed of particles and to distribute the flow uniformly over the entire cross-section.

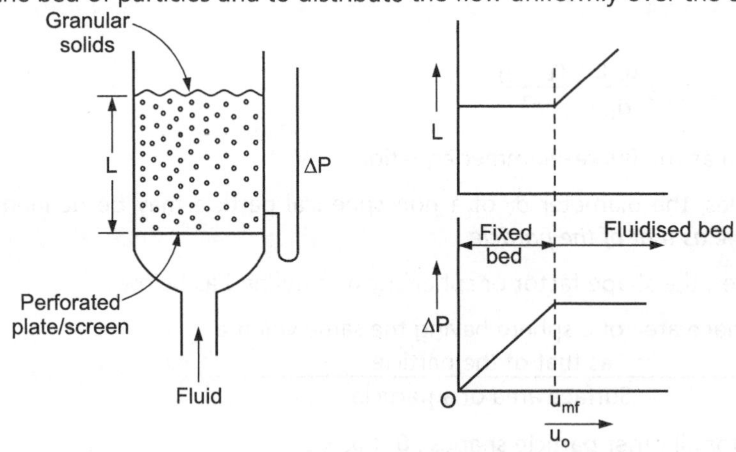

Fig. 7.25 : Pressure drop and bed heights v/s superficial velocity for bed of solids

Further, consider the upward flow of a fluid through a vertical bed of fine particles. At a low flow rate, the fluid simply passes through the void spaces between stationary particles without causing any particle motion. The flow will be laminar and the pressure drop across the bed will be proportional to the superficial velocity u_0.

As we go on increasing steadily the flow rate of the fluid, a stage will be reached when all the particles are just suspended in the flowing fluid. At this stage, the pressure drop across the bed equalizes the weight of the bed. The bed in this condition is considered to be just fluidised and is termed as an incipiently fluidised bed.

Thus, for a bed at minimum fluidisation, we have

$$\Delta P\, A = W = A\, L_{mf}\,(1 - e_{mf})\,(\rho_p - \rho)\, g \qquad \text{... (7.159)}$$

where A = cross-sectional area of the tube and ρ_p = density of the particle.

Rearranging Equation (7.159) gives

$$\frac{\Delta P}{L_{mf}} = (1 - e_{mf})\,(\rho_p - \rho)\, g \qquad \text{... (7.160)}$$

In many practical applications of fluidisation, since the particles are very small and the fluid velocity is low, the flow is streamline/laminar. Therefore, Equation (7.154) becomes

$$\frac{\Delta P}{L_{mf}} = \frac{150\, \mu\, u_{mf}}{d_p^2} \times \frac{(1 - e_{mf})^2}{e_{mf}^3} \qquad \text{... (7.161)}$$

The superficial fluid velocity at minimum fluidisation conditions is obtained by equating the right hand sides of Equations (7.160) and 7.161). Therefore, we get u_{mf} as

$$u_{mf} = \frac{d_p^2}{150} \times \frac{\rho_p - \rho}{\mu} g \left(\frac{e_{mf}^3}{1 - e_{mf}}\right) \quad \text{... (7.162)}$$

Once the bed is fluidised, the pressure drop across the bed remains constant but the bed height increases continuously with increasing velocity (i.e., flow).

Equations (7.160) and (7.161) for a given fluid-solid system predict that

$$\frac{\Delta P}{L(1-e)} = k_1 = k_2 u_0 \frac{(1-e)}{e^3} \quad \text{... (7.163)}$$

where k_1 and k_2 are constants.

$$k_1 = k_2 u_0 \frac{1-e}{e^3}$$

$$u_0 = \frac{k_1}{k_2} \frac{e^3}{1-e}, \text{ Let } k = k_1/k_2$$

$$u_0 = k \frac{e^3}{1-e} \quad \text{... (7.164)}$$

where k is a constant for the system.

Equation (7.164) gives us idea regarding the variation of the bed porosity (e) with superficial fluid velocity (u_0) in a fluidised bed. This equation has been found to apply equally well for a liquid-solid system for e (bed voidage) less than 0.80.

The relationship between bed heights and bed voidages is given by

$$\frac{L}{L_{mf}} = \frac{1-e_{mf}}{1-e} \quad \text{... (7.165)}$$

Equation (7.165) is applicable for particle Reynolds numbers less than 1.0 $N_{Re,p} = d_p u_t \rho/\mu$.

7.46 MEASUREMENT OF FLUID FLOW

In the chemical process industry, it is desirable to know the amount of a fluid flowing to or from the process equipment. Many different types of flow meters are used industrially to measure the rate at which a fluid is flowing through a pipe or a duct. The *flow measuring devices* or *flow meters* are classified as :

1. Variable head meters, e.g., orifice meter, venturi meter.
2. Variable area meters, e.g., rotameter.
3. Current meters, e.g., cup type current meter.
4. Positive displacement meters, e.g., wet gas meter.
5. Electromagnetic meters, e.g., magnetic meter.

7.46.1 Variable Head Meter and Variable Flow Meter

Variable head meters such as venturi meters, orifice meters and pitot tubes and variable area meters such as rotameters of various designs are widely used for flow measurements.

In variable head meters the area of flow is constant and the pressure drop varies with the flow rate, while in variable area meters, the pressure drop is constant and the flow area varies with the flow rate.

7.47 VENTURI METER

A venturi meter is a flow measuring device that uses a converging section of pipe to give an increase in the flow velocity and a corresponding pressure drop which is used to calculate the flow rate.

A venturi meter is a variable head meter which is used for measuring the flow rate of a fluid through a pipe. In this meter, the fluid is gradually accelerated to a throat and then gradually retarded in a diverging section where the flow expands to the pipe size. A large portion of the kinetic energy is thus recovered (converted back into pressure energy).

In this meter the fluid is gradually accelerated to the throat, the velocity of the fluid increases. The increase in velocity results in decreasing the pressure at the throat and the pressure difference between the upstream side of the meter and the throat is used to obtain the flow rate.

Principle :

- When the cross-sectional area of flow decreases, some pressure head is converted to velocity head. When the velocity increases, the pressure decreases.

OR

- The basic principle on which a venturi meter works is that *by reducing the cross-sectional area of the flow passage, a pressure difference is created and the measurement of the pressure difference (between the inlet of the meter and at a point of reduced pressure) enables the estimation of the discharge/flow rate through the pipe.*

Construction

A venturi meter consists of

(i) **An inlet section followed by a convergent cone/section**. The inlet section of the venturi meter is of the same diameter as that of the pipe line in which it is installed which is followed by the short convergent section with a converging cone angle of 15-20° and its length parallel to the axis is approximately equal to 2.7 ($D - D_T$), where D is the pipe diameter and D_T is the throat diameter. In the convergent section, the fluid is accelerated.

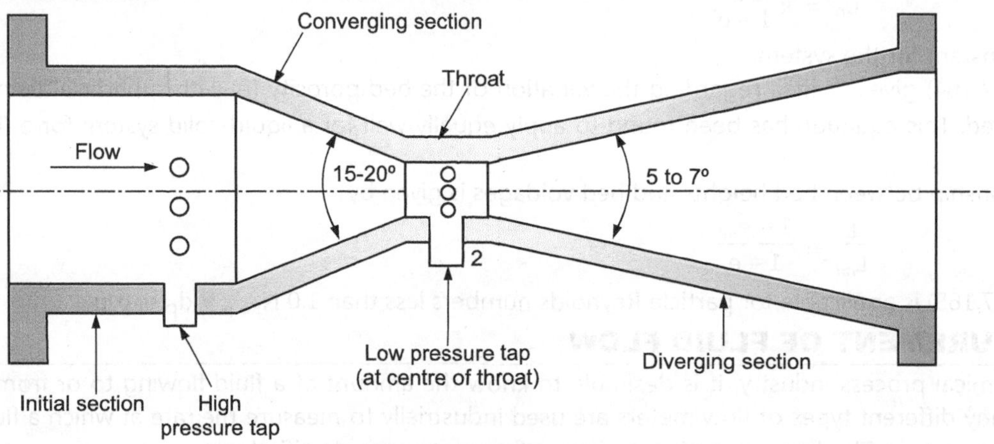

Fig. 7.26 : Venturi meter

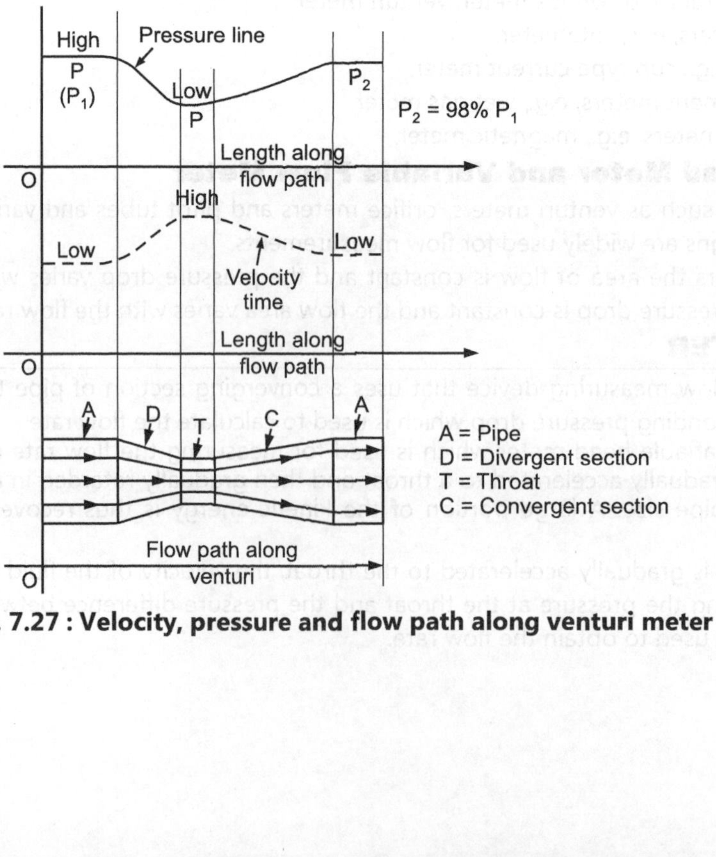

Fig. 7.27 : Velocity, pressure and flow path along venturi meter

(ii) **A cylindrical throat** – the section of constant cross-section with its length equal to diameter. The flow area is minimum at the throat. In general, the diameter of throat may vary from 1/3 to 3/4 of the pipe diameter and more commonly the diameter of throat is 1/2 the pipe diameter.

(iii) **A long diverging section/a gradual divergent cone with a cone angle of about 5 - 7°** wherein the fluid is retarded and a large portion of the kinetic energy is converted back into the pressure energy.

The convergent section/cone is a short pipe that tapers from the original pipe size to that of the throat of the meter. The divergent cone/section of the venturi meter is a gradually diverging pipe with its cross-sectional area gradually increasing from that of the throat to the original pipe size. At the inlet section and the throat (centre of throat), i.e., at stations 1 and 2 of the venturi meter, pressure taps are provided through the pressure rings/piezometer rings.

A piezometer ring is an annular chamber provided at the pressure taps with small holes drilled from the inside of the tube and is used for averaging out the individual pressures transmitted through the several small holes to a pressure measuring device.

Working :

A venturi meter of known coefficient is installed in the pipeline and the pressure taps are connected to a pressure measuring device. Air pockets, if any, are removed from connecting tubings, measuring device, etc. after starting the flow of fluid through the pipeline in which it is installed for flow measurement. In the meter, the fluid is accelerated in the converging cone and then retarded in the diverging cone gradually. An increase in the flow velocity at the throat (a section of minimum flow area) results in a decrease in the pressure at the throat. Due to this a pressure difference is developed between the inlet section and the throat section which is measured/noted with the help of manometer or pressure gauges (connected to the pressure taps) after the steady state is attained. This pressure difference/drop is then related to the flow rate by a mathematical flow equation for the meter.

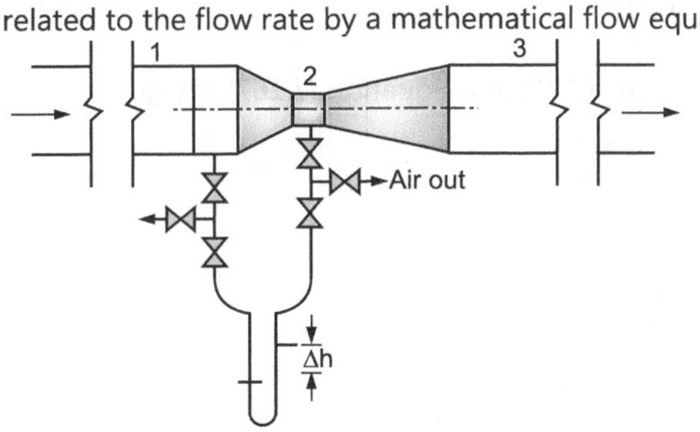

Fig. 7.28 : Venturi meter with U-tube manometer

In the venturi meter, fluid is accelerated in the convergent cone from the inlet section 1 to the throat section 2 and in the divergent cone, it is retarded from the throat section 2 to the end section 3 of the venturi meter. In order to avoid the possibility of flow separation and consequent energy loss, the divergent cone of a venturi meter is made longer with a gradual divergence. Since the separation of flow may occur in the divergent cone of a venturi meter, this portion is not used for measuring the flow rate.

Since there is a gradual reduction in the area of flow, there is no vena contracta and the flow area is minimum at the throat so that the coefficient of contraction is unity.

7.48 FLOW EQUATION FOR A VENTURI METER

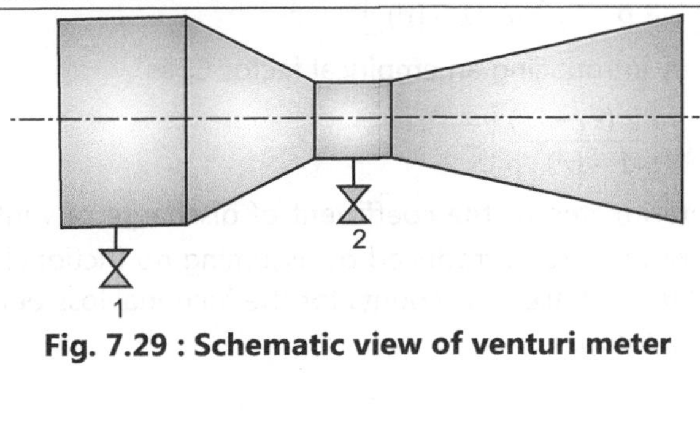

Fig. 7.29 : Schematic view of venturi meter

Let P_1, P_2 and u_1, u_2 be the pressures and velocities at section/stations 1 and 2, respectively. Let A_1 and A_T be the flow areas at stations 1 and 2, respectively. Section/ station 1 is at the upstream side of the convergent cone (inlet section) and station 2 is at the throat.

Assume that the flowing fluid is incompressible and there are no frictional losses between station 1 and station 2.

The Bernoulli equation in terms of energy units (J/kg) between stations 1 and 2 is

$$\frac{P_1}{\rho} + \frac{\alpha_1 u_1^2}{2} + g Z_1 = \frac{P_2}{\rho} + \frac{\alpha_2 u_2^2}{2} + g Z_2 \quad \text{... (7.166)}$$

The venturi is connected in a horizontal pipe (Fig. 7.29), so $Z_1 = Z_2$ (or if we assume that a datum is passing through the axis of the venturi meter, then $Z_1 = Z_2 = 0$) and the above equation then reduces to

$$\frac{P_1}{\rho} + \frac{\alpha_1 u_1^2}{2} = \frac{P_2}{\rho} + \frac{\alpha_2 u_2^2}{2} \quad \text{... (7.167)}$$

From the equation of continuity, we have

$$\dot{m} = \rho\, u_1 A_1 = \rho\, u_2 A_T \quad \text{... (7.168)}$$

where $A_1 = \pi/4.D^2$ and $A_T = \pi/4.D_T^2$

in which D = pipe diameter and D_T = throat diameter

$$\therefore \quad u_1 (\pi/4.D^2) = u_2(\pi/4.D_T^2)$$

Let $D_T/D = \beta$ = diameter ratio, diameter of throat/diameter of pipe

$$u_1 = (D_T/D)^2\, u_2 \quad \text{... (7.169)}$$

$$u_1 = \beta^2.u_2 \quad \text{... (7.170)}$$

Substituting for u_1 from Equation (7.170) into Equation (7.167), we get

$$\frac{P_1}{\rho} + \frac{\alpha_1 (\beta^2 u_2)^2}{2} = \frac{P_2}{\rho} + \frac{\alpha_2 u_2^2}{2} \quad \text{... (7.171)}$$

Rearranging, we get

$$\frac{\alpha_2 u_2^2}{2} - \frac{\alpha_1 \beta^4 u_2^2}{2} = \frac{P_1 - P_2}{\rho}$$

$$\alpha_2 u_2^2 - \alpha_1 \beta^4 u_2^2 = \frac{2 (P_1 - P_2)}{\rho}$$

$$\alpha_1 \left[\frac{\alpha_2}{\alpha_1} u_2^2 - \beta^4 u_2^2\right] = \frac{2 (P_1 - P_2)}{\rho}$$

Usually $\alpha_2/\alpha_1 = 1.0$. Therefore,

$$\alpha_1 \left[u_2^2 - \beta^4 u_2^2\right] = 2 (P_1 - P_2) / \rho$$

$$u_2^2 (1 - \beta^4) = \frac{2 (P_1 - P_2)}{\alpha_1 . \rho}$$

$$u_2 = \left[\frac{2 (P_1 - P_2)}{\rho} \times \frac{1}{\alpha_1 (1 - \beta^4)}\right]^{1/2} \quad \text{... (7.172)}$$

Equation (7.172) is corrected by introducing an empirical factor C_v as

$$u_2 = C_v \left[\frac{2 (P_1 - P_2)}{(1 - \beta^4) . \rho}\right]^{1/2} \quad \text{... (7.173)}$$

where C_v is the coefficient of venturi meter, or the coefficient of discharge of venturi meter, or simply the venturi coefficient and it takes into account the error introduced by assuming no frictional losses between stations 1 and 2 as well as assuming $\alpha_1/\alpha_2 = 1$ and $\alpha_1 = 1.0$ [i.e., it accounts for the frictional loss between stations 1 and 2 and small effects of kinetic energy factors α_1 and α_2].

Volumetric flow rate, Q, is given by

$$Q = u_2 A_T \qquad ... (7.174)$$

where A_T = cross-sectional area of the throat.

Combining Equations (7.173) and (7.174), we get

$$Q = C_v A_T \left[\frac{2(P_1 - P_2)}{(1-\beta^4)\rho}\right]^{1/2}$$

$$Q = \frac{C_v A_T}{\sqrt{1-\beta^4}} \cdot \sqrt{\frac{2(P_1 - P_2)}{\rho}} \qquad ... (7.175)$$

where Q is the actual discharge / volumetric flow rate

Q_{th} = theoretical discharge and is given by

$$Q_{th} = A_T \left[\frac{2(P_1 - P_2)}{(1-\beta^4) \cdot \rho}\right]^{1/2} \qquad ... (7.176)$$

$$\therefore \quad C_v = Q / Q_{th}$$

The coefficient of discharge of a venturi meter is the *ratio of the actual discharge to the theoretical discharge of the venturi meter.*

Equation (7.176) gives the value of theoretical discharge through the venturi as it has been obtained by considering no frictional losses in the system.

Equation (1.175) is the desired flow equation for a venturi meter.

For a well designed venturimeter, the coefficient of venturi, C_v is about **0.98** for pipe diameters ranging from 50 to 200 mm and about **0.99** for larger sizes.

If U-tube manometer is used for measuring the pressure difference between station 1 and station 2 and Δh is the manometer reading obtained in terms of meters of the manometric fluid, then pressure drop across the meter is given by

$$(P_1 - P_2) = \Delta h (\rho_M - \rho) g \qquad ... (7.177)$$

where ρ_M = density of manometric fluid in kg/m^3

ρ = density of flowing fluid in kg/m^3

and g = 9.81 m/s^2

Combining Equations (7.176) and (7.177), we get

$$Q = \frac{C_v A_T}{\sqrt{1-\beta^4}} \cdot \sqrt{\frac{2 \Delta h (\rho_M - \rho) g}{\rho}} \qquad ... (7.178)$$

If ΔH is the pressure difference across the venturi expressed in terms of meters of the flowing fluid, then

$$\Delta H = \Delta h \frac{(\rho_M - \rho)}{\rho} \qquad ... (7.179)$$

Combining Equations (7.175) and (7.179), we get

$$Q = \frac{C_v A_T}{\sqrt{1-\beta^4}} \cdot \sqrt{2 g \Delta H} \qquad ... (7.180)$$

Equation (7.178) is the desired flow equation for a venturi meter where the pressure difference is expressed in terms of manometer reading and Equation (7.180) is the desired flow equation where the pressure difference is expressed in terms of meters of the flowing fluid.

7.49 PRESSURE RECOVERY IN VENTURI METER

Since the angle of divergence in the recovery cone (i.e., in the divergent cone) is small, the permanent pressure loss from a venturi meter is relatively small. In a well designed meter, the permanent pressure loss is about 10 percent of the venturi differential $P_1 - P_2$ (i.e., 10 percent of the pressure drop across the meter) which means that about 90 percent of the venturi differential is recovered.

The pressure recovery in a venturi meter is very high and thus may be used where only a small pressure head is available though it is expensive.

7.50 ADVANTAGES OF VENTURI METER

1. Low permanent pressure loss and hence high pressure recovery.
2. High accuracy over wide flow ranges.
3. It can be used for flow measurement of compressible and incompressible fluids.
4. It can also be used where only a small pressure head is available.
5. High reproducibility.
6. Less power loss.

7.51 DISADVANTAGES OF VENTURI METER

1. It is expensive and bulky.
2. It occupies considerable space.
3. Relatively complex in construction.
4. The ratio of throat diameter to pipe diameter cannot be changed.
5. Not suitable for flow measurement of highly viscous slurries.
6. Used only for permanent installations.
7. It cannot be altered once it is installed.

7.52 APPLICATIONS OF VENTURI METER

Venturi meters are used :

- For flow measurements of gases, liquids, water, slurries and dirty liquids.
- Whenever high pressure recovery is required.
- For high flow rate measurement in pipes having diameters in a few meter.
- In medicine - for measuring the rate of blood flow in the arteries.
- In aviation – to measure the speed of the air around the plane.
- In automobile – to measure the fuel and air distribution in carburettor.

7.53 ORIFICE METER

An orifice meter is a differential pressure flow meter which reduces the area for fluid flow using an orifice plate.

It is a variable head meter used for measuring the discharge / flow rate through a pipe. In this meter, the fluid is accelerated by causing it to flow through a sudden constriction (orifice), the kinetic energy of the fluid increases and the pressure energy therefore decreases. With this meter, the overall pressure drop across the meter is high - a large percentage of the pressure drop across the meter is not recoverable, but is relatively cheap and reliable instrument and its installation requires a small length as compared to the venturi meter. Because of this where the space is limited, the orifice meter may be used for the measurement of discharge (flow rate) through pipes.

Principle :

- Increase in fluid velocity results in decrease in fluid pressure.

OR

- The basic principle on which an orifice meter works is that *by reducing the cross-sectional area of the flow passage, the fluid is accelerated and a pressure difference is created/developed, and the measurement of the pressure difference (between the inlet of the meter and a point of reduced pressure) enables the determination of the discharge (flow rate) through the pipe.*

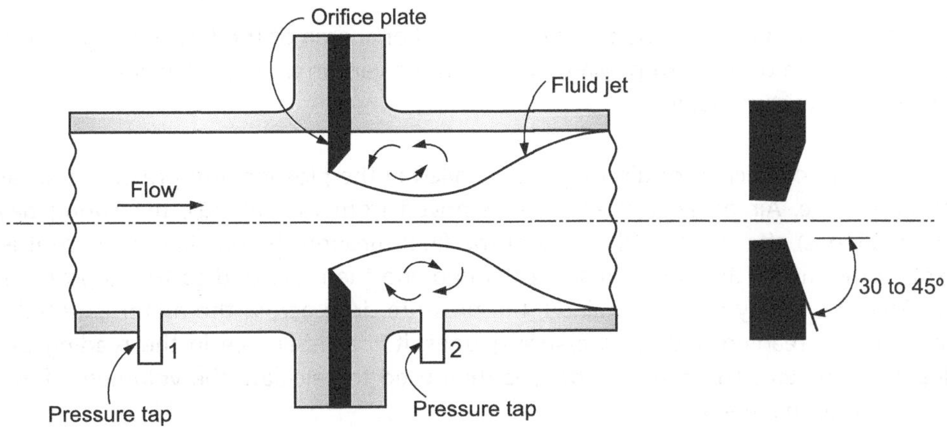

Fig. 7.30 : Orifice meter

Construction :

An orifice meter consists of a flat circular plate with an accurately machined and drilled circular hole called an orifice. The orifice is concentric with the pipe axis and drilled with sharp edges. The thickness of the plate is less than or equal to 0.05 times the diameter of the pipe. From the upstream of the plate, the edge of the orifice is made flat for a thickness less than or equal to 0.02 times the diameter of the pipe and the remaining portion of the plate is bevelled with the bevel angle of 30 to 45° (preferably 45°). The plate is usually made of stainless steel [SS – 304 or SS – 316] to resist corrosion.

The orifice plate is clamped between two pipe flanges with the bevelled surface facing the downstream (Fig. 7.30). The orifice diameter may vary from 0.2 to 0.85 times the pipe diameter, but generally the orifice diameter of 0.5 times the diameter of pipe is used. Pressure taps are provided, one each on the upstream side (station 1) and downstream side (station 2) of the orifice plate. On the downstream side of the orifice plate, the pressure tap is provided at the section where a converging jet of the fluid has a minimum cross-sectional area (This point of minimum cross-section of flow is known as the vena contracta) resulting in almost a maximum velocity and consequently a minimum pressure at this section. Hence, a maximum possible pressure difference exists between the section/station 1 (at a point short distance from the plate on the upstream where flow is undisturbed) and the section/station 2 (at a point where the flow area is minimum on the downstream side). This pressure difference is measured by connecting a manometer or other pressure measuring device, e.g., pressure gauges to the pressure taps.

7.54 FLOW PATTERN IN ORIFICE METER

The flow pattern of flowing fluid through an orifice plate is shown in Fig. 7.31. The flow lines run parallel and widely spaced on the upstream side of the plate but as the flow approaches the orifice plate, the flow lines slowly converge at the orifice and then diverge out to pipe area smoothly. Since the flow lines converge and then diverge, the flow area becomes minimum at a short distance on the downstream side of the orifice plate which is known as the vena-contracta. The jet of fluid issuing from the orifice plate gradually expands from the vena-contracta to again fill the entire cross-section of the pipe.

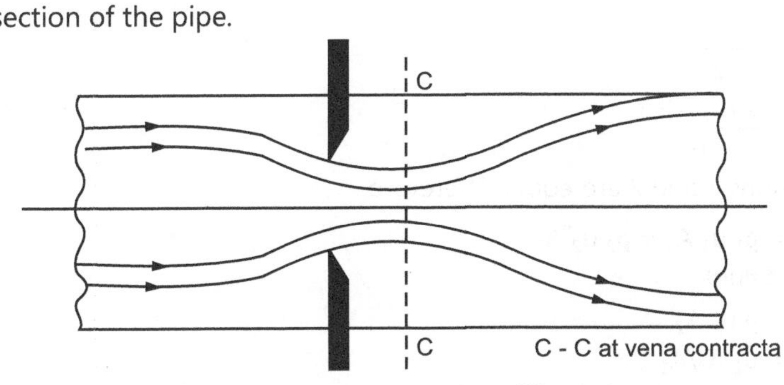

Fig. 7.31 : Flow pattern in orifice meter

In this meter, since there is a sudden change in the cross-sectional area of the flow passage and not a gradual change in the cross-sectional area of the flow passage as that in the venturi meter, there is a greater loss of energy in the orifice meter than in the venturi meter.

Working :

An orifice meter of known coefficient of discharge is installed in the pipeline and pressure taps are connected to a pressure measuring device. Air pockets, if any, are removed from the tubings, measuring device (may be manometer, or pressure gauges) after starting the flow of the fluid through the pipeline in which it is installed for the flow measurement. In the meter, the fluid is first accelerated and then retarded so that a pressure drop across the meter is created. Once the steady state is attained, the pressure drop across the meter is noted (for a U-tube manometer, it is a manometer reading and for pressure gauges, it is a difference in the reading of the pressure gauges on both sides of the meter). This pressure drop is then used to calculate the volumetric flow rate using a mathematical flow equation for the meter.

Venturi and orifice meters require straight runs of pipe before and after the point of installation. While installing this meter, a sufficiently long run of the straight pipe should be provided on both the sides of the meter so as to establish normal velocity distributions. There should not be any pipe fittings or a valve close to the meter (within a distance of about 1 to 1.5 m from the meter on both the sides).

7.55 FLOW EQUATION FOR AN ORIFICE METER

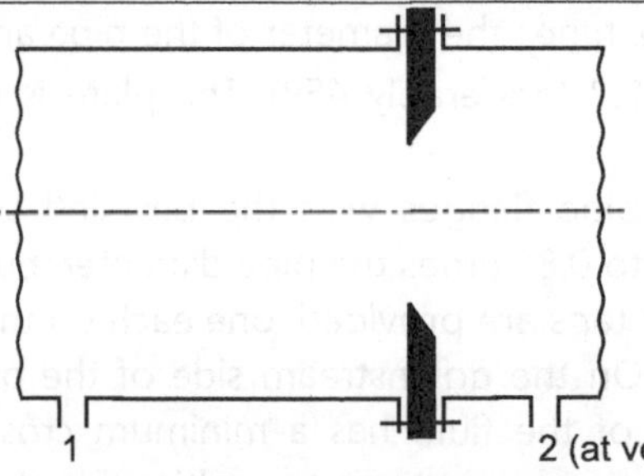

Fig. 7.32 : Orifice plate for flow equation

Let P_1, P_2 and u_1, u_2 be the pressures and velocities at stations 1 and 2 respectively. Applying the Bernoulli's equation between stations 1 (upstream of the meter) and 2 (at the vena contracta) for compressible fluids and assuming negligible frictional loss and $\alpha_1 = \alpha_2 = 1.0$, we get

$$\frac{P_1}{\rho} + g Z_1 + \frac{u_1^2}{2} = \frac{P_2}{\rho} + g Z_2 + \frac{u_2^2}{2} \quad \text{... (7.181)}$$

Equation (3.30) is the Bernoulli's equation across the orifice meter in terms of energy units (J/kg).

For the orifice meter installed in a horizontal pipeline, we have : $Z_1 = Z_2$

Therefore, Equation (7.181) reduces to

$$\frac{P_1}{\rho} + \frac{u_1^2}{2} = \frac{P_2}{\rho} + \frac{u_2^2}{2} \quad \text{... (7.182)}$$

$$\frac{P_1 - P_2}{\rho} = \frac{u_2^2 - u_1^2}{2}$$

$$u_2^2 - u_1^2 = \frac{2 (P_1 - P_2)}{\rho} \quad \text{... (7.183)}$$

The mass flow rates at stations 1 and 2 are equal. Therefore,

$$\dot{m} = \rho_1 u_1 A_1 = \rho_2 u_2 A_2$$

but $\rho = \rho_1 = \rho_2$ = density of fluid

$$\therefore \quad \rho u_1 A_1 = \rho u_2 A_2$$

$$u_1 = u_2 A_2/A_1 \quad \text{... (7.184)}$$

where A_1 = area of the pipe and A_2 = area of flow at the vena contracta.

Substituting the value of u_1 from Equation (7.184) into Equation (7.183) gives

$$u_2^2 - u_2^2 (A_2/A_1)^2 = \frac{2 (P_1 - P_2)}{\rho}$$

$$u_2^2 [1 - (A_2/A_1)^2] = 2 (P_1 - P_2)/\rho$$

$$u_2 = \left[\frac{2 (P_1 - P_2)}{\rho [1 - (A_2/A_1)^2]}\right]^{1/2} \quad \text{... (7.185)}$$

From the continuity equation, we have

$$\dot{m} = \rho u_2 A_2 = \rho u_o A_o$$

where u_o = velocity through the orifice

A_o = area of the orifice

$$\therefore \quad u_2 = u_o (A_o/A_2) \quad \text{... (7.186)}$$

The area of the vena contracta (A_2) can be related to the area of the orifice (A_o) by the coefficient of contraction.

$$\text{Coefficient of contraction} = C_c = A_2/A_o \quad \text{... (7.187)}$$

$$A_2 = C_c A_o \quad \text{... (7.188)}$$

Substituting the value of A_2 from Equation (7.188) into Equation (7.186), we get

$$u_2 = \frac{u_o A_o}{C_c . A_o} = \frac{u_o}{C_c} \quad \text{... (7.189)}$$

Substituting the value of u_2 from Equation (7.189) and A_2 from Equation (7.188) into Equation (7.188), we get

$$\frac{u_o}{C_c} = \left[\frac{2 (P_1 - P_2)}{\rho [1 - (C_c.A_o/A_1)^2]}\right]^{1/2}$$

$$u_o = C_c\left[\frac{2 (P_1 - P_2)}{\rho [1 - (C_c.A_o/A_1)^2]}\right]^{1/2} \quad \text{... (7.190)}$$

For taking into account the frictional losses in the meter and the parameters C_c, α_1 and α_2 (kinetic energy correction factors), the coefficient of discharge of orifice meter (also termed as the coefficient of a orifice meter), C_o can be introduced into the above equation and thus Equation (7.190) can be written as

$$u_o = C_o\left[\frac{2 (P_1 - P_2)}{\rho [1 - (A_o/A_1)^2]}\right]^{1/2} \quad \text{... (7.191)}$$

where A_o = area of the orifice

$A_o = \pi/4 . D_o^2$, where D_o is the orifice diameter

$A_1 = \pi/4\ D^2$, where D is the diameter of the pipe

$$A_o/A_1 = (\pi/4 . D_o^2) / (\pi/4 . D^2) = (D_o/D)^2$$

Let β = diameter of orifice/diameter of pipe = D_o/D

Then, $(A_o/A_1)^2 = (D_o/D)^4 = \beta^4$

Substituting for (A_o/A_1), Equation (7.191) becomes

$$u_o = C_o \sqrt{\frac{2 (P_1 - P_2)}{\rho [1 - \beta^4]}}$$

$$u_o = \frac{C_o}{\sqrt{1 - \beta^4}} . \sqrt{\frac{2(P_1 - P_2)}{\rho}} \quad \text{... (7.192)}$$

Equation (7.192) is used to calculate the velocity through the orifice meter.

The volumetric flow rate through the orifice meter is obtained by multiplying the velocity by the area of the orifice.

$$Q = u_o A_o \qquad \text{... (7.193)}$$

Combining Equations (7.192) and (7.193) gives

$$Q = \frac{C_o A_o}{\sqrt{1-\beta^4}} \cdot \sqrt{\frac{2(P_1 - P_2)}{\rho}} \qquad \text{... (7.194)}$$

In the above equation, Q is in m³/s, D_0 and D are in m, ρ is in kg/m³ and P_1 and P_2 are in N/m².

$(P_1 - P_2)$ = pressure drop across the meter in N/m² (Pa).

Q calculated from Equation (7.194) is the actual discharge / flow rate.

$$Q_{th} = \text{Theoretical discharge}$$

$$\therefore \quad Q = Q_{actual} = C_0 \cdot Q_{th} \qquad \text{... (7.195)}$$

$$Q = \frac{A_0}{\sqrt{1-\beta^4}} \cdot \sqrt{\frac{2(P_1 - P_2)}{\rho}} \qquad \text{... (7.196)}$$

Equation (7.194) is the simple working equation for evaluating the discharge (Q) from the meter knowing the pressure drop over the meter (i.e., for evaluating the flow rate through the pipe line in which the meter is installed).

The value of the coefficient of orifice meter is about **0.61** (for Reynolds numbers more than 10^4) indicating that there is a substantial loss in pressure over the instrument (i.e., the pressure recovery in this meter is very poor) which is the most serious disadvantage of the meter.

The coefficient of discharge of orifice meter is much lower than that of venturi meter. This is because of the fact in the orifice meter, there are no gradual converging and diverging flow passages as those exist in the venturi meter, which results in a greater loss of energy and consequent reduction of the coefficient of the discharge of the orifice meter. The pressure drop in the orifice meter for the same conditions as in the venturi meter is many times greater than that in the venturi meter.

If a mercury manometer is used to measure the pressure difference over the meter and if Δh is the difference in the levels in two limbs of the manometer in terms of the manometric fluid, then

$$(P_1 - P_2) = \Delta h (\rho_M - \rho) g \qquad \text{... (7.197)}$$

For a mercury manometer it becomes

$$(P_1 - P_2) = \Delta h (\rho_{Hg} - \rho) g \qquad \text{... (7.198)}$$

Substituting $(P_1 - P_2)$ from Equation (7.197) in Equation (7.194), we get

$$Q = \frac{C_0 A_0}{\sqrt{1-\beta^4}} \cdot \sqrt{\frac{2\Delta h (\rho_M - \rho) g}{\rho}} \qquad \text{... (7.199)}$$

or

$$Q = \frac{C_0 A_0}{\sqrt{1-\beta^4}} \cdot \sqrt{\frac{2\Delta h (\rho_{Hg} - \rho) g}{\rho}} \qquad \text{... when the mercury manometer is used ... (7.200)}$$

where ρ_M = density of manometric fluid, ρ_{Hg} = density of mercury (manometric fluid), ρ = density of flowing fluid and Δh = difference in the levels in the two limbs of manometer in m of the manometric fluid.

$$\Delta H = \frac{\Delta h (\rho_M - \rho)}{\rho} \qquad \text{... (7.201)}$$

where ρ_M = density of manometric fluid in kg/m³

ρ = density of flowing fluid in kg/m³

ΔH = pressure difference across meter in terms of m of the flowing fluid.

Combining Equations (7.200) and (7.201), we get

$$Q = \frac{C_0 A_0}{\sqrt{1-\beta^4}} \cdot \sqrt{2 g \Delta H} \qquad \text{... (7.202)}$$

where A_0 is in m², g is in m/s², ΔH is in m and Q is in m³/s. Equation (7.202) is used to calculate the volumetric flow rate through the pipe when the pressure drop across the meter is expressed in terms of m of the flowing fluid.

7.56 PRESSURE RECOVERY IN ORIFICE METER

The orifice meter has a low discharge coefficient of about 0.61 which means that there is a substantial loss in pressure over the meter (the pressure recovery is poor) because of the large friction losses from eddies generated by the re-expanding jet below the vena contracta. This results in power loss. The permanent pressure loss from an orifice meter depends on the value of β (Fig. 7.33). For β = 0.5, the permanent pressure loss is about 73 percent of the orifice differential.

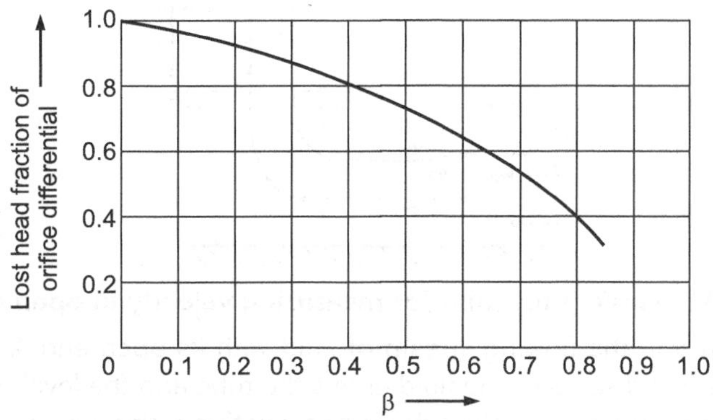

Fig. 7.33: Overall pressure loss in orifice meters

Orifice meters are used for testing purposes and in steam lines, where the power loss is of less importance.

7.57 ADVANTAGES OF ORIFICE METER

1. Simple in construction.
2. It is easy to install.
3. Installing a new orifice plate with different opening is a simple job. (Greater flexibility).
4. It can be used with all types of differential pressure transmitters.
5. It is cheap.
6. The ratio of orifice diameter to pipe diameter can be changed.

7.58 DISADVANTAGES OF ORIFICE METER

1. Substantial loss of pressure over the instrument and thus the pressure recovery is poor.
2. Considerable power loss.
3. It cannot be used for slurries that may clog the orifice opening.
4. For accurate readings, it requires a certain straight run of pipe without any pipe fittings or a valve on both sides of the meter.

7.59 APPLICATIONS OF ORIFICE METERS

There are used for flow measurement of water, air, steam and gases in petrochemical, chemical, power generation, water treatments and gas generation plants.

7.60 PITOT TUBE

It is a device used to measure the local or point velocity by measuring the difference between impact pressure and static pressure. The orifice meter and venturi meter measure the average velocity of the entire fluid stream, whereas the pitot tube measures the velocity at one point only.

Principle

The basic principle is that *if the velocity of flow at a particular point is reduced to zero (known as the stagnation point), the pressure at that point increases due to conversion of the kinetic energy into the pressure energy, and by measuring the increase in the pressure energy at this point the velocity may be determined.*

Construction and Working :

A simple form of pitot tube is shown in Fig. 7.34. It consists of a glass tube, sufficient large for capillary effects to be negligible and bent at right angles. A single tube of this kind may be used for measuring the flow velocity in an open channel.

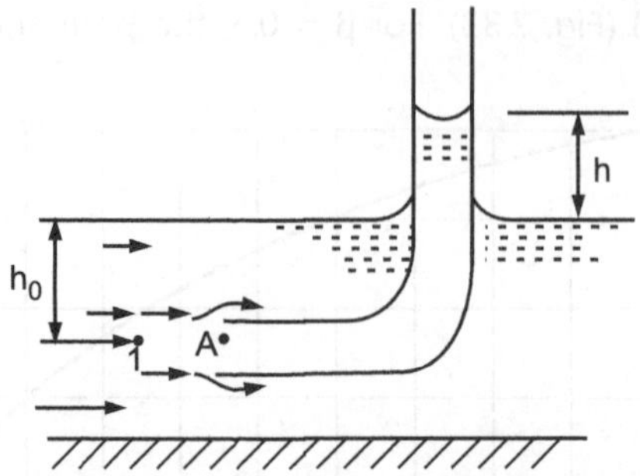

Fig. 7.34 : Simple pitot tube for measuring velocity in open channel

The tube is dipped vertically in the flowing stream of fluid with its open end A facing the flow and the other open end projecting above the liquid surface. The fluid enters the tube and the level of the fluid in the tube exceeds that of the fluid surface because the end A of the tube is a stagnation point where the fluid is at rest and the fluid approaching end A divides at this point and passes around the tube. At the stagnation point, the kinetic energy is converted into the pressure energy and hence the fluid in the tube rises above the surrounding fluid surface by a height that corresponds to the velocity of fluid approaching the end A of the tube.

Consider a point 1 at the upstream of end A lying in a horizontal plane in the flowing stream where the flow velocity is u.

Applying the Bernoulli's equation between points 1 and A, we get

$$h_o + u^2/2g = h_o + h \qquad \text{... (7.203)}$$

$$\therefore \quad u = \sqrt{2gh} \qquad \text{... (7.204)}$$

where h is the height of the fluid raised in the tube above the free surface and h_o is the depth of the tube in the fluid, (h_o + h) is the stagnation pressure head at point A which consists of a static pressure head h_o and dynamic pressure head/impact pressure head h.

Hence, the velocity at any point in the flowing stream can be determined by dipping the pitot tube to the required point and measuring the height h of the fluid raised in the tube above the free surface, i.e., by measuring the rise of fluid in the tube (i.e., h), we can find the velocity of the flowing fluid.

Taking into account the loss of energy, Equation (7.204) is modified to yield the actual flow velocity as

$$u = C\sqrt{2gh} \qquad \text{... (7.205)}$$

where C is the coefficient of pitot tube and it is generally taken as 0.98.

7.61 PITOT TUBE FOR MEASUREMENT OF VELOCITY OF FLOW IN PIPES

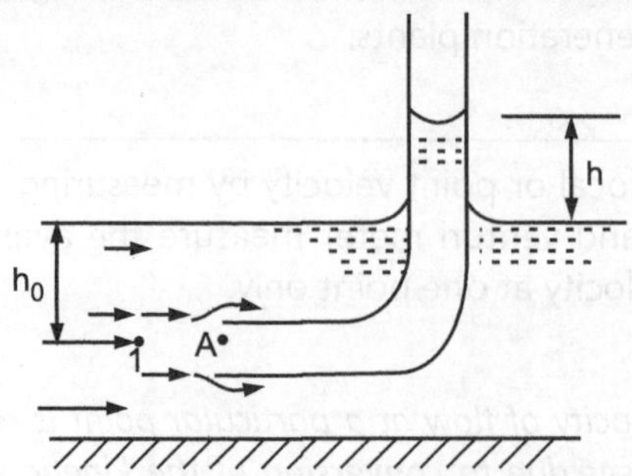

Fig. 7.35 : Pitot tube used for measuring velocity in pipes

The pitot tube shown in Fig. 7.35 consists of an impact tube, the opening of which facing to the direction of flow and a static tube, the opening of which is perpendicular to the direction of flow. The impact tube measures the impact/dynamic pressure and the static tube measures the static pressure. The two tubes may be connected to the arms of a manometer for measuring the pressure difference.

Applying Bernoulli's equation between stations 1 and 2, we get

$$\frac{P_1}{\rho} + \frac{u^2}{2} = \frac{P_2}{\rho} \qquad \text{... (7.206)}$$

$$\therefore \qquad u = 2\left(\frac{P_2 - P_1}{\rho}\right)^{1/2} \qquad \text{... (7.207)}$$

P_2 is greater than P_1, where P_1 = static pressure, P_2 = stagnation / impact pressure. $(P_2 - P_1)$ can be expressed in terms of a manometer reading Δh.

$$(P_2 - P_1) = \Delta h\,(\rho_M - \rho)\,g \qquad \text{... (7.208)}$$

$$u = \left[2\,\Delta h\,\frac{(\rho_M - \rho)}{\rho}\,g\right]^{1/2} \qquad \text{... (7.209)}$$

If ΔH is the pressure head expressed in terms of m of flowing fluid, we get

$$\Delta H = \Delta h\left(\frac{\rho_M - \rho}{\rho}\right) \qquad \text{... (7.210)}$$

$$\therefore \qquad u = [2\,\Delta H\,g]^{1/2}$$

$$u = \sqrt{2\,g\,\Delta H} \qquad \text{... (7.211)}$$

By taking into account the loss of energy due to friction, the actual point/local velocity is given by

$$u = C\sqrt{2\,g\,\Delta H} \qquad \text{... (7.212)}$$

where C is the coefficient of pitot tube.

7.62 ROTAMETER (VARIABLE AREA METER)

In the orifice meter or venturi meter (variable head meters), the area of constriction / area of flow is constant and the differential pressure / pressure drop across the meter varies with the flow rate / discharge, while in the variable area meter, the pressure drop across the meter is constant and the flow rate is a function of the area of constriction / area of flow. Thus, any change in the flow rate can be measured in terms of the change in the area of flow.

The most important area meter is the rotameter.

A **rotameter** is a device used to measure flow rate, in which a float rises in a tapered vertical tube to a height that depends on the flow rate through the tube.

Principle :

The flow raises the float in the tapered tube that increases the passage area for the fluid flow. The greater the flow, the higher the float is raised (variable area principle).

OR

Rotameter operates on the principle that *there is a different constriction / orifice area (flow area) for each flow rate and the pressure drop across the meter is constant. In other words, the pressure drop across the meter is constant and the area through which the fluid flows varies with the flow rate. This area is related, through a proper calibration, to the flow rate.*

Construction :

It consists of a tapered glass tube mounted vertically in a frame with the large end up. The tube contains a freely moving solid float. The diameter of the float is smaller than the diameter of the bottom of the tapered tube. The density of a float material is higher than that of the liquid. A perforated plate is provided at both the ends of the tapered tube for arresting the float in the tube. A guide is provided for the float so it always remains at the centre in the tube along the axis of the tube. The float is usually constructed out of corrosion resistance materials such as stainless steel (SS – 304 or SS – 316), aluminium, monel, bronze, nickel and plastics. The float material decides the flow range of the meter.

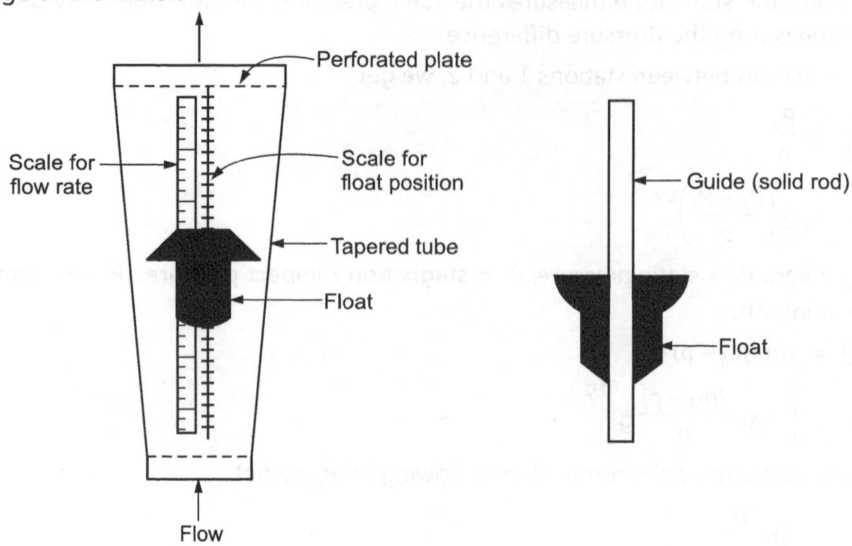

Fig. 7.36 : Rotameter

A nearly linear flow scale is etched/marked on the glass tube or it is mounted close to the tube so that the position of the float can be marked and the flow rate is then obtained from a calibration curve or a direct scale of the flow rate can also be provided over the tapered tube or near the tube. *The float is the indicating element and the reading edge of the float is taken at the largest cross-section of the float.* The either ends of the meter may be screwed or flanged. Rotameters do not require straight runs of pipe before and after the point of installation.

For low pressure and temperature applications, glass tubes are used but for opaque liquids, high temperature and pressure applications or for other conditions where glass is impracticable (not suitable), metal tubes are used. When a metal tube is used (the float in this case is invisible) or when a liquid is very dark or dirty, an external indicator is required (for the meter reading).

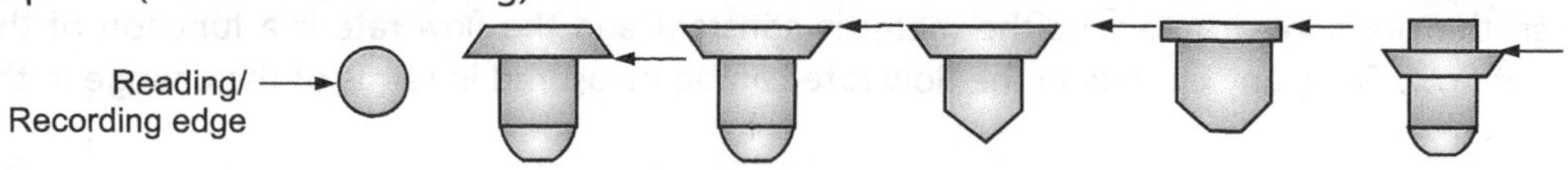

Fig. 7.37 : Float shapes

Working :

In rotameter as the flow varies, the float rises or falls, thus changing the area of the annular space between the float and the wall of the tube. The area available for flow is the annular space/annulus between the float and the wall of the tube.

When no fluid flows through the meter, the rotameter float rests at the bottom of the tube. But as the fluid begins to flow from the lower side of the tube, the float rises until the differential pressure just balance the weight of the float and the fluid flows through the meter through the annular space. As the flow rate increases, the float rises further in the tube, thus, increasing the area available for flow (flow area) keeping a differential pressure across it constant. On the other hand, as the flow rate decreases, the float falls in the tube, thus decreasing the flow area with constant pressure drop across it. At a given flow rate, float stabilises at a certain fixed position in the tube and at steady state it is recorded as the rotameter reading on the scale provided.

The variation of the flow area with flow rate can be measured in terms of change in the float position. A rotating motion of the float helps to keep it steady.

Rotameters are widely used in the chemical industry for the measurement of flow rates of compressible as well as incompressible fluids. They provide a directing reading of the flow rate.

7.63 FLOW EQUATION FOR ROTAMETER

For a given flow rate, let P_1 and P_2 be the pressures across the float at its equilibrium condition.

At equilibrium condition, the forces acting on the float must be balanced so that no net force acts to move the float.

For dynamic equilibrium of the float, the force balance can be written as :

Net upward force = Net downward force

The forces acting on the float are : (i) a buoyant force, (ii) a differential pressure force and (iii) a gravity force.

∴ Buoyant force + Differential pressure force = Gravity force

(i) Buoyant force of the liquid acting so as to lift the float = $V_f \rho g$.

(ii) Differential pressure force (drag force on the float resulting from friction for flow around the float) = $(P_1 - P_2) A_f$.

(iii) Gravity force (weight of the float) = $V_f \rho_f g$. ... A gravity force is acting downward on the float

$$\therefore \quad V_f \rho g + (P_1 - P_2) A_f = V_f \rho_f g \qquad \text{... (7.213)}$$

$$(P_1 - P_2) A_f = V_f (\rho_f - \rho) g \qquad \text{... (7.214)}$$

where m_f = mass of float, V_f = volume of float, ρ_f = density of float, ρ = density of flowing fluid, A_f = the cross-sectional area of the largest part of the float.

For a given float and fluid of fixed density, the right hand side of Equation (7.214) is constant indicating that ΔP is constant.

$$(P_1 - P_2) = \Delta P = \frac{V_f (\rho_f - \rho).g}{A_f} \qquad \text{... (7.215)}$$

Let the area of the annulus between the float and the wall of the tube be A_2 at a certain location of the float and the cross-sectional area of the tube at that location be A_1.

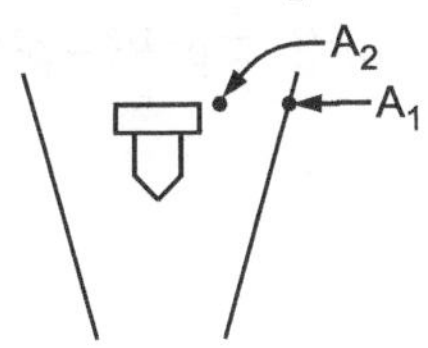

Equation (7.215) can be written as

$$\frac{P_1 - P_2}{\rho} = g\,\Delta H = \frac{V_f (\rho_f - \rho)}{A_f \rho} g \qquad \text{... (7.216)}$$

Annulus diameter = $D_1 - D_f$ (at float position)

where D_f = diameter of the largest portion of the float

D_1 = diameter of the tube at the float position

Considering the annulus flow area between the tube and float as an orifice of diameter $D_1 - D_f$ having a differential pressure of $(P_1 - P_2)$ across it, the flow equations derived for the orifice meter can be used to express the flow rate through the rotameter.

$$Q = \frac{C_D A_2}{\sqrt{1 - \beta^4}} \cdot \sqrt{\frac{2 (P_1 - P_2)}{\rho}} \qquad \text{... (7.217)}$$

where $A_2 = \pi/4 \, (D_1^2 - D_f^2)$

Combining Equations (7.216) and (7.217), we get

$$Q = \frac{C_D A_2}{\sqrt{1 - \beta^4}} \cdot \sqrt{\frac{2 g V_f}{A_f}\left(\frac{\rho_f - \rho}{\rho}\right)} \qquad \text{... (7.218)}$$

Here, $$\beta = \frac{D_1 - D_f}{D_1} = (A_2/A_1)^{1/2} \qquad \text{... (7.219)}$$

$$\therefore \quad Q = \frac{C_D A_2}{\sqrt{1-(A_2/A_1)^2}} \times \sqrt{\frac{2\, gV_f}{A_f}\left(\frac{\rho_f - \rho}{\rho}\right)} \qquad \text{... (7.220)}$$

The velocity is given by

$$u_2 = Q/A_2 = \frac{C_D}{\sqrt{1-(A_2/A_1)^2}} \cdot \sqrt{\frac{2\, g\, V_f\, (\rho_f - \rho)}{A_f\, .\, \rho}} \qquad \text{... (7.221)}$$

The mass flow rate is given by

$$\dot{m} = \rho\, u_2\, A_2 = \rho\, Q$$

Substituting for Q from Equation (7.220) in the above equation, we get

$$\dot{m} = \frac{\rho\, C_D\, A_2}{\sqrt{1-(A_2/A_1)^2}} \sqrt{\frac{2\, g\, V_f\, (\rho_f - \rho)}{A_f\, .\, \rho}}$$

$$\dot{m} = \frac{C_D\, A_2}{\sqrt{1-(A_2/A_1)^2}} \sqrt{\frac{2\, g\, V_f\, (\rho_f - \rho)\, \rho}{A_f}} \qquad \text{... (7.222)}$$

The coefficient of discharge of rotameter (C_D) depends on the shape of the float and the Reynolds number (based on the velocity in the annular space and the equivalent diameter of the annular space/annulus).

This meter can be made relatively insensitive to the density changes of the fluid by selection of ρ_f. Therefore, when $d\dot{m}/d\rho = 0$, the flow rate for a given rotameter is independent of ρ.

From Equation (7.222),

$$\frac{d\dot{m}}{d\rho} = \frac{C_D A_2 \sqrt{2\, gV_f}}{\sqrt{1-(A_2/A_1)^2}} \left[(\rho_f - \rho)^{1/2} \,.\, \frac{1}{2}\, \rho^{-1/2} - \rho^{1/2} \,.\, \frac{1}{2}\, (\rho_f - \rho)^{1/2} \right]$$

when $d\dot{m}/d\rho = 0$

$$(\rho_f - \rho)^{1/2} \,.\, \frac{1}{2}\, \rho^{-1/2} = \rho^{1/2}\, \frac{1}{2}\, (\rho_f - \rho)^{-1/2}$$

$$(\rho_f - \rho)^{1/2}\, (\rho_f - \rho)^{1/2} = \rho^{1/2} \,.\, \rho^{1/2}$$

$$(\rho_f - \rho) = \rho$$

$$\therefore \quad \rho_f = 2\, \rho \qquad \text{... (7.223)}$$

Therefore, if the density of the float is twice that of the fluid, then the position of the float for a given flow rate is independent of the density of the fluid.

7.64 ADVANTAGES OF ROTAMETERS

(i) It provides linear scale.
(ii) Simple to install and maintain.
(iii) Low pressure drop.
(iv) No external power is required.
(iv) Cost effective.
(v) Repeatability.
(vi) Inexpensive.
(vii) Can be used with corrosive fluid.
(viii) Reliable and accurate.

7.65 DISADVANTAGES OF ROTAMETERS

(i) It must be installed in a vertical position only.

(ii) When an opaque fluid is used, float may not be visible.

(iii) Glass tube types are subjected to breakage.

7.66 APPLICATIONS OF ROTAMETERS

- Rotameters are used in process industries for flow measurement of liquids and gases. It is also used for high pressures and high temperature flow measurement.
- It is used to monitor gas and water flow in plants and laboratories.
- It is used to monitor filtration loading.

7.67 CALIBRATION OF A ROTAMETER

The calibration chart for a given rotameter prepared over its entire range is a relationship between the rotameter reading (i.e., height of float, float position) and the volumetric flow rate. A typical calibration chart is shown in Fig. 7.38. The calibration chart as well as a proper correction for the density and the viscosity of a fluid when the chart of one fluid is to be used for the other fluid may usually be supplied by a manufacturer.

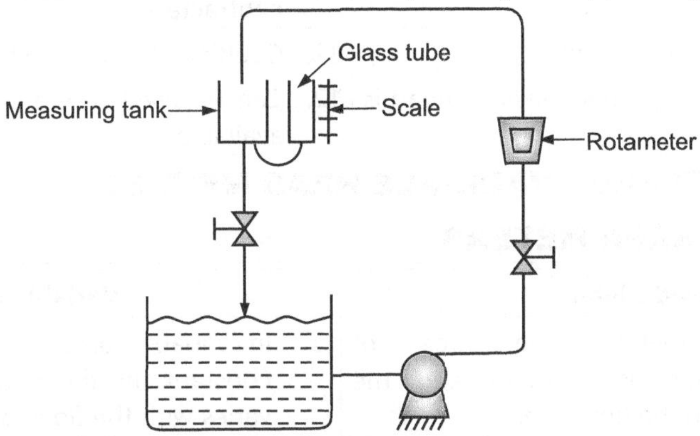

Fig. 7.38 : Set up for calibration of a given rotameter

For calibrating a given rotameter, the flow of the fluid (liquid e.g., water) through the meter is started by slightly opening a valve at the inlet to the meter. Time is allowed to attain a steady state and for this valve opening, the float position is noted and the liquid is collected in a measuring tank over a known period of time. The volumetric flow rate is obtained from the volume collected and time noted. This procedure is repeated for several valve positions to cover the entire range of the meter and the calibration chart is prepared.

On the calibration chart, we should provide information such as : name, density and temperature of the fluid handled.

$$\text{Volumetric flow rate} = \frac{\text{Volume of liquid collected}}{\text{time}}$$

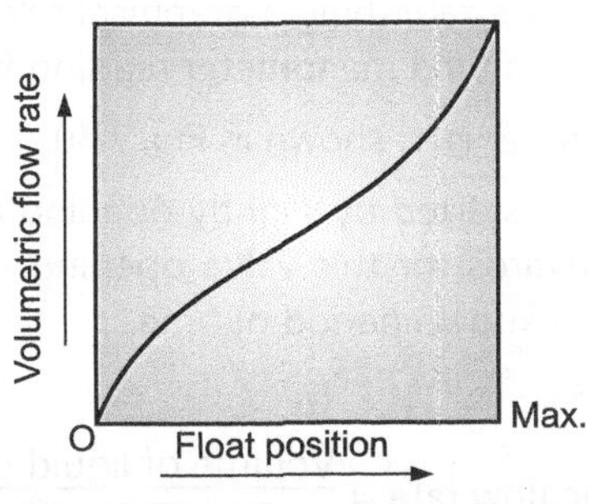

Fig. 7.39 : Calibration chart of rotameter

7.68 DIFFERENCE BETWEEN ORIFICE METER AND VENTURI METER

Orifice Meter	Venturi Meter
1. Simple in construction.	1. Relatively complex in construction.
2. Low space requirement.	2. Occupies considerable space.
3. Relatively cheap.	3. Expensive.
4. Ratio of orifice diameter to pipe diameter can be changed simply by installing a new orifice plate.	4. Ratio of throat diameter to pipe diameter cannot be changed.
5. Pressure recovery is very poor.	5. Pressure recovery is very high.
6. Coefficient of discharge is about 0.61.	6. Coefficient of discharge is about 0.98.
7. Larger power loss.	7. Smaller power loss.
8. Fluid is retarded at once.	8. Fluid is retarded gradually.
9. Flow separates from the wall, i.e., flow separation is there.	9. No flow separation (possibility very less).
10. Area is minimum at the vena contracta.	10. Area is minimum at the throat and there is no vena contracta.
11. Coefficient of contraction is not unity.	11. Coefficient of contraction is unity.
12. Cannot be used where only small pressure head is available.	12. Can be used where only a small pressure head is available.

7.69 DIFFERENCE BETWEEN VARIABLE HEAD METERS AND VARIABLE AREA METERS

Variable Head Meter	Variable Area Meter
1. In variable head meters, the area of constriction/area of flow is constant and the pressure drop varies with the flow rate.	1. In variable area meters, the pressure drop is constant and the area of flow/area of constriction varies with the flow rate.
2. Variables head meters include orifice meter and venturi meter.	2. Variable area meters include rotameters of various designs.
3. Simple in construction.	3. Complex in construction.
4. Relatively cheaper.	4. Relatively costly.
5. Needs straight runs of pipe before and after the meter (i.e., without fittings, valve).	5. Does not need straight runs of pipe before and after the meter.
6. It cannot give direct visual reading of the flow rate.	6. It can give direct visual reading of the flow rate.

7.70 CALIBRATION OF AN ORIFICE METER (OR A VENTURI METER)

Calibration of a given orifice meter means establishing a graphical relationship between the volumetric flow rate and the pressure drop across the meter (may be the manometer reading if a manometer is used).

A set up for calibrating a given orifice meter is as shown in Fig. 7.40.

The flow of the fluid (liquid e.g., water) is started by slightly opening a valve at the exit of the pipe line. Time is allowed to attain a steady state and afterwards for this valve opening the manometer reading is noted and the liquid is collected in a measuring tank over a known period of time.

The volumetric flow rate is obtained as

$$\text{Volumetric flow rate} = \frac{\text{Volume of liquid collected}}{\text{Time}}$$

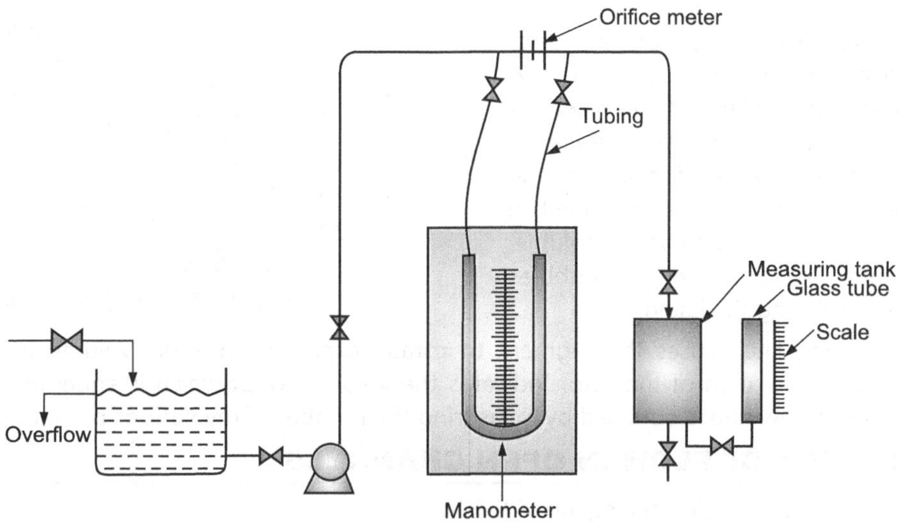

Fig. 7.40 : Set up for calibration of an orifice meter

This procedure is repeated for several valve positions to cover the entire range of flow rate and the calibration chart is prepared [it is a plot of volumetric flow rate v/s manometer reading (on an ordinary graph paper)].

This plot is used to obtain the volumetric flow rate knowing a manometer reading across the meter (i.e., for any known/observed manometer reading).

7.71 MAGNETIC FLOW METER

It consists of a non-magnetic pipe enclosed by an electromagnet (which generates a uniform magnetic field) as shown in Fig. 7.41. Two or more metal electrodes are mounted flush with the inner wall of the pipe. A fluid flows through the non-magnetic pipe.

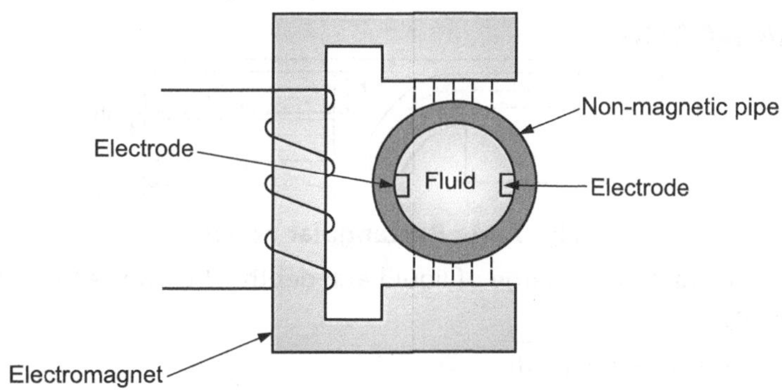

Fig. 7.41 : Magnetic flow meter

In this meter, the fluid flows through a magnetic field and generates a voltage which is directly proportional to the velocity of the fluid through the meter. The use of this meter is limited to fluids having some electrical conductivity, such as liquid metal coolants. They are now developed to handle liquids having poor electrical conductivity, such as tap water. Special features of the meter include – a very negligible drop in pressure and ability to handle fluids containing a high percentage of solids.

7.72 QUANTITY METERS

These meters measure directly the total quantity of fluid which has passed in a given time. This quantity of fluid divided by the time of passage gives the average flow rate. The wet and dry gas meters are examples of flow meters falling in this class for the measurement of the flow of gas through a pipe.

A wet gas meter (Fig. 7.42) consists of a segmented cylindrical rotor in four equal parts. A narrow slot is provided in the axis of the rotor for gas inlet in the meter and a narrow slot parallel to the axis of rotor is provided for gas discharge at the outer edge of the rotor. The rotor is immersed in water or suitable liquid in which the gas is insoluble. Gas is introduced into the hollow axis and bubbles into the segment by displacing the liquid.

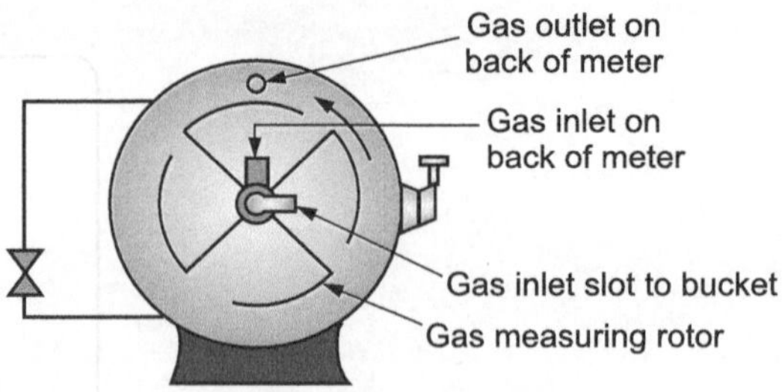

Fig. 7.42 : Wet gas meter

Gas fills the segment, and causes the segment to rotate, expelling an equal volume of gas from another segment. One complete revolution of the rotor indicates the volume swept which is equal to the volume of four segments. The volumetric flow rate is obtained by measuring the number of revolutions per minute.

7.73 MEASUREMENT OF FLOW IN OPEN CHANNELS

A notch may be defined as an opening provided in the side of a tank/vessel/reservoir such that the liquid surface in the tank (i.e., the upstream liquid level) is below the top edge of the opening (See Fig. 7.43).

A notch may have only the bottom edge and sides because its top edge above the liquid surface/level serves no purpose. The bottom edge of a notch, over which the liquid flows, is known as sill or crest of the notch.

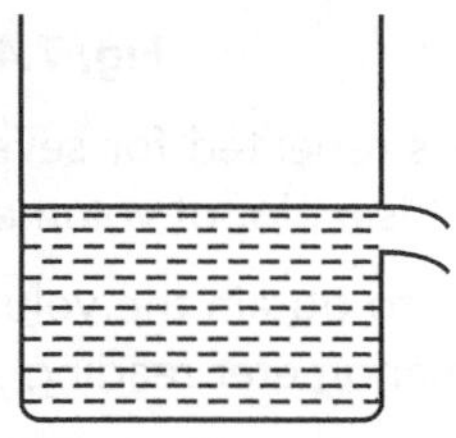

Fig. 7.43 : A notch

Usually, a notch is made of a metallic plate. It is used to measure the rate of discharge of liquids/the rate of flow of liquids flowing in open channels, not in closed pipelines.

The notches are usually classified according to their shapes :

(i) Rectangular notch. (ii) Triangular notch/V-notch and (iii) Trapezoidal notch.

7.74 RECTANGULAR NOTCH

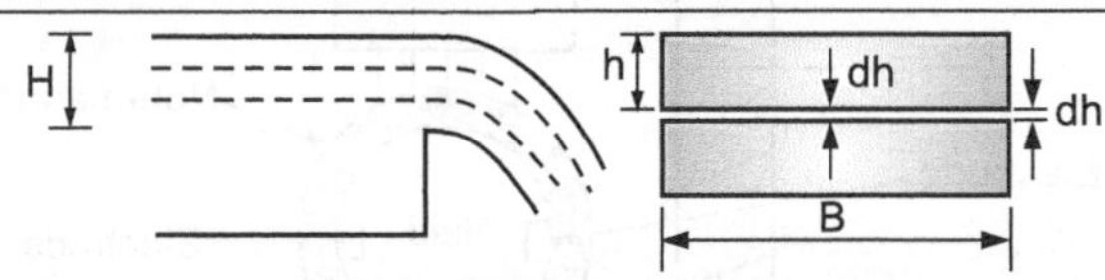

Fig. 7.44 : Rectangular notch

For a rectangular notch, the rate of discharge of liquid at a depth h below the free surface through a horizontal strip of thickness dh is given by

$$dQ = C_d \, B \, dh \sqrt{2gh} \quad \text{... (7.224)}$$

where B is the width or length of the notch and C_d is the coefficient of discharge (usually about 0.60).

The theoretical velocity of liquid through the strip at a depth h from the liquid-free surface is

$$V = \sqrt{2gh} \quad \text{... (7.225)}$$

$$A = \text{Area of the strip} = B \cdot dh$$

$$\text{Theoretical discharge} = VA = B \cdot dh \sqrt{2gh} \quad \text{... (7.226)}$$

$$\text{Coefficient of discharge} = C_d = \frac{\text{Actual discharge}}{\text{Theoretical discharge}}$$

$$\therefore \quad \text{Actual discharge} = C_d \text{ (Actual discharge)}$$

∴ Actual discharge through the strip, i.e., the rate of discharge or rate of flow through the strip is

$$dQ = C_d \, B \, dh \sqrt{2gh} \quad \text{... Equation (7.224)}$$

The total discharge/the total rate of flow through the whole notch is found out by integrating the above equation within the limits 0 and H.

$$\therefore \quad Q = C_d B \sqrt{2g} \int_0^H \sqrt{h} \cdot dh = C_d B \sqrt{2g} \left[\frac{h^{3/2}}{3/2}\right]_0^H$$

$$\therefore \quad Q = \frac{2}{3} C_d B \sqrt{2g} (H)^{2/3} = \frac{2}{3} C_d B \sqrt{2g} (H)^{1.5} \qquad \text{... (7.227)}$$

where H is the height of liquid sill/crest of the notch.

7.75 TRIANGULAR NOTCH

A triangular notch is also called a V-notch.

Consider a V-notch/triangular notch as shown in Fig. 7.45. Let H be the height of the liquid above the apex of the notch, θ be the angle of the notch and C_d is the coefficient of discharge. Let us find the width of the notch at the liquid surface.

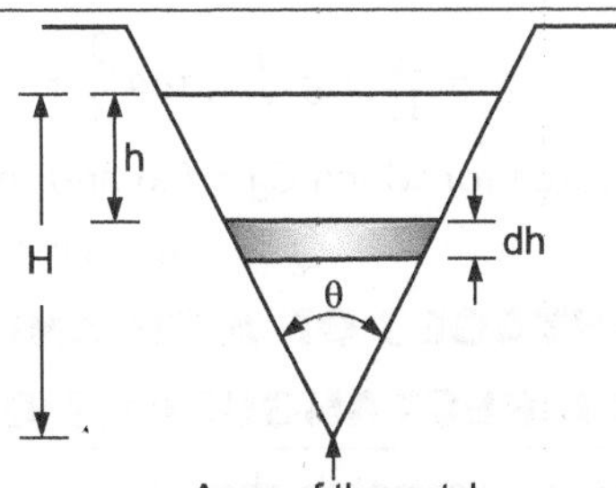

Fig. 7.45 : Triangular notch/V-notch

From the geometry of Fig. 7.46,

$$\frac{x}{H} = \tan\frac{\theta}{2}$$

$$x = H \tan\frac{\theta}{2}$$

Width of the notch at the liquid surface = $2x = 2H \tan\frac{\theta}{2}$.

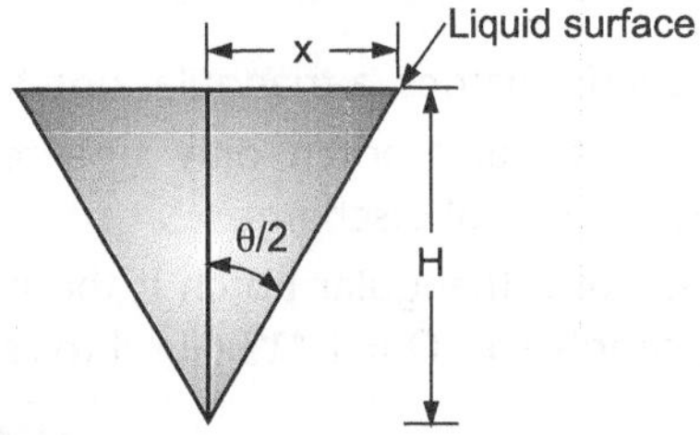

Fig. 7.46 : Triangular notch

For a V-notch, the rate of discharge of liquid at a depth h below the liquid free surface through a horizontal strip of thickness dh (of liquid) is given by

$$dQ = C_d \cdot 2 (H - h) \tan\frac{\theta}{2} \, dh \sqrt{2gh} \qquad \text{... (7.228)}$$

The velocity of liquid through the strip at a depth h from the liquid free surface is

$$V = \sqrt{2gh}$$

$$\text{Width of the strip} = 2 (H - h) \tan\frac{\theta}{2} \text{ (from the geometry of the figure)}$$

$$A = \text{area of the strip} = \text{width} \times \text{thickness} = 2 (H - h) \tan\frac{\theta}{2} \cdot dh$$

$$\text{Theoretical discharge through the strip} = A \cdot V = 2 (H - h) \tan\frac{\theta}{2} \cdot dh \cdot \sqrt{2gh}$$

$$C_d = \frac{\text{Theoretical discharge}}{\text{Actual discharge}}$$

Actual discharge through the strip = dQ.

$$dQ = C_d \times \text{theoretical discharge}$$

$$dQ = C_d \cdot 2 (H - h) \tan\frac{\theta}{2} \cdot dh \cdot \sqrt{2gh}$$

The total discharge/the total rate of flow through the notch may be obtained by integrating the above equation within the limits 0 and H.

$$\therefore \quad Q = \int_0^H C_d \cdot 2\,(H-h)\tan\frac{\theta}{2}\cdot dh\cdot\sqrt{2gh}$$

$$= 2\,C_d\sqrt{2g}\,\tan\frac{\theta}{2}\int_0^H (H-h)\sqrt{h}\,dh = 2\,C_d\sqrt{2g}\cdot\tan\frac{\theta}{2}\int_0^H [H\,h^{1/2}-h^{3/2}]\,dh$$

$$= 2\,C_d\sqrt{2g}\,\tan\frac{\theta}{2}\left[\frac{H\cdot h^{3/2}}{3/2}-\frac{h^{5/2}}{5/2}\right]_0^H = 2\,C_d\sqrt{2g}\,\tan\frac{\theta}{2}\left[\frac{H^{5/2}}{3/2}-\frac{H^{5/2}}{5/2}-0+0\right]$$

$$Q = \frac{8}{15}\,C_d\sqrt{2g}\,\tan\frac{\theta}{2}\,H^{5/2} = \frac{8}{15}\,C_d\sqrt{2g}\,\tan\frac{\theta}{2}\,(H)^{25} \quad \text{... (7.229)}$$

For a 90° notch for which $C_d = 0.6$ and for $g = 9.81\ m/s^2$,

$$Q = 1.417\ H^{5/2} = 1.417\ (H)^{25} \text{ ... in SI units} \quad \text{... (7.230)}$$

7.76 ADVANTAGES OF A TRIANGULAR NOTCH/V-NOTCH OVER A RECTANGULAR NOTCH

1. For a triangular notch, the rate of discharge is proportional to the liquid depth raised to a power of 2.5 and for a rectangular notch to a power of 1.5. Therefore, a triangular notch can handle a wider range of flow rates.
2. For low discharges, a triangular notch gives more accurate results than a rectangular notch.
3. For a triangular notch only one reading, i.e., head/liquid depth (H) is needed to be taken for the measurement of discharge.
4. In case of a triangular notch if the angle of the notch is 90°, then the total discharge equation becomes very simple [i.e., $Q = 1.417\ (H)^{2.5}$] to remember.

SOLVED EXAMPLES

Example 7.1 : *A pressure gauge on a tank reads 50 psig on a day when a barometer reads a pressure of 28 inches of mercury. Find the absolute pressure in the tank in psi.*

Solution : Gauge pressure = 50 psig

Pressure recorded by the barometer = 28 in Hg (inches Hg)

Atmospheric pressure = Barometric pressure = 28 in Hg

We have : 29.92 in Hg = 1 atm = 14.7 psi

$$\therefore \quad \text{Atmospheric pressure} = 28 \text{ in Hg} \times \frac{14.7 \text{ psi}}{29.92 \text{ in Hg}} = \mathbf{13.76\ psi}$$

We have : Absolute pressure = Gauge pressure + Atmospheric pressure

= 50 + 13.76 = 63.76 psi

Absolute pressure in the rank = **63.76 psi** ... **Ans.**

(psi ⇒ poundal per square inch ⇒ pound force per square inch).

Example 7.2 : *A U-tube manometer is used to measure the pressure of water flowing in a pipeline. The right arm of the manometer is open to the atmosphere and the left arm is connected to the pipe. The centre of the pipe is in level with that of mercury in the right arm. Find the pressure in the pipe, if the difference of mercury levels in the two arms is 100 mm.*

Solution : In this case, h_m = Height of the manometric fluid above the datum z-z

$= 100 \text{ mm} = 100 \times 10^{-3}$ m

h_1 = Height of the flowing fluid above the datum z-z

$= 100 \text{ mm} = 100 \times 10^{-3}$ m

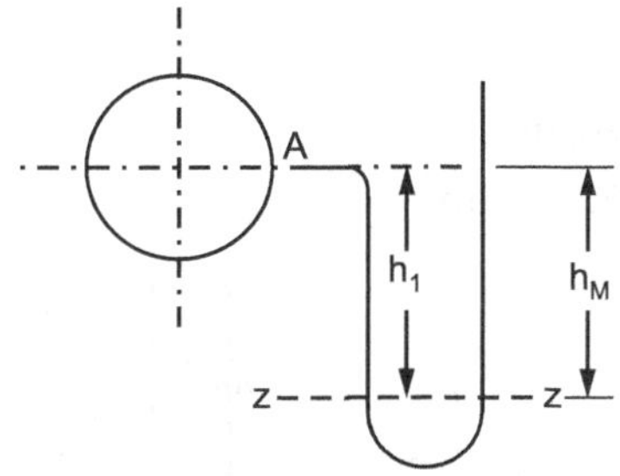

Fig. Ex. 7.2

$$\rho_m = 13600 \text{ kg/m}^3 = \text{Density of the manometric fluid}$$
$$\rho = 1000 \text{ kg/m}^3 = \text{Density of the flowing fluid}$$

Pressure in the pipe/gauge pressure is

$$P_A = h_m\rho_m g - h_1\rho g$$
$$= 100 \times 10^{-3} \times 13600 \times 9.81 - 100 \times 10^{-3} \times 1000 \times 9.81$$
$$= 12360 \text{ N/m} = 12.36 \text{ kN/m}^2 = \mathbf{12.36 \text{ kPa}}$$ **... Ans.**

Example 7.3 : *A U-tube mercury manometer is used to measure the pressure of water flowing in a pipeline. The mercury level in the open arm is 60 mm higher than the level of mercury in the arm connected to the pipeline. If the height/level of water in the arm connected to the pipeline is 50 mm, find the pressure in the pipeline in N/m² and in terms of head of water.*

Data : Density of water = 1000 kg/m³, Density of mercury = 13600 kg/m³.

Solution : Density of water (flowing fluid) = ρ = 1000 kg/m³.

Density of the manometric fluid = ρ_m = 13600 kg/m³

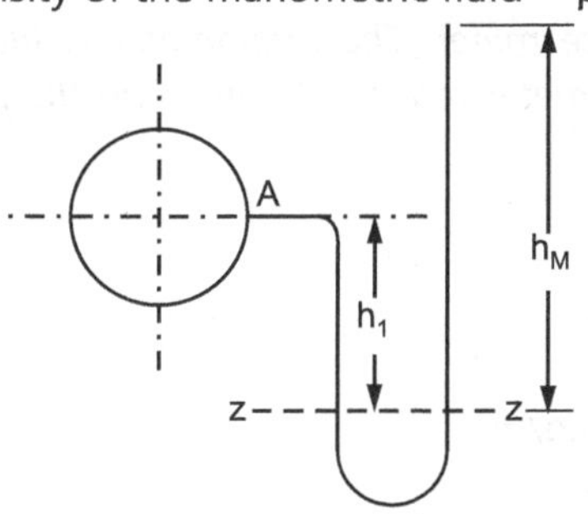

$$h_m = 60 \text{ mm} = 60 \times 10^{-3} \text{ m}$$
$$h_1 = 50 \text{ mm} = 50 \times 10^{-3} \text{ m}$$

Fig. Ex. 7.3

The gauge pressure or pressure of water in the pipeline is

$$P_A = h_m\rho_m g - h_1\rho g$$
$$= 60 \times 10^{-3} \times 13600 \times 9.81 - 50 \times 10^{-3} \times 1000 \times 9.81$$
$$= \mathbf{7514.46 \text{ N/m}^2}$$ **... Ans.**
$$= \mathbf{7.5145 \text{ kN/m}^2}$$ **... Ans.**

Pressure head in terms of m of water is

$$h_A = \frac{P_A}{\rho g} = \frac{7514.46}{1000 \times 9.81} = \mathbf{0.766 \text{ m of water or } 766 \text{ mm of water}}$$ **... Ans.**

Example 7.4 : *A U-tube manometer is used to measure the pressure of oil, having a specific gravity of 0.80, flowing through a pipeline. The right arm is open to the atmosphere and the left arm of the manometer is connected to the pipeline. The centre of the pipe is 90 mm below the level of mercury in the right arm. If the difference in mercury levels in the two arms is 150 mm, find the pressure of oil in the pipeline.*

Density of mercury = 13600 kg/m³ and density of water = 1000 kg/m³.

Solution : Specific gravity of oil = 0.80.

Density of oil = ρ = 0.80 × 1000 = 800 kg/m^3

Density of manometric fluid = ρ_m = 13600 kg/m^3

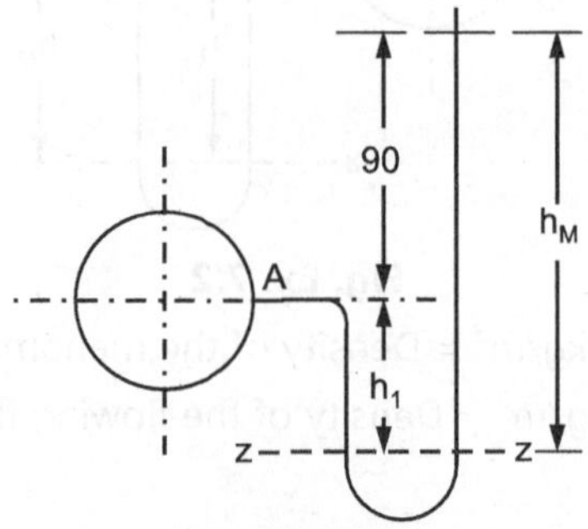

Fig. Ex. 7.4

h_m = height of manometric fluid above the datum line

= 150 mm = 150 × 10^{-3} m

h_1 = height of flowing fluid (oil) above the z-z datum

= 150 – 90 = 60 mm = 60 × 10^{-3} m

Pressure of oil in the pipeline or gauge pressure at A is

$$P_A = h_m \rho_m g - h_1 \rho g$$

$$= 150 \times 10^{-3} \times 13600 \times 9.81 - 60 \times 10^{-3} \times 800 \times 9.81$$

$$= 19541.5 \text{ N/m}^2 = \mathbf{19.54\ kN/m^2}$$ ... **Ans.**

Example 7.5 : *A simple U-tube manometer is installed across an orifice meter. The manometer is filled with mercury and the liquid above the mercury is carbon tetrachloride. The manometer reads 200 mm. Find the pressure difference over the manometer in N/m^2.*

Density of mercury = 13600 kg/m^3.

Density of carbon tetrachloride = 1600 kg/m^3.

Solution : Manometer reading = 200 mm = 0.20 m

ρ_m = density of the manometric fluid = 13600 kg/m^3

ρ = density of the flowing fluid = 1600 kg/m^3

The pressure difference over the manometer is

$$\Delta P = h_m (\rho_m - \rho)\, g$$

$$= 0.20\,(13600 - 1600) \times 9.81$$

$$= 23544 \text{ N/m}^2$$

$$= 23.544 \text{ kN/m}^2 = \mathbf{23.544\ kPa}$$ ... **Ans.**

Example 7.6 : *A flate plate having an area of 1.6 m^2 is pulled with a velocity of 0.40 m/s relative to another plate located at a distance of 0.25 mm from it. Find the force required to maintain this velocity, if the oil separating this plate has a viscosity of 2P (poise).*

Solution : A = area of the plate = 1.6 m^2

y = thickness of the oil layer = 0.25 mm = 0.25 × 10^{-3} m

μ = viscosity of oil = 2P = 0.20 (N.s)/m^2

u = velocity of the upper plate relative to the lower plate

= 0.40 m/s

The Newton's law of viscosity is

$$\tau = \mu \cdot \frac{u}{y}$$

$$\tau = \text{shear stress} = \text{force/area}$$

$$= \frac{\text{force required to maintain the velocity/speed}}{\text{area of the plate}}$$

$$\therefore \quad \frac{\text{Force}}{\text{area}} = \mu \cdot \frac{u}{y}$$

$$\text{Force} = \mu \cdot \frac{u}{y} \cdot \text{area} = 0.20 \times \frac{0.4}{0.25 \times 10^{-3}} \times 1.6 = 512 \text{ N}$$

Force required to maintain the speed/velocity of the plate = **512 N** ... **Ans.**

Example 7.7 : *Two horizontal plates are placed 12.5 mm apart. The space between these plates is filled with an oil, having a viscosity of 14 P (poise). If the upper plate is moved with a velocity of 2.5 m/s, compute the shear stress in the oil.*

Solution : The shear stress in the oil is given by

$$\tau = \mu \cdot \frac{u}{y}$$

where
μ = viscosity of the oil = 14 P = 1.4 (N.s)/m^2
u = velocity of the upper plate = 2.5 m/s
y = thickness of the oil layer between plates
= 12.5 mm = 12.5×10^{-3} m

$$\tau = 1.4 \times \frac{2.5}{12.5 \times 10^{-3}} = 280 \text{ N/m}^2$$

Shear stress in the oil = **280 N/m²** ... **Ans.**

Example 7.8 : *A plate which is 2.5 mm from another plate (fixed plate) moves with a velocity of 0.8 m/s. A force of 40 N/m² is required to maintain this speed of the moving plate. The space between the plates is filled by an oil. Find the viscosity of the oil.*

Solution : The Newton's law of viscosity is

$$\tau = \mu \cdot \frac{u}{y}$$

where
τ = shear stress = 40 N/m^2 = force/area
u = velocity of the moving plate = 0.80 m/s
y = thickness of the oil layer between the plates
= 2.5 mm = 2.5×10^{-3} m
μ = viscosity of the oil

$$\mu = \tau \cdot (y/u) = 40 \times \frac{2.5 \times 10^{-3}}{0.80}$$

$$= 0.125 \text{ (N.s)/m}^2 = 0.125 \text{ Pa·s} = 1.25 \text{ P}$$

Viscosity of the oil = 0.125 (N.s)/m^2 = 0.125 Pa.s = **1.25 P** ... **Ans.**

Example 7.9 : *An oil has a viscosity of 5 poise and a specific gravity of 0.80. Calculate the kinematic viscosity of the oil in stokes.*

Solution : Viscosity of the oil = μ = 5 P = 0.5 (N.s)/m^2 = 0.5 Pa.s.

Specific gravity of the oil = 0.80.

Density of the oil = ρ = 0.80 × density of water
= 0.80 × 1000 = 800 kg/m^3

Kinematic viscosity of the oil = $\upsilon = \mu/\rho$

$$= \frac{0.50}{800} = 0.000625 = 6.25 \times 10^{-4}\ m^2/s$$

We have : 1 stoke = $10^{-4}\ m^2/s$

$\therefore$ Kinematic viscosity of the oil = $6.25 \times 10^{-4}\ m^2/s \times \frac{1\ \text{stoke}}{10^{-4}\ m^2/s}$ = **6.25 stokes** ... **Ans.**

Example 7.10 : *An oil has a kinematic viscosity of 8 stokes and a specific gravity of 0.90. Find the absolute viscosity of the oil in poise.*

Solution : Density of water = $1000\ kg/m^3$

Specific gravity of the oil = 0.90

Density of the oil = $0.90 \times 1000 = 900\ kg/m^3$

Kinematic viscosity of the oil = 8 stokes = $8 \times 10^{-4}\ m^2/s$

[since 1 stoke = $10^{-4}\ m^2/s$]

We have : $\upsilon = \frac{\mu}{\rho}$

$\therefore$ $\mu = \upsilon \cdot \rho$

$= 8 \times 10^{-4} \times 900$

$= 0.72\ (N.s)/m^2 = 7.2\ P$

[since 1 P (poise) = $0.10\ (N.s)/m^2$]

Viscosity of the oil = **7.2 P** ... **Ans.**

Example 7.11 : *Find the kinematic viscosity of an oil in stokes. The specific gravity of the oil is 0.95 and the viscosity of the oil is 1 poise.*

Solution : Density of water = $1000\ kg/m^3$

μ = viscosity of the oil = 1 poise = $0.10\ (N.s)/m^2$

ρ = density of the oil = specific gravity × density of water

$= 0.95 \times 1000 = 950\ kg/m^3$

The kinematic viscosity of the oil is

$$\upsilon = \frac{\mu}{\rho} = \frac{0.10}{950} = 0.000105\ m^2/s$$

We have : 1 stoke = 10^{-4} m/s

We have : $\upsilon = 0.000105\ m^2/s = 1.05 \times 10^{-4}\ m^2/s$

$$\upsilon = 1.05 \times 10^{-4}\ m^2/s \times \frac{1\ \text{stoke}}{10^{-4}\ m^2/s} = 1.05\ \text{stoke}$$

Kinematic viscosity of the oil = **1.05 stoke** ... **Ans.**

$\upsilon = \mu/\rho$

Units of $\upsilon \Rightarrow \frac{\text{units of } \mu}{\text{units of } \rho}$

$$\Rightarrow \frac{(N.s)/m^2}{kg/m^3} \Rightarrow \left(\frac{kg.m}{s^2}\right) \cdot \frac{1}{m^2} \cdot \frac{m^3}{kg}$$

$\Rightarrow m^2/s$ [since 1 N = 1 (kg.m)/s²]

Example 7.12 : *An open reservoir contains a liquid having a density of 1250 kg/m³. At a certain point in the reservoir, the gauge pressure is 32.424 kN/m².g. What height above the given point is the liquid level ?*

Solution :

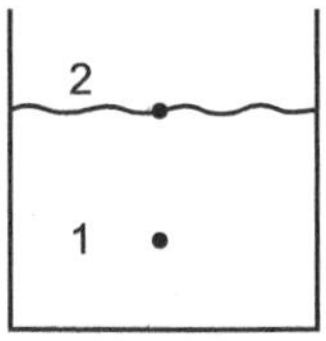

Fig. Ex. 7.12

Density of liquid = 1250 kg/m³

At point 2 the pressure is atmospheric as the reservoir is open to the atmosphere, i.e., 101.325 kN/m²

At point 1 the pressure is 32.424 kN/m²g. Let h be the height of liquid above the point 1.

$$\Delta P = (P_1 - P_2) = h\rho g \quad \text{... (1)}$$

We know that :

Absolute pressure = gauge pressure + atmospheric pressure.

$\therefore$ $P_1 = 32.424 + 101.325 = 133.749 \text{ kN/m}^2 = 133749 \text{ N/m}^2$

Pressure at point 2 = $P_2 = 101.325 \text{ kN/m}^2 = 101325 \text{ N/m}^2$

$g = 9.81 \text{ m/s}^2$

ρ = density of liquid = 1250 kg/m³

Substituting the values of ρ, g, P_1 and P_2 in Equation (1), we get

$$(P_1 - P_2) = h\rho g$$

$$(133.749 - 101.325) \times 10^3 = h \times 1250 \times 9.81$$

$$h = 2.644 \text{ m} = \mathbf{264.4 \text{ cm}}$$ **... Ans.**

Example 7.13 : *Water is flowing through a pipe at a rate of 90 l/s. Find the mass flow rate of water through the pipe.*

Solution : Volumetric flow rate of water = Q = 90 *l*/s

Density of water = ρ = 1000 kg/m³ = 1 kg/*l*

Mass flow rate of water = $\dot{m}$ kg/s

The mass flow rate of water is $\dot{m} = Q \cdot \rho$

$= 90 \; l/s \times 1 \; kg/l$

$= \mathbf{90 \; kg/s}$ **... Ans.**

Example 7.14 : *Water is flowing through a pipe at a rate of 20000 kg/h. Find the flow rate of water in l/s.*

Solution : Mass flow rate of water = $\dot{m} = 20000 \text{ kg/h} = \frac{20000}{3600} = 5.56 \text{ kg/s}$

Density of water = 1000 kg/m³ = 1 kg/*l*

The volumetric flow rate of water is

$$Q = \frac{\dot{m}}{\rho} = \frac{5.56 \text{ (kg/s)}}{1 \text{ (kg/}l)} = \mathbf{5.56 \; l/s}$$ **... Ans.**

Example 7.15 : *Water is flowing through a pipe of 100 mm diameter with an average velocity of 10 m/s. Find the volumetric flow rate in l/s and mass flow rate in kg/s.*

Solution : Diameter of the pipe = 100 mm = 0.10 m

Area of the pipe (cross-sectional area) = $A = \pi/4 D^2 = \pi/4 \, (0.10)^2 = 7.854 \times 10^{-3} \text{ m}^2$

Average velocity of water = u = 10 m/s

The volumetric flow rate of water is

$$Q = u \cdot A$$
$$= 10 \text{ m/s} \times 7.854 \times 10^{-3} \text{ m}^2 = 7.854 \times 10^{-2} \text{ m}^3\text{/s}$$

We have : $1000\, l = 1 \text{ m}^3$

$\therefore$ $$Q = 7.854 \times 10^{-2} \text{ m}^3\text{/s} \times \frac{1000\, l}{1 \text{ m}^3} = \mathbf{78.54\ l/s}$$ ... **Ans.**

The mass flow rate of water is $\dot{m} = Q \cdot \rho$

$$= 78.54\ l\text{/s} \times 1 \text{ kg}/l$$
$$= \mathbf{78.54\ kg/s}$$ ... **Ans.**

Example 7.16 : *Water is flowing through a pipe of 50 mm diameter with an average velocity of 5 m/s. Find the mass flow rate of water in kg/s.*

Solution : Diameter of the pipe = D = 50 mm = 0.05 m

Cross-sectional area of the pipe = $A = \pi/4 \cdot D^2 = \pi/4 \times (0.05)^2 = 0.001963 \text{ m}^2$

Average velocity of water = u = 5 m/s

Density of water = ρ = 1000 kg/m^3

The mass flow rate of water is $\dot{m} = \rho u A$

$$= 1000 \text{ (kg/m}^3) \times 5 \text{ (m/s)} \times 0.001963 \text{ (m}^2)$$
$$= \mathbf{9.815\ kg/s}$$ ... **Ans.**

Example 7.17 : *Water is flowing through a pipe of 50 mm diameter at a rate of 12 kg/s. Find the average velocity of water through the pipe.*

Solution : Diameter of the pipe = D = 50 mm = 0.05 m

Cross-sectional area of the pipe = $A = \pi/4 \cdot D^2$

$$= \pi/4 \times (0.05)^2 = 0.001963 \text{ m}^2$$

Mass flow rate of water = $\dot{m}$ = 12 kg/s

We have : $\dot{m} = \rho u A$

where ρ is the density of water (1000 kg/m^3).

The average velocity of water is $u = \dfrac{\dot{m}}{\rho A} = \dfrac{12}{1000 \times 0.001963} = \mathbf{6.113\ m/s}$ ... **Ans.**

Example 7.18 : *A simple U-tube manometer is installed across an orifice meter. The manometric fluid is mercury (sp. gr. 13.6) and flowing fluid through piping is carbon tetrachloride (sp. gr. 1.6). The manometer reads 200 mm. What is the pressure difference over the manometer in N/m² ?*

Solution : ρ_A = density of manometric fluid = 13.6 g/cm³ = 13600 kg/m³

ρ_B = density of flowing fluid = 1.6 g/cm³ = 1600 kg/m³

h = manometer reading = 200 mm = 0.2 m

Pressure difference over the manometer = $\Delta P = h.(\rho_A - \rho_B) \cdot g$

$$\Delta P = 0.2\,(13600 - 1600)\,9.81$$
$$\Delta P = \mathbf{23544\ N/m^2}$$ ... **Ans.**

Example 7.19 : *The pressure difference over a manometer is 2452 N/m². If the manometric fluid is carbon tetrachloride (sp. gr. 1.6) and water is flowing through the pipeline and fills the manometer leads, what will be the manometer reading ?*

Solution : ΔP = pressure difference over the manometer = 2452 N/m²

ρ_A = density of the manometric fluid = 1600 kg/m³

ρ_B = density of the flowing fluid = 1000 kg/m³

$$h = \text{manometer reading, m}$$

$$\Delta P = h \,.\, (\rho_A - \rho_B)\, g$$

$$2452 = h\,(1600 - 1000)\, 9.81$$

$$h = \text{manometer reading} = \mathbf{0.416\ m = 41.6\ cm}$$ **... Ans.**

Example 7.20 : *A U-tube manometer is used to measure pressure drop across an orifice meter. The manometric fluid is mercury (sp. gr. 13.6) and fluid flowing through the pipeline and filling manometric leads is brine (sp. gr. 1.26).*

When the pressure at taps are equal, the level of mercury in the manometer is one meter below the taps. In operating conditions, the pressure at the upstream tap is 115.324 kN/m² absolute and that at the downstream tap is 33.864 kN/m³ below the atmospheric pressure. What is the reading of manometer in centimeters ?

Solution : P_1 = pressure at the upstream tap (absolute pressure) = 115.324 kN/m²

$$P_1 = 115324\ \text{N/m}^2$$

P_2 = pressure at the downstream tap (absolute pressure)

$$= 101.325 - 33.864 = 67.461\ \text{kN/m}^2$$

$$P_2 = 67461\ \text{N/m}^2$$

$$\Delta P = P_1 - P_2 = 115324 - 67461 = 47863\ \text{N/m}^2$$

We have : $$\Delta P = h \cdot (\rho_A - \rho_B)\, g$$

where ρ_A = 13600 kg/m³

ρ_B = 1260 kg/m³

h = manometer reading

$$47863 = h\,(13600 - 1260) \times 9.81$$

$$h = 0.3954\ \text{m}$$

Manometric reading = **39.54 cm** **... Ans.**

Example 7.21 : *The wall shear stress, τ_w = 0.981 N/m² and average shear rate in circular pipe, γ = 981 (s)⁻¹ is obtained for the fluid flow. Calculate the viscosity of the fluid.*

Solution : Relationship between shear rate and shear stress is given by

$$\tau = \mu\,(du/dy)$$

$$\tau = \mu\,(\gamma)$$

τ = shear stress = 0.981 N/m²

γ = shear rate = 981 $(s)^{-1}$

μ = viscosity

$$0.981 = \mu\,(981)$$

$\therefore$ $$\mu = \mathbf{1.0 \times 10^{-3}\ (N.s)/m^2 = 1.0 \times 10^{-3}\ Pa.s}$$ **... Ans.**

Example 7.22 : *The viscosity of a certain oil sample is given as 2500 Redwood seconds. Find its kinematic viscosity.*

Solution : Here $t = 2500$ s

The kinematic viscosity is given by

$$\upsilon = At - \frac{B}{t}$$

For $t > 100$ s : A = 0.00247 and B = 0.50.

$\therefore$ $$\upsilon = 0.00247 \times 2500 - \frac{0.5}{2500} = 6.175 - 0.0002$$

$$\upsilon = \mathbf{6.1748\ centistokes}$$ **... Ans.**

Example 7.23 : *The absolute viscosity of a flowing fluid is 0.67 N.s/m². If the density is 10 kg/m³, calculate the kinematic viscosity of the fluid.*

Solution : Given : Absolute viscosity = μ = 0.67 N.s/m², Density = ρ = 10 kg/m³.

The kinematic viscosity is given by

$$\upsilon = \frac{\mu}{\rho} = \frac{0.67 \text{ N.s/m}^2}{10 \text{ kg/m}^3}$$

$$\upsilon = 0.067 \frac{\text{N.s}}{\text{kg/m}}$$

But 1 N = 1 kg.m/s², therefore :

$$\upsilon = 0.067 \frac{\text{kg.m}}{\text{s}^2} \times \frac{\text{s}}{\text{kg/m}}$$

$$\upsilon = \mathbf{0.067\ m^2/s}$$

The kinematic viscosity of the fluid is 0.067 m²/s. **... Ans.**

Example 7.24 : *Calculate the density of a fluid having an absolute viscosity of 0.89 N.s/m² and kinematic viscosity of 2 m²/s.*

Solution : The density is given by

$$\rho = \frac{\mu}{\upsilon}$$

where, μ = viscosity = 0.89 N.s/m²

υ = kinematic viscosity = 2 m²/s

$$\therefore \quad \rho = \frac{0.89 \text{ N.s/m}^2}{2 \text{ m}^2\text{/s}} = 0.445 \left(\frac{\text{kg.m}}{\text{s}^2}\right)\frac{(\text{s/m}^2)}{(\text{m}^2\text{/s})}$$

$$= \mathbf{0.445\ kg/m^3}$$

The density of the fluid is 0.445 kg/m³.

Example 7.25 : *A pressure gauge on a tank reads 50 psig on a day when the barometer reads a pressure of 28 in Hg (28 inches of mercury). Find the absolute pressure in the tank in psi.*

Solution : Given : Pressure recorded by the pressure gauge = 50 psig.

(as the letter g follows the unit, it is a gauge pressure, i.e., pressure recorded by the pressure gauge)

P = 50 psig

We know that :

Atmospheric pressure = Barometric pressure ⇒ Pressure recorded by the barometer

= 28 in Hg

Let us convert the atmospheric pressure from Hg to psi.

We have for pressure :

29.92 in Hg = 1 atm

1 atm = 14.7 psi

$$\therefore \quad \text{Atmospheric pressure} = 28 \text{ in Hg} \times \frac{1 \text{ atm}}{29.92 \text{ in Hg}} \times \frac{14.7 \text{ psi}}{1 \text{ atm}}$$

$$= 13.75668 \simeq 13.76 \text{ psi}$$

The relationship between absolute pressure and gauge pressure is

Absolute pressure = Gauge pressure + Atmospheric pressure = 50 + 13.76

Absolute pressure in the tank = **63.76 psi** **... Ans.**

Example 7.26 : *A pressure gauge on a tower indicates a vacuum of 3.53 in Hg. The barometric pressure is 29.31 in Hg. Find the absolute pressure in the tower in mmHg.*

Solution : Vacuum = 3.53 in Hg

We know that :

$$\text{Atmospheric pressure} = \text{Barometric pressure} = 29.31 \text{ in Hg}$$

The relationship between absolute pressure and vacuum is

$$\text{Absolute pressure} = \text{Atmospheric pressure} - \text{Vacuum}$$

$$= 29.31 - 3.53 = 25.78 \text{ in Hg}$$

We have for pressure :

$$29.92 \text{ in Hg} = 760 \text{ mmHg}$$

$$\therefore \quad \begin{bmatrix}\text{Absolute pressure} \\ \text{in the tower}\end{bmatrix} = 25.78 \text{ in Hg} \times \frac{760 \text{ mmHg}}{29.92 \text{ in Hg}}$$

$$= 654.84 \approx \mathbf{655\ mmHg}$$ ... **Ans.**

Example 7.27 : *Convert a pressure of 800 mmHg to the following units : (i) kPa, (ii) atm, (iii) psi and (iv) bar.*

Solution : (i) To find P in kPa.

Given : P = 800 mmHg

We have for pressure :

$$760 \text{ mm Hg} = 101.325 \text{ kPa}$$

$$\text{Pressure} = 800 \text{ (mmHg)} \times \left(\frac{101.325}{760}\right)\left(\frac{\text{kPa}}{\text{mmHg}}\right) = 106.5789 = 106.6 \text{ kPa}$$ ... **Ans. (i)**

(ii) To find P in atm :

Given : P = 800 mmHg

We have for pressure :

$$760 \text{ mmHg} = 1 \text{ atm}$$

$$\text{Pressure} = 800 \text{ (mmHg)} \times \left(\frac{1}{760}\right)\left(\frac{\text{atm}}{\text{mmHg}}\right) = 1.0526 \approx 1.053 \text{ atm}$$ ... **Ans. (ii)**

(iii) To find P in psi :

Given : P = 800 mmHg

We have for pressure :

$$760 \text{ mmHg} = 14.7 \text{ psi}$$

$$\text{Pressure} = 800 \text{ (mmHg)} \times \left(\frac{14.7}{760}\right)\left(\frac{\text{psi}}{\text{mmHg}}\right) = 15.47 \text{ psi}$$ ... **Ans. (iii)**

(iv) To find P in bar : **Given :** P = 800 mmHg

We have for pressure :

$$760 \text{ mmHg} = 1.01325 \text{ bar}$$

$$\text{Pressure} = 800 \text{ (mmHg)} \times \left(\frac{1.01325}{760}\right)\left(\frac{\text{bar}}{\text{mmHg}}\right)$$

$$= 1.06657 \approx 1.07 \text{ bar}$$ ... **Ans. (iv)**

We have for pressure :

$$760 \text{ mmHg} = 1.01325 \text{ bar}$$

$$\therefore \quad 1 \text{ mmHg} = 1.333 \times 10^{-3} \text{ bar}$$

$$\text{Pressure} = 800 \text{ (mmHg)} \times \left(\frac{1.333 \times 10^{-3}}{1}\right)\left(\frac{\text{bar}}{\text{mmHg}}\right)$$

$$= 1.0664 \approx 1.07 \text{ bar}$$ ... **Ans. (iv)**

Example 7.28 : *Convert a pressure of 2 atm to the following units : (i) mmHg, (ii) kPa and (iii) N/m².*

Solution : (i) To find P in mmHg :

Given : P = 2 atm

We have for pressure :

$$1 \text{ atm} = 760 \text{ mmHg}$$

$$\text{Pressure} = 2 \text{ atm} \times \frac{760 \text{ mmHg}}{1 \text{ atm}} = \mathbf{1520 \text{ mmHg}} \quad \text{... Ans. (i)}$$

(ii) To find P in kPa :

We have for pressure :

$$1 \text{ atm} = 101.325 \text{ kPa}$$

$$\text{Pressure} = 2 \text{ atm} \times \frac{101.325 \text{ kPa}}{1 \text{ atm}} = \mathbf{202.65 \text{ kPa}} \quad \text{... Ans. (ii)}$$

(iii) To find P in N/m² :

We have for pressure :

$$1 \text{ atm} = 101325 \text{ N/m}^2$$

$$\text{Pressure} = 2 \text{ atm} \times \frac{101325 \text{ N/m}^2}{1 \text{ atm}} = \mathbf{202650 \text{ N/m}^2} \quad \text{... Ans. (iii)}$$

Example 7.29 : *A process equipment is maintained under a vacuum of 475 mmHg. Find the absolute pressure in the equipment in kPa and bar.*

Solution : Given : Vacuum in the equipment = 475 mmHg.

We know that

Atmospheric pressure = 760 mmHg

We have, Absolute pressure = atmospheric pressure – vacuum

= 760 – 475

= 285 mmHg

(i) To find the absolute pressure in the equipment in kPa :

We have, P = 285 mmHg

We have for pressure :

$$760 \text{ mmHg} = 101.325 \text{ kPa}$$

$$\text{Absolute pressure in the equipment} = 285 \text{ mmHg} \times \frac{101.325 \text{ kPa}}{760 \text{ mmHg}} = 37.9968 \approx \mathbf{38 \text{ kPa}} \quad \text{... Ans. (i)}$$

(ii) To find P in bar :

We have for pressure :

$$760 \text{ mmHg} = 1.01325 \text{ bar}$$

$$\text{Absolute pressure in the equipment} = 285 \text{ mmHg} \times \frac{1.01325 \text{ bar}}{760 \text{ mmHg}} = 0.379968 \approx \mathbf{0.38 \text{ bar}} \quad \text{... Ans. (ii)}$$

Example 7.30 : *A force of 19.635 kgf is applied on a piston having a of diameter 5 cm. Find the pressure exerted on the piston in kPa and psi.*

Solution : Given : F = 19.635 kgf

D = 5 cm ... (find area using D)

$$\text{Area of the piston} = \frac{\pi}{4} D^2 = \frac{\pi}{4} (5)^2 = 19.635 \text{ cm}^2$$

We have, Pressure $= \frac{\text{Force}}{\text{Area}} = \frac{19.635 \text{ kgf}}{19.635 \text{ cm}^2} = 1 \text{ kgf/cm}^2$

Pressure exerted on the piston $= 1 \text{ kgf/cm}^2$

Let us now convert this pressure in kPa and psi in turn.

(i) To find P in kPa :

We have, $P = 1 \text{ kgf/cm}^2$

We have for pressure :

$$1.033 \text{ kgf/cm}^2 = 101.325 \text{ kPa}$$

Pressure exerted on the piston $= 1 \text{ kgf/cm}^2 \times \frac{101.325 \text{ kPa}}{1.033 \text{ kgf/cm}^2} = 98.088 \approx$ **98.09 kPa** ... **Ans. (i)**

(ii) To find P in psi : $P = 1 \text{ kgf/cm}^2$

We have for pressure :

$$1.033 \text{ kgf/cm}^2 = 14.7 \text{ psi}$$

Pressure exerted on the piston $= 1 \text{ kgf/cm}^2 \times \frac{14.7 \text{ psi}}{1.033 \text{ kgf/cm}^2} =$ **14.23 psi** ... **Ans. (ii)**

Example 7.31 : *Convert a viscosity of 1 poise (P) to SI units.*

Solution : Given : Viscosity = 1 P = 1 poise

The conversion factor between poise and (N.s)/m^2 is :

$$1 \text{ P} = 0.1 \text{ (N.s)/m}^2$$

$\therefore$ Viscosity $= 1 \text{ P} \times \frac{0.1 \text{ (N.s)/m}^2}{1 \text{ P}} =$ **0.10 (N.s/m² or 0.1 Pa.s** ... **Ans.**

[Since 1 Pa = 1 N/m^2]

Example 7.32 : *Water is flowing through a pipe of 100 mm diameter with an average velocity of 10 m/s. Determine the velocity of water at the other end of the pipe, if the diameter of the pipe is gradually changed to 200 mm.*

Solution : Average velocity of water through the pipe where the diameter is 100 mm is 10 m/s.

Therefore, $u_1 = 10 \text{ m/s}$

$D_1 = 100 \text{ mm} = 0.10 \text{ m}$

Average velocity of water through the pipe where the diameter is 200 mm = u_2

$D_2 = 200 = 0.20 \text{ m}$

We have : $\dot{m} = \rho u A = \text{constant}$

$$\rho_1 u_1 A_1 = \rho_2 u_2 A_2$$

$$u_1 A_1 = u_2 A_2 \quad \text{... since } \rho_1 = \rho_2 = \rho = \text{density of water}$$

$$u_2 = \frac{u_1 A_1}{A_2} = \frac{u_1 (\pi/4 \cdot D_1^2)}{\pi/4 \cdot D_2^2} = \frac{u_1 D_1^2}{D_2^2} = \frac{10\,(0.10)^2}{(0.2)^2} = 2.5 \text{ m/s}$$

Average velocity of water at the end of pipe where the diameter is 200 mm = **2.5 m/s** ... **Ans.**

Example 7.33 : *Water is flowing through a pipe of 50 mm diameter with a velocity of 3 m/s. Find the discharge through the pipe in l/s.*

Solution : Diameter of the pipe = 50 mm = 0.05 m

Area of pipe $= A = \pi/4 \cdot D^2$

$= \pi/4\,(0.05)^2$

$= 0.001963 \text{ m}^2$

Velocity of water $= u = 3 \text{ m/s}$

The discharge, i.e., volumetric flow rate of water through the pipe is

$$Q = u \cdot A$$
$$= 3 \times 0.001963 = 0.00589 \text{ m}^3/\text{s}$$

We have : $1000\ l = 1 \text{ m}^3$

$\therefore$ $Q = 0.00589 \times 1000 = \mathbf{5.89\ l/s}$ ... **Ans.**

Example 7.34 : *Find the size of a pipe in mm through which an oil is flowing with a velocity of 3 m/s.*

Given : Specific gravity of the oil is 0.80 and the volumetric flow rate of the oil is 2 m^3/s.

Solution : Volumetric flow rate of the oil = Q = 2 m^3/s

Specific gravity of the oil = 0.80

Density of the oil = ρ = Specific gravity of oil × Density of water

$= 0.80 \times 1000 = 800 \text{ kg/m}^3$

Velocity of the oil = u = 3 m/s

Cross-sectional area of the pipe = $A = \pi/4 \cdot D^2$

Diameter of the pipe = D, m

The volumetric flow rate of oil is given by

$$Q = u \cdot A$$

$\therefore$ $A = Q/u$

$$\pi/4 \cdot D^2 = Q/u$$

$\therefore$ $D^2 = 4Q/\pi u$

$\therefore$ $D = [4Q/\pi \cdot u]^{1/2} = \left[\dfrac{4 \times 2}{\pi \cdot 3}\right]^{1/2}$

$$D = 0.921 \text{ m} \approx 0.92 \text{ m}$$

Diameter of the pipe = **0.92 m = 920 mm** ... **Ans.**

Example 7.35 : *Water (density = 1000 kg/m³) flows through the piping system shown in Fig. 7.35. An equal quantity of water flows through each of the pipes C. The flow through pipe A is 10 m³/h. Calculate (a) mass-flow rate in each pipe, (b) the average velocity in each pipe and (c) mass velocity in pipes A and B.*

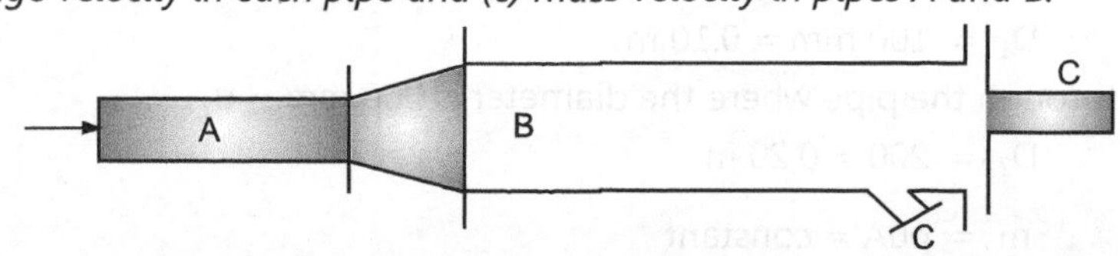

Fig. Ex. 7.35

Pipe A = 50 mm i.d. pipe = D_1

Pipe B = 75 mm i.d. pipe = D_2

Pipe C = 40 mm i.d. pipe = D_3

Solution :

Volumetric flow rate of water through the pipe A = Q_A = 10 m³/h

Cross-sectional area of the pipe A = $\pi/4\ D_1^2$

Density of water = ρ = 1000 kg/m³

Mass flow rate through the pipe 'A' = $\dot{m}_1 = Q_A \cdot \rho$

$= 10 \times 1000 = \mathbf{10000\ kg/h}$ ... **Ans. (a)**

Mass flow rate through the pipe 'B' $= \dot{m}_2$ = mass flow rate through the pipe A = **10,000 kg/h** ... **Ans. (a)**

Mass flow rate through the pipe 'C' $= \dot{m}_3 = \frac{1}{2}$ (mass flow rate through the pipe A)

$$= \frac{1}{2} \times 10000 = \mathbf{5000\ kg/h}$$ ... **Ans. (a)**

We have : Mass flow rate $= \dot{m} = \rho \cdot u \cdot A$

For pipe A :

u_1 = Velocity of water through the pipe A

A_1 = Cross - sectional area of the pipe A

$$A_1 = \pi/4\ D_1^2, \quad \text{where } D_1 = 50 \text{ mm} = 50 \times 10^{-3} \text{ m}$$

$$= \pi/4\ (50 \times 10^{-3})^2$$

$$A_1 = 1.963 \times 10^{-3} \text{ m}^2$$

We have for pipe A : $\dot{m}_1 = \rho u_1 A_1$

$\therefore \quad 10{,}000 = 1000 \times u_1 \times 1.963 \times 10^{-3}$

$$u_1 = 5092 \text{ m/h}$$

Average velocity through the pipe A $= \frac{5092}{3600} = \mathbf{1.41\ m/s}$ ... **Ans. (b)**

For pipe B :

Velocity through B is obtained by the relation :

$$\frac{u_1}{u_2} = \left(\frac{D_2}{D_1}\right)^2$$

$$u_2 = u_1 \left(\frac{D_1}{D_2}\right)^2$$

where u_2 = velocity through the pipe B

u_1 = velocity through the pipe A = 1.41 m/s

D_1 = diameter of the pipe A $= 50 \times 10^{-3}$ m

D_2 = diameter of the pipe B $= 75 \times 10^{-3}$ m

$$u_2 = 1.41 \left(\frac{50 \times 10^{-3}}{75 \times 10^{-3}}\right)^2$$

u_2 = Velocity through pipe B = **0.63 m/s** ... **Ans. (b)**

For pipe C :

Mass flow rate through the pipe C = 5000 kg/h

We know, $\dot{m} = \rho \cdot u \cdot A$

$$\dot{m}_3 = \rho\ u_3\ A_3$$

where $\dot{m}_3$ = 5000 kg/h

ρ = 1000 kg/m³

A_3 = area of flow of pipe C

$$A_3 = \frac{\pi}{4}(D_3)^2, \quad \text{where } D_3 = 40 \text{ mm} = 40 \times 10^{-3} \text{ m}$$

$$= \pi/4\,(40 \times 10^{-3})^2 = 1.25 \times 10^{-3} \text{ m}^2$$

$$u_3 = \text{Average velocity through the pipe C} = \frac{\dot{m}}{\rho A_3}$$

$$= \frac{5000}{1000 \times 1.25 \times 10^{-3}} = 3978.8 \text{ m/h} = \mathbf{1.1 \text{ m/s}}$$ ... **Ans. (b)**

$$\text{Mass velocity through the pipe A} = G_1 = \frac{\text{mass flow rate}}{\text{cross-sectional area of pipe A}}$$

$$= \frac{10000}{1.963 \times 10^{-3}} = 5094243.5 \text{ kg/(m}^2\text{.h)} = \mathbf{1415 \text{ kg/(m}^2\text{.s)}}$$ ... **Ans. (c)**

$$\text{Mass velocity through the pipe B} = \frac{\text{mass flow rate through the pipe B}}{\text{cross-sectional area of the pipe B}}$$

$$\text{Mass velocity through the pipe B} = \frac{1000}{\pi/4\,(75 \times 10^{-3})} = 226337 \text{ kg/(m}^2\text{ . h)}$$

$$= \mathbf{628.76 \text{ kg/(m}^2\text{.s)}}$$ ... **Ans. (c)**

Example 7.36 : *A 75 mm diameter pipe discharges liquid at the rate of 30 l/s. Determine the type of flow, if the kinematic viscosity of the liquid is 1.2×10^{-4} m²/s.*

Solution : D = diameter of the pipe = 75 mm = 0.075 m

A = cross-sectional area of the pipe $= \pi/4\,(0.075)^2 = 0.004418 \text{ m}^2$

$$Q = 30 \text{ l/s} = 30 \times 10^{-3} \text{ m}^3\text{/s}$$

$$\upsilon \text{ of water} = 1.2 \times 10^{-4} \text{ m}^2\text{/s}$$

$$u = \text{velocity of water} = Q/A$$

$$= \frac{30 \times 10^{-3}}{0.004418} = 6.79 \text{ m/s}$$

$$N_{Re} = \text{Reynolds number} = \frac{Du\rho}{\mu} = \frac{Du}{\upsilon} = \frac{0.075 \times 6.79}{1.2 \times 10^{-4}} = 4243.75$$

Since N_{Re} is greater than 4000, the flow is turbulent. ... **Ans.**

Example 7.37 : *Water is flowing through a pipe of 25 mm diameter with a velocity of 2 m/s. Determine the type of flow, if the viscosity of water is 1 mPa·s.*

Solution : Diameter of the pipe = D = 25 mm = 0.025 m

$$\text{Velocity of water} = u = 2 \text{ m/s}$$

$$\text{Density of water} = \rho = 1000 \text{ kg/m}^3$$

$$\text{Viscosity of water} = \mu = 1 \text{ mPa·s} = 1 \times 10^{-3} \text{ Pa·s}$$

$$\mu = 1 \times 10^{-3} \text{ Pa·s} = 1 \times 10^{-3} \text{ kg/(m.s)}$$

$$\text{The Reynolds number is} \quad N_{Re} = \frac{Du\rho}{\mu} = \frac{0.025 \times 2 \times 1000}{1 \times 10^{-3}} = 50000$$

Since $N_{Re} > 4000$, **the flow is turbulent**. ... **Ans.**

Example 7.38 : *An oil of kinematic viscosity 20 stokes is flowing through a pipe of 300 mm diameter. If the flow rate of oil through the pipe is 25 l/s, determine the type of flow.*

Solution : D = diameter of pipe = 300 mm = 0.30 m

$$A = \text{area of pipe} = \pi/4 \times (0.30)^2 = 0.0707 \text{ m}^2$$

$$Q = \text{flow rate of oil} = 25 \text{ l/s} = 25 \times 10^{-3} \text{ m}^3\text{/s}$$

$$u = \text{velocity of oil} = \frac{Q}{A} = \frac{25 \times 10^{-3}}{0.0707} = 0.3536 \approx 0.354 \text{ m/s}$$

$$\upsilon = \text{kinematic viscosity of oil} = \frac{\mu}{\rho} = 20 \text{ stokes} = 20 \times 10^{-4} \text{ m}^2/\text{s}$$

$$[\text{since 1 stoke} = 10^{-4} \text{ m}^2/\text{s}]$$

$$\text{Reynolds number} = N_{Re} = \frac{Du\rho}{\mu} = \frac{Du}{\upsilon}$$

$$= \frac{0.30 \times 0.354}{20 \times 10^{-4}} = 53.10$$

Since N_{Re} is less than 2100, the flow is laminar. **... Ans.**

Example 7.39 : *An oil of specific gravity 0.95 is flowing at a rate of 60 l/s through a pipeline of 200 mm diameter. Determine the type of flow, if the viscosity of the oil is 1 poise.*

Solution : Density of oil = ρ = specific gravity = density of water = 0.95 × 1000 = 950 kg/m^3

$$D = \text{diameter of pipeline} = 200 \text{ mm} = 0.20 \text{ m}$$

$$A = \text{cross-sectional area} = \pi/4 \cdot D^2 = \pi/4\,(0.20)^2$$

$$= 0.0314 \text{ m}^2$$

$$Q = \text{volumetric flow rate of oil} = 60 \text{ l/s} = 60 \times 10^{-3} \text{ m}^3/\text{s}$$

$$\text{Viscosity of oil} = 1 \text{ P} = 0.10 \text{ kg/(m·s)}$$

$$u = \text{velocity of oil} = Q/A = 60 \times 10^{-3}/0.0314$$

$$= 1.911 \text{ m/s}$$

$$\text{Reynolds number} = N_{Re} = \frac{Du\rho}{\mu} = \frac{0.20 \times 1.911 \times 950}{0.10} = 3630.9 \approx 3631$$

Since N_{Re} is less than 4000 and > 2100 the flow is in the transient region but may be considered as turbulent.

Example 7.40 : *In a laboratory experiment, an oil flowing through a pipe of 50 mm diameter at a velocity of 1.5 m/s results in a pressure drop of 18 kPa over two locations that are 8 m apart. Find the viscosity of the oil.*

Solution : The Hagen-Poiseullie equation is

$$\Delta P = \frac{32\, L\, \mu u}{D^2}$$

where

$$\Delta P = 18 \text{ kPa} = 18 \times 10^3 \text{ Pa} = 18 \times 10^3 \text{ N/m}^2$$

$$L = \text{length of the pipe} = 8 \text{ m}$$

$$D = \text{diameter of the pipe} = 50 \text{ mm} = 0.05 \text{ m}$$

$$u = \text{velocity of the oil} = 1.5 \text{ m/s}$$

$$\mu = \text{viscosity of the oil}$$

The viscosity of the oil is

$$\mu = \frac{\Delta P \cdot D^2}{32\, Lu}$$

$$= \frac{18 \times 10^3 \times (0.05)}{32 \times 8 \times 1.5} = 0.117 \text{ (N·s)/m}^2 = \mathbf{1.17\ P}$$ **... Ans.**

Example 7.41 : *Water is flowing through a pipeline of 200 mm diameter at a rate of 65 l/s. Find the velocity of water at a point P, 50 mm away from the centre of the pipeline.*

Solution :

$$D = \text{diameter of the pipeline} = 200 \text{ mm} = 0.20 \text{ m}$$

$$r_w = \text{radius of the pipeline} = 0.2/1 = 0.10 \text{ m}$$

$$r = \text{distance of the point P from the centre of the pipeline}$$

$$= 50 \text{ mm} = 0.05 \text{ m}$$

$$A = \text{area of the pipeline}$$

$$= \pi/4 \cdot D^2 = \pi/4\,(0.2)^2 = 0.0314 \text{ m}^2$$

u = average velocity of water in the pipeline

Q = volumetric flow rate of water = 65 l/s = 65×10^{-3} m^3/s

$$u = \frac{Q}{A} = \frac{65 \times 10^{-3}}{0.0314} = 2.07 \text{ m/s}$$

u_{max} = maximum velocity of water = 2 (average velocity)

$= 2.u = 2 \times 2.07 = 4.14$ m/s

The velocity of water at any point P from the centre is

$$\frac{u}{u_{max}} = \left[1 - \left(\frac{r}{r_w}\right)^2\right]$$

$$u = u_{max}\left[1 - \left(\frac{r}{r_w}\right)^2\right] = 4.14\left[1 - \left(\frac{0.05}{0.1}\right)^2\right] = \mathbf{3.11\ m/s}$$ ... **Ans.**

Example 7.42 : *An oil of specific gravity 0.91 and kinematic viscosity 0.002 m²/s flows through a pipe of 150 mm diameter at a rate of 60 l/s. Find the loss of head due to viscosity of the oil over a length of 30 m.*

Solution : Specific gravity of oil = 0.91

Density of water = 1000 kg/m^3

Density of oil = $0.91 \times 1000 = 910$ kg/m^3

Kinematic viscosity of oil = $\upsilon = 0.002$ m^2/s = μ/ρ

Viscosity of oil = $\mu = \upsilon \cdot \rho$

$= 0.002 \times 910 = 1.82$ (N·s)/m^2

Diameter of pipe = 150 mm = 0.150 m

A = cross-sectional area of pipe = $\pi/4\ (0.150)^2$

$= 0.01767$ m^2

Q = volumetric flow rate of oil = 60 l/s = 60×10^{-3} m^3/s

$$u = \text{velocity of oil} = \frac{Q}{A} = \frac{60 \times 10^{-3}}{0.01767} = 3.395 \text{ m/s}$$

Given : L = 30 m

We have :

$$\Delta P = \frac{32\, L\, \mu u}{D^2}$$

$$= \frac{32 \times 30 \times 1.82 \times 3.395}{(0.150)^2} = 263633 \text{ N/m}^2$$

The loss of head due to viscosity in terms of m of oil is

$$\text{Loss head} = \frac{\Delta P}{\rho g} = \frac{263633}{910 \times 9.81} = \mathbf{29.53\ m}$$ ... **Ans.**

Example 7.43 : *Calculate the friction factor when the Reynolds number is 1600 for flow of fluid through the pipe.*

Solution : Reynolds number = N_{Re} = 1600

As N_{Re} given is less than 2100, the flow is laminar.

For laminar flow, the relationship between the friction factor (f) and the Reynolds number (N_{Re}) is

$$f = \frac{16}{N_{Re}} = \frac{16}{1600}$$

$\therefore$ $f = \mathbf{0.01}$ ... **Ans.**

Example 7.44 : *Calculate the critical velocity of water flowing through 25 mm i.d. pipe.*

Data : *Density of water* = *1000 kg/m³*

Viscosity of water = *0.0008 (N.s)/m²*

Solution : The velocity at which flow changes from laminar to turbulent is called as the critical velocity. It occurs at the Reynolds number of 2100.

Let u_c = critical velocity

D = diameter of water = 25 mm = 25×10^{-3} m.

ρ = density of water = 1000 kg/m³

μ = viscosity of water = 0.0008 (N.s)/m² = 0.0008 kg/(m.s)

$$N_{Re,c} = 2100 = \frac{Du_c \cdot \rho}{\mu}$$

$$2100 = \frac{25 \times 10^{-3}\, u_c \times 1000}{0.0008}$$

$$\therefore \quad u_c = 0.0672 \text{ m/s}$$

Critical velocity of water flowing through 25 i.d. pipe = **0.0672 m/s** ... **Ans.**

Example 7.45 : *Water of density 1000 kg/m³ and viscosity 0.0008 (N.s)/m² is pumped at a rate of 1000 cm³/s through a 25 mm i.d. pipe. Calculate the Reynolds number.*

Solution : Density of water = ρ = 1000 kg/m³

Viscosity of water = μ = 0.0008 (N.s)/m² = 0.0008 kg/(m.s)

I.D. of pipe = D = 25 mm

= 25×10^{-3} m

Volumetric flow rate = Q = 1000 cm³/s

= 1000×10^{-6} m³/s

$$\text{Area of flow} = \frac{\pi}{4} D^2$$

$$\text{Area of flow} = A = \frac{\pi}{4}(25 \times 10^{-3})^2 = 4.9 \times 10^{-4} \text{ m}^2$$

$$\text{Average velocity} = u = Q/A = \frac{\text{Volumetric flow rate}}{\text{Cross-sectional area}} = \frac{1000 \times 10^{-6}}{4.9 \times 10^{-4}} = 2.041 \text{ m/s}$$

$$\text{Reynolds number} = N_{Re} = \frac{D\,u\,\rho}{\mu} = \frac{25 \times 10^{-3} \times 2.041 \times 1000}{0.0008} = 63{,}662$$

Reynolds number = **63,662.** ... **Ans.**

Example 7.46: *Acetic acid is to be pumped at a rate of 0.02 m³/s through a 75 mm i.d. pipe line. What is the pressure drop in the pipe line over a length of 70 m ?*

Date : *Density of acetic acid* = *1060 kg/m³*

Viscosity of acetic acid = *0.0025 (N.s)/m²*

Solution : Volumetric flow rate of acetic acid = Q = 0.02 m³/s

D = inside diameter of pipe = 75 mm = 0.075 m

$$A = \text{Cross-sectional area of the pipe} = \frac{\pi}{4} \times D^2 = \frac{\pi}{4}(0.075)^2$$

$$A = 4.418 \times 10^{-3} \text{ m}^2$$

u = Average velocity through the pipe = Q/A

$$= \frac{0.02}{4.418 \times 10^{-3}} = 4.53 \text{ m/s}$$

$$N_{Re} = \text{Reynolds number} = \frac{D\,u\,\rho}{\mu}$$

$$\text{where } \rho = 1060 \text{ kg/m}^3$$

$$\mu = 0.0025 \text{ (N.s)/m}^2 \equiv (0.0025 \text{ kg/(m.s)})$$

$$N_{Re} = \frac{0.075 \times 4.53 \times 1060}{0.0025} = 144054$$

As the Reynolds number is greater than 4000, the flow is turbulent.

The relationship between f and N_{Re} for the turbulent flow is

$$f = \frac{0.078}{(N_{Re})^{0.25}}$$

$$f = \frac{0.078}{(144054)^{0.25}}$$

$$f = 0.004$$

$$\frac{\Delta P}{\rho} = \frac{4 f L u^2}{2 D}$$

$$\Delta P = \text{pressure drop over length L}$$

$$\Delta P = \rho \left[\frac{4f Lu^2}{2 D}\right] \quad \text{... (1)}$$

$$\text{where } \rho = 1060 \text{ kg/m}^3$$

$$f = 0.004$$

$$u = 4.53 \text{ m/s}$$

$$L = 70 \text{ m}$$

$$D = 0.075 \text{ m}$$

Substituting the values of various terms in Equation (1) gives

$$\Delta P = 1060 \left[\frac{4 \times 0.004 \times 60 \times (4.53)^2}{2 \times 0.075}\right] = 139213.8 \text{ N/m}^2 = \mathbf{139.214 \text{ kN/m}^2} \quad \textbf{... Ans.}$$

Example 7.47 : *Water is flowing at a velocity of 2.5 m/s through 25 mm internal diameter pipe. Find out the friction factor ' f '.*

Data : *Density of water = 1000 kg/m³*

Viscosity of water = 0.0008 (N.s)/m²

Solution :

$$N_{Re} = \frac{Du\rho}{\mu}$$

$$\text{where } D = \text{diameter of pipe} = 25 \text{ mm} = 0.025 \text{ m}$$

$$u = \text{velocity} = 2.5 \text{ m/s}$$

$$\rho = \text{density} = 1000 \text{ kg/m}^3$$

$$\mu = \text{viscosity} = 0.0008 \text{ (N.s)/m}^2 = 0.0008 \text{ kg/(m.s)}$$

$$\therefore \quad N_{Re} = \frac{0.025 \times 2.5 \times 1000}{0.0008}$$

= 78125 Therefore, the flow is turbulent

For turbulent flow, the relationship between N_{Re} and 'f ', friction factor is

$$\frac{1}{\sqrt{f}} = 4.0 \log \left(N_{Re} \sqrt{f}\right) - 0.4 = 4 \log N_{Re} + 4 \log \sqrt{f} - 0.4$$

$$\therefore \quad \frac{1}{f} - 4 \log \sqrt{f} = 4 \log N_{Re} - 0.4 = 4 \log (78125) - 0.4$$

$$\therefore \quad \frac{1}{\sqrt{f}} - 4 \log \sqrt{f} = 19.17$$

Here we have to adopt a trial and error procedure to find f.

Assume a certain value of f, calculate the L.H.S. of the above equation and see that whether the L.H.S. calculated is equal to 19.17 or not. If not, assume a new value of f and repeat the procedure till we do not get :

$$\text{L.H.S.} \cong \text{R.H.S.}$$

$$\text{with } f = 0.004$$

$$\text{L.H.S.} = 20.60$$

But $\text{R.H.S.} = 19.17$

$\therefore \quad f = 0.0045$

$$\frac{1}{\sqrt{0.0045}} - 4\log\sqrt{0.0045} = 19.17$$

$$19.60 = 19.17$$

Since there is a considerable difference between L.H.S. and R.H.S., assume a new value for f, and repeat the procedure till we get the L.H.S. approximately equal to the R.H.S. (19.17). Doing this, we get

$$f = \text{friction factor} = \mathbf{0.00475}$$ **... Ans.**

Example 7.48 : *Water is flowing at a rate of 5 m³/h through a pipe line of 78 mm i.d. The viscosity of water is 8.0×10^{-4} Pa.s. Calculate the pressure drop and frictional loss over a length of 50 metres of the pipeline.*

Solution : Volumetric flow rate = 5 m^3/h

$$\text{Diameter of pipe} = 78 \text{ mm} = 0.078 \text{ m}$$

$$\text{Viscosity of water} = 8.0 \times 10^{-4} \text{ Pa.s}$$

$$\text{Length} = 50 \text{ m}$$

$$\text{Reynolds number, } N_{Re} = \frac{D \times u \times \rho}{\mu}$$

$$D = 0.078 \text{ m}$$

$$\rho = 1000 \text{ kg/m}^3$$

$$\mu = 8 \times 10^{-4} \text{ kg/(m.s)} = \text{Pa.s}$$

$$Q = 5 \text{ m}^3/\text{h} = 1.39 \times 10^{-3} \text{ m}^3/\text{s}$$

$$A = \frac{\pi D^2}{4} = \frac{\pi \times (0.078)^2}{4} = 4.778 \times 10^{-3} \text{ m}^2$$

$$u = Q/A$$

$$\therefore \quad u = \frac{1.39 \times 10^{-3}}{4.778 \times 10^{-3}} = 0.3 \text{ m/s}$$

$$\therefore \quad N_{Re} = \frac{0.078 \times 0.3 \times 1000}{8 \times 10^{-4}} = 29250$$

Since $N_{Re} > 4000$, the flow is turbulent.

The friction factor for turbulent flow is given by

$$\frac{1}{\sqrt{f}} = 4.0\log\left(N_{Re}\sqrt{f}\right) - 0.4 = 4\log N_{Re} + 4\log\sqrt{f} - 0.4$$

$$\frac{1}{\sqrt{f}} - 4\log\sqrt{f} = 4\log N_{Re} - 0.4 = 4\log(29250) - 0.4$$

$$\frac{1}{\sqrt{f}} - 4\log\sqrt{f} = 17.46$$

To evaluate f we have to adopt a trial and error procedure. Assume a value of f, calculate the LHS, check for LHS ≈ RHS. If not, repeat the procedure such that LHS ≈ RHS.

We have : R.H.S. = 17.46

With f = 0.005

L.H.S. = 18.74

for f = 0.006 , L.H.S. = 17.35

f = 0.0059 , L.H.S. = 17.478

Since for f = 0.0059,

L.H.S. (17.478) ≈ R.H.S. (17.46), f = 0.0059

OR

$$f = 0.078/(N_{Re})^{0.25} = 0.078/(29250)^{0.25} = 0.00596$$

$$\therefore \quad f = 0.0059$$

Pressure drop is given by the relation :

$$\frac{\Delta P}{\rho} = \frac{4\,f\,L\,u^2}{2\,D}$$

f = 0.0059, L = 50 m

u = 0.3 m/s, D = 0.078 m

$$\frac{\Delta P}{1000} = \frac{4 \times 0.0059 \times 50 \times (0.30)^2}{2 \times 0.078}$$

$\therefore \quad \Delta P =$ **681 N/m²** ... **Ans.**

Example 7.49 : *An oil is contained between two identical parallel plates of 2.0 m² area each. The top plate is pulled to the left (–x-direction) with a force of 0.33 N at a velocity of 0.3 m/s. The bottom plate is pulled in the opposite direction with a force of 0.11 N at a velocity of 0.10 m/s. Find the viscosity of the oil if the plates are 5 mm apart.*

Solution : (i) Top plate pulled : Given : F = 0.33 N, u = 0.3 m/s, A = 2 m², τ = F/A = 0.33/2 = 0.165 N/m² or Pa

Assuming linear velocity distribution, we can write

$$\frac{du}{dy} = \frac{\Delta u}{\Delta y} = \frac{0.3}{5 \times 10^{-3}} = 60\ s^{-1}$$

Therefore, the viscosity of oil is given by

$$\mu = \tau / (\Delta u/\Delta y) = 0.165/60 = 2.75 \times 10^{-3}\ Pa\,.\,s$$

Viscosity of the oil = **2.75 mPa . s** ... **Ans.**

(ii) Bottom plate pulled :

Given : F = 0.11 N, u = 0.1 m/s, A = 2 m²

τ = F/A = 0.11/2 = 0.55 N/m² or Pa

$$du/dy = \Delta u/\Delta y = \frac{0.10}{5 \times 10^{-3}} = 20\ s^{-1}$$

$$\mu = \tau / (\Delta u/\Delta y) = 0.055/20 = 2.75 \times 10^{-3}\ Pa\,.s$$

Viscosity of the oil = **2.75 mPa . s** ... **Ans.**

Example 7.50 : *Water flows through a 25 mm diameter pipeline at a rate of 0.8 l/s. If the temperature of water is 303 K (30°C), determine the type of flow. At 303 K (30°C) the density of water is 996 kg/m³ and the viscosity of water is 0.8 mPa.s.*

Solution : D = 25 mm = 0.025 m, ρ = 996 kg/m³, μ = 0.8 mPa.s = 0.8×10^{-3} Pa.s

Cross-sectional area of pipeline = $\frac{\pi}{4}(0.025)^2 = 4.909 \times 10^{-4}\ m^2$

Volumetric flow rate of water = 0.8 l/s = $0.8 \times 10^{-3}\ m^3/s$

$$\text{Velocity of water, u} = \frac{0.8 \times 10^{-3}}{4.909 \times 10^{-4}} = 1.63 \text{ m/s}$$

$$\text{Reynolds number, } N_{Re} = \frac{Du\rho}{\mu} = \frac{0.025 \times 1.63 \times 996}{0.8 \times 10^{-3}} = \mathbf{50734}$$

As the Reynolds number is greater than 4000, the flow is **turbulent.** ... **Ans.**

Example 7.51 : *In an air pipeline, the flow has following conditions at station 1 :*

Temperature = 298 K (25°C), pressure = 1.8 bar, velocity = 15 m/s and pipe inside the diameter = 50 mm.

The conditions at station 2 are :

Temperature = 298 K (25°C), pressure = 1.3 bar and pipe inside the diameter = 75 mm.

Density of air at 298 K and 1.8 bar pressure is 2.1 kg/m³.

Estimate the mass flow rate of air and the velocity at station 2.

Solution :

At station 1 : D_1 = 50 mm = 0.05 m, T_1 = 298 K, P_1 = 1.8 bar, u_1 = 15 m/s and ρ_1 = 2.1 kg/m^3

At station 2 : D_2 = 75 mm = 0.075 m, T_2 = 298 K, P_2 = 1.3 bar, u_2 = ?, ρ_2 = ?

Cross-sectional area of pipe at station 1

$$= \pi/4\ (0.05)^2 = 0.00196 \text{ m}^2$$

$$\text{Mass flow rate of air} = \rho_1 A_1 u_1 = 2.1 \times 0.00196 \times 15 = \mathbf{0.062 \text{ kg/s}}$$ ... **Ans.**

Density of air at station 2 :

We know that for an ideal gas :

$$\rho \propto P$$

When applied to two stations, this becomes

$$\frac{\rho_1}{\rho_2} = \frac{P_1}{P_2}$$

$$\rho_2 = \rho_1 \cdot \frac{P_2}{P_1} = \frac{2.1 \times 1.3}{1.8} = 1.517 \text{ kg/m}^3$$

The continuity equation is

$$\rho_1 u_1 A_1 = \rho_2 u_2 A_2$$

$$\rho_1 u_1 (\pi/4) D_1^2 = \rho_2 u_2 (\pi/4) D_2^2$$

$$u_2 = \frac{\rho_1 u_1 D_1^2}{\rho_2 D_2^2} = u_1 \left(\frac{\rho_1}{\rho_2}\right)\left(\frac{D_1}{D_2}\right)^2 = 15\left(\frac{2.10}{1.517}\right)\left(\frac{0.05}{0.075}\right) = \mathbf{9.23 \text{ m/s}}$$ ... **Ans.**

Example 7.52 : *For laminar flow, the velocity profile in a circular pipe is given by*

$$u = u_{max}\left[1 - \left(\frac{r}{R}\right)^2\right]$$

where u_{max} *is the velocity at the centre line of the pipe*

r *is the radial distance from the centre line of the pipe*

R *is the radius of the pipe.*

(i) What is the average velocity ?

(ii) Show that the velocity gradient varies linearly with radius.

(iii) Find the velocity gradient at the wall and at the centre line.

Solution : Consider a differential element of a fluid of radius r and thickness dr concentric with the pipe. This fluid element moves with a velocity u.

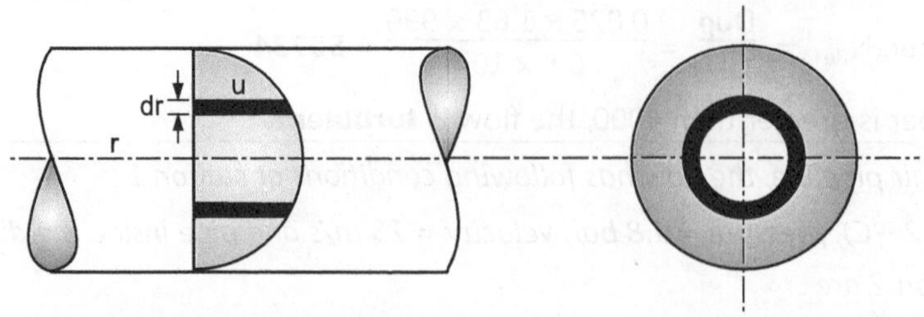

Fig. Ex. 7.52 : Fluid element at radius r of thickness dr moving with velocity u

Cross-sectional area of the fluid element = $2 \pi r \, dr$

Volumetric flow rate of the fluid through the differential area = $(2 \pi r \, dr) . u$

The average velocity is given by

$$u_{avg.} = \frac{\text{Volumetric flow rate}}{\text{Cross-sectional area of pipe}}$$

$$= \frac{2\pi \int_0^R r.u \, dr}{2\pi \int_0^R r \, dr} = \frac{\int_0^R u_{max} \left[1 - (r/R)^2\right] r \, dr}{\int_0^R r \, dr}$$

Since u_{max} is constant, we can write

$$u_{avg} = \frac{u_{max} \int_0^R \left[1 - (r/R)^2\right] dr}{R^2/2} = \frac{2 \, u_{max}}{R^2} \int_0^R (r \, dr - r^3/R^2) \, r \, dr = \frac{2 \, u_{max}}{R^2} \left[\frac{R^2}{2} - \frac{R^4}{4 \, R^2}\right]$$

$$u_{avg.} = \frac{2 \, u_{max}}{R^2} \cdot \frac{R^2}{4} = \frac{1}{2} u_{max} = \mathbf{0.5 \, u_{max}}$$ **... Ans. (i)**

We have, $\mathbf{u = u_{max} (1 - r^2/R^2)}$

Differentiating with respect to r gives

$$\frac{du}{dr} = u_{max} \left[0 - \frac{1}{R^2} \cdot 2 \, r\right]$$

$$\frac{du}{dr} = -\frac{2r \, . \, u_{max}}{R^2} = k.r$$

where $\mathbf{k = -2 \, u_{max}/R^2}$

$\therefore$ $du/dr \propto r$

Therefore, the velocity gradient varies linearly with r. **... Ans. (ii)**

We have $$\frac{du}{dr} = \frac{-r \, u_{max}}{R^2}$$

At the wall of the pipe, we have

$$r = R$$

$\therefore$ $$\frac{du}{dr} = \frac{-R \, u_{max}}{R^2} = \mathbf{\frac{-u_{max}}{R}}$$ **... Ans. (iii)**

At the centre line of the pipe, we have : r = 0

$$\therefore \quad \frac{du}{dr} = \frac{-r\,u_{max}}{R^2} = \frac{-0\,.\,u_{max}}{R^2} = 0$$

$$\mathbf{\frac{du}{dr} = 0} \qquad \textbf{... Ans. (iv)}$$

Example 7.53 : *For turbulent flow in a smooth pipe, the velocity profile is given by*

$$u = u_{max}\left(1 - \frac{r}{R}\right)^n$$

where $n = 1/7$ *for* N_{Re} *upto* 10^5

$n = 1/8$ *for* N_{Re} *more than* 10^5 *upto* 4×10^5

For n = 1/7, determine :

(i) The average velocity.

(ii) The velocity gradient at the wall and at the centre line of the pipe.

Solution : The average velocity is given by

$$u_{avg.} = \frac{\text{Volumetric flow rate}}{\text{Cross-sectional area of pipe}} = \frac{2\pi \int_0^R ur\,dr}{2\pi \int_0^R r\,dr} = \frac{\int_0^R u\,r\,dr}{R^2/2}$$

Substituting for u by $u_{max}(1 - r/R)^n$ gives

$$u_{avg} = \frac{2\int_0^R u_{max}(1 - r/R)^n\, r\,dr}{R^2}$$

Integrating, we get $u_{avg.} = \mathbf{\dfrac{2\,u_{max}}{(n + 2)\,(n + 1)}}$ **... Ans.**

Integration is done as given below :

Let $1 - \frac{r}{R} = y \quad \therefore \quad r = (1 - y)\,R$

$dr = -R\,dy$

At r = 0, y = 1 and at r = R, y = 0

$$u_{avg.} = \frac{2\,u_{max}\int_1^0 y^n (1 - y)\,R\,(-R\,dy)}{R^2} = -2\,u_{max}\int_1^0 y^n(1 - y)\,dy = 2\,u_{max}\int_0^1 y^n(1 - y)\,dy$$

$$= 2\,u_{max}\left[\int_0^1 y^n\,dy - \int_0^1 y^{n+1}\,dy\right] = 2\,u_{max}\left[\left(\frac{y^{n+1}}{n+1}\right)_0^1 - \left(\frac{y^{n+2}}{n+2}\right)_0^1\right]$$

$$= 2\,u_{max}\left[\frac{1}{n+1} - \frac{1}{n+2}\right] = 2\,u_{max}\frac{[(n + 2) - (n + 1)]}{(n + 1)\,(n + 2)}$$

$$u_{avg.} = \mathbf{2\,u_{max} / (n + 1)\,(n + 2)} \qquad \textbf{... Ans. (i)}$$

We have : $u = u_{max}\left(1-\frac{r}{R}\right)^n$

Let $1 - r/R = y$

$\therefore \quad r = (1-y)\,R$

$u = u_{max} \cdot y^n$

$\therefore \quad \frac{du}{dy} = n \cdot u_{max}\, y^{n-1}$

$r = (1-y)\,R$

$dr = -R\,dy$

$\therefore \quad dy/dr = -1/R$

$$\frac{du}{dr} = \frac{du}{dy}\cdot\frac{dy}{dr}$$

$$= n \cdot u_{max} \cdot y^{n-1}\left(\frac{-1}{R}\right) = -\frac{u_{max}}{R} \cdot n \cdot \left(1-\frac{r}{R}\right)^{n-1}$$

$$\boxed{\frac{du}{dr} = \frac{-u_{max} \cdot n}{R}\left(\frac{R-r}{R}\right)^{n-1}}$$

At the wall, r = R $\quad \mathbf{\frac{du}{dr} = \infty}$ **... Ans. (ii)**

This implies that the wall shear stress becomes infinitely large. This impossibility can be clarified by the fact that the given velocity distribution ceases to be valid near the wall of the pipe.

$$\frac{du}{dr} = \frac{-u_{max} \cdot n}{R}\left[\frac{R-r}{R}\right]^{n-1}$$

At the centreline of pipe, r = 0

$$\therefore \quad \frac{du}{dr} = \frac{-u_{max} \cdot n}{R}$$

For n = 1/7

$$\frac{du}{dr} = \frac{\mathbf{-u_{max}}}{\mathbf{7R}}$$ **... Ans. (ii)**

Example 7.54 : *By doubling the flow in a pipe it is found that the frictional losses increase by 3.5 times, how do the frictional losses vary with velocity and what is the nature of the flow ?*

Solution : Pressure drop due to friction varies with u as

$$\Delta P \propto u^n$$

where u is the average velocity.

Let at $u = u_1$, $\Delta P = \Delta P_1$

and at $u = u_2$, $\Delta P = \Delta P_2$

$$\Delta P = c \cdot u^n$$

$$\Delta P_1 = c\,u_1^n \text{ and } \Delta P_2 = c \cdot u_2^n$$

$$\left(\frac{\Delta P_2}{\Delta P_1}\right) = \left(\frac{u_2}{u_1}\right)^n$$

Given : $\Delta P_2 = 3.5\ \Delta P_1$ at $u_2 = 2u_1$

$$\therefore \quad 3.6 = (2)^n$$

$$\text{or} \quad n = 1.81$$

$$\therefore \quad \Delta P \propto \mathbf{u^{1.81}}$$ **... Ans.**

We know that the frictional losses in the turbulent flow region vary about 1.7 to 2 power of the velocity, i.e., $\Delta P \propto u^{1.7-2}$ and the frictional losses in the laminar flow region vary directly as the flow velocity, i.e., $\Delta P \propto u$.

Since n = 1.81 which is between 1.7 and 2, **the flow is turbulent.** **... Ans.**

Example 7.55 : *Estimate the size of a pipe required to convey 15 l/s of an oil to a distance of 3 km with a frictional head loss of 24 m.*

Data : *Kinematic viscosity of oil is 10^{-5} m²/s.*

Solution : For solving this problem a trial and error procedure is to be adopted.

The frictional head loss is given by

$$h_f = \frac{\Delta P}{\rho g} = \frac{4 f L u^2}{2 g D}$$

Volumetric flow rate = 15 l/s = 15×10^{-3} m³/s

Let the diameter of pipe be 'D' m.

$$\text{Velocity of oil through the pipe} = \frac{15 \times 10^{-3}}{\pi/4 \, . \, D^2} = \frac{0.0191}{D^2}$$

L = 3 km = 3000 m, $\upsilon = 10^{-5}$ m²/s.

$$h_f = 24 = \frac{4 \times f \times 3000}{2 \times 9.81 \times D}\left(\frac{0.0191}{D^2}\right)^2$$

$$\therefore \quad D^5 = 9.3 \times 10^{-3} f$$

$$D = (9.3 \times 10^{-3} f)^{1/5} \quad \text{... (1)}$$

$$N_{Re} = \frac{Du\rho}{\mu} = \frac{Du}{\nu} = \frac{D \times (0.0191/D^2)}{10^{-5}}$$

$$N_{Re} = 1910/D \quad \text{... (2)}$$

We have, for turbulent flow

$$f = \frac{0.078}{(N_{Re})^{0.25}} \quad \text{... (3)}$$

Now assume f, calculate D by Equation (1), calculate N_{Re} by Equation (2) and then calculate f by Equation (3). Repeat this procedure till 'f ' assumed is approximately equal to 'f ' calculated.

f , assumed	**D , m**	**N_{Re}**	**f , calculated**
0.005	0.136	14044	0.0071
0.007	0.145	13172	0.0073
0.0072	0.146	13082	0.00729
0.0073	0.147	12993	0.00731

Therefore, f = 0.0073 and D = 0.147 m, since $f_{\text{, assumed (0.0073)}} \approx f_{\text{, calculated (0.00731)}}$

Check : u = 0.0191/D² = 0.0191/ (0.0147)² = 0.884 m/s, f = 0.0073

$$\therefore \quad h_f = \frac{4 f L u^2}{2 g D} = \frac{4 \times 0.0073 \times 3000 \times (0.884)^2}{2 \times 9.81 \times 0.147}$$

$$= 23.74 \text{ m} \simeq 24 \text{ m}$$

∴ Size of the pipe, i.e., diameter of the pipe = 0.147 m = **147 mm.** **... Ans.**

Example 7.56 : *Two pipes, each 300 m long, are available for connecting to a reservoir from which the flow at a rate of 0.085 m³/s is required. The diameters of the two pipes are 300 mm and 150 mm respectively. Determine the ratio of the head lost when pipes are connected in series to the head lost when they are connected in parallel. Neglect minor losses. Assume the flow to be turbulent.*

Solution : When pipes are connected in series, the head lost due to friction is given by

$$h_f\text{, series} = h_{f_1} + h_{f_2}$$

$$= \frac{4\,f\,L\,u_1^2}{2\,g\,D_1} + \frac{4\,f\,L\,u_2^2}{2\,g\,D_2} = \frac{2\,f\,L}{g}\left[\frac{u_1^2}{D_1} + \frac{u_2^2}{D_2}\right]$$

u_1 is the velocity through the pipe of diameter D_1 and u_2 is the velocity through the pipe of diameter D_2.

D_1 = 300 mm = 0.30 m, D_1 = 150 mm = 0.15 m, L = 300 m, g = 9.81 m/s².

Volumetric flow rate through the pipes in series = 0.085 m³/s

Cross-sectional area of pipe of diameter $D_1 = \pi/4\,(0.30)^2 = 0.0707$ m²

$$\therefore \quad u_1 = 0.085/0.0707 = 1.2 \text{ m/s}$$

Cross-sectional area of pipe of diameter $D_2 = \pi/4\,(0.15)^2 = 0.01767$ m²

$u_2 = 0.085/0.01767 = 4.81$ m/s

$$h_f\text{, series} = \frac{2 \times f \times 300}{9.81}\left[\frac{(1.2)^2}{0.30} + \frac{(4.81)^2}{0.15}\right] = 9727\,f$$

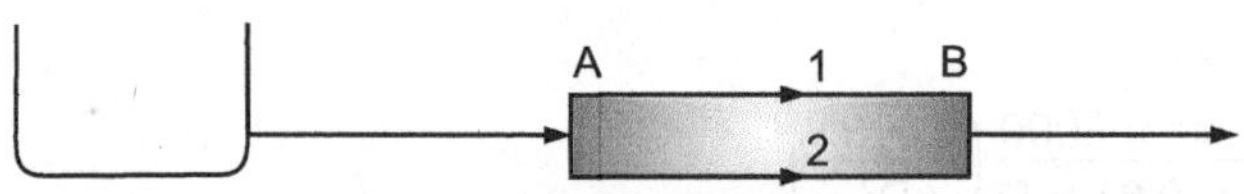

Fig. Ex. 7.56

When the pipes are connected in parallel, flow takes place under the difference of head between sections A and B and hence the loss of head between sections A and B will be the same whether the liquid flows through the pipe 1 or the pipe 2.

The rate of discharge in the main line is equal to the sum of discharges in each of the parallel pipes.

$$Q = Q_1 + Q_2 = 0.085 \text{ m}^3/\text{s}$$

Q_1 is the volumetric flow rate through the pipe of diameter D_1.

Q_2 is the volumetric flow rate through the pipe of diameter D_2.

The head loss due to friction is given by

$$h_f\text{, parallel} = \frac{4\,f\,L\,u_1^2}{2\,g\,D_1} = \frac{4\,f\,L\,u_2^2}{2\,g\,D_2}$$

$$\therefore \quad \frac{4\,f\,L\,(Q_1/A_1)^2}{2\,g\,D_1} = \frac{4\,f\,L\,(Q_2/A_2)^2}{2\,g\,D_2}$$

$$\therefore \quad \frac{(Q_1/A_1)^2}{D_1} = \frac{(Q_2/A_2)^2}{D_2}$$

$$\frac{Q_1^2}{D_1^5} = \frac{Q_2^2}{D_2^5}$$

$$Q_1^2 = \left(\frac{D_1}{D_2}\right)^2 \cdot Q_2^2 = \left(\frac{0.30}{0.15}\right)^5 Q_2^2 = 32\,Q_2^2$$

$$Q_1 = 5.66\,Q_2$$

We have,

$$Q_1 + Q_2 = 0.085$$

$$5.66\, Q_2 + Q_2 = 0.085$$

$$\therefore \quad Q_2 = 0.0128 \text{ m}^3/\text{s}$$

$$Q_1 = 5.66 \times 0.0128 = 0.0724 \text{ m}^3/\text{s}$$

$$h_f\text{, parallel} = \frac{4\, f\, L\, (Q_2 / A_2)^2}{2\, g\, D_2}$$

where $A_2 = \pi/4\, D_2^2 = \pi/4\, (0.15)^2 = 0.01767 \text{ m}^2$

$$h_f\text{, parallel} = \frac{4 \times f \times 300 \times (0.0128/0.01767)^2}{2 \times 9.81 \times 0.15} = 214.2\, f$$

$$\frac{\text{Head loss when pipes in series}}{\text{Head loss when pipes in parallel}} = \frac{h_f\text{, series}}{h_f\text{, parallel}}$$

$$\frac{h_f\text{, series}}{h_f\text{, parallel}} = \frac{9727\, f}{214.2\, f} = \mathbf{45.41}$$ **... Ans.**

Example 7.57 : *Crude oil having a specific gravity of 0.91 and a viscosity of 0.124 Pa.s is pumped at a rate of 7 l/s through a pipe line 75 mm diameter having a length of 62 m and whose outlet is 3 m higher than its inlet. Calculate the power required for the pump if its efficiency is 60%.*

Solution : D = 75 mm = 0.075 m, L = 62 m

$\rho = 0.91 \times 1000 = 910 \text{ kg/m}^3$, $\mu = 0.124$ Pa.s

Volumetric flow rate of the oil $= 7\ l/s = 7 \times 10^{-3} \text{ m}^3/\text{s}$

Cross-sectional area of the pipe $= \pi/4\, (0.075)^2 = 4.418 \times 10^{-3} \text{ m}^2$

Velocity of oil through the pipe, $u = 7 \times 10^{-3} / 4.418 \times 10^{-3}$

$$u = 1.584 \text{ m/s}$$

Mass flow rate of the oil, $\dot{m} = 7 \times 10^{-3} \times 910 = 6.37$ kg/s

$$\text{Reynolds number, } N_{Re} = \frac{Du\rho}{\mu} = \frac{0.075 \times 1.584 \times 910}{0.124} = 872$$

Since $N_{Re} < 2100$, the flow is laminar.

The head loss due to friction is given by

$$h_f = \frac{32\, \mu\, u\, L}{\rho \,.\, g\, D^2} = \frac{32 \times 0.124 \times 1.584 \times 62}{910 \times 9.81 \times (0.075)^2} = 7.76 \text{ m}$$

Total head in this case is

$$\text{Total head} = Z + \frac{u^2}{2g} + h_f = 3 + (1.584)^2 / (2 \times 9.81) + 7.76 = 10.9 \text{ m}$$

Efficiency of the pump is 60%

$$\therefore \quad \eta = 0.6$$

Power required for the pump $= (\dot{m} \times \text{head} \times g) / \eta$

$= 6.37 \times 10.9 \times 9.81 / 0.60$

$= 1135 \text{ W} = \mathbf{1.13\ kW}$ **... Ans.**

Example 7.58 : *Find the pressure drop due to friction in a 60 m long and 25 mm i.d. pipe when water is flowing at a rate of 3 kg/s. If the pressure drop falls by one half, what will the new flow rate be ?*

Data : Density of water = 1000 kg/m³, viscosity of water = 0.0008 Pa.s

Solution : Mass flow rate of water $= \dot{m} = 3$ kg/s

D = 25 mm = 0.025 m, $\rho = 1000 \text{ kg/m}^3$, $\mu = 0.0008$ Pa.s, L = 60 m

Cross-sectional area of the pipe $= \pi/4 \times (0.025)^2 = 4.909 \times 10^{-4} \text{ m}^2$

We have : $\dot{m} = \rho u A$

$u = \dot{m}/\rho A = 3 / (1000 \times 4.909 \times 10^{-4}) = 6.11$ m/s

Reynolds number, $N_{Re} = Du\rho/u$

$N_{Re} = 0.025 \times 6.11 \times 1000 / 0.0008 = 190937$

As the Reynolds number is greater than 4000, the flow is turbulent. For turbulent flow, we have :

$$f = 0.078 / (N_{Re})^{0.25}$$

$$= 0.078 / (190937)^{0.25} = 0.00373$$

Pressure drop due to friction is given by

$$\frac{\Delta P}{\rho} = \frac{4 f L u^2}{2 D}$$

$$\Delta P = 1000 \times 4 \times 0.00373 \times 60 \times (6.11)^2/2 \times 0.025$$

$$= 668394 \text{ N/m}^2 = \mathbf{668.394 \text{ kN/m}^2} \quad \text{... Ans.}$$

The new pressure drop is half of the pressure drop calculated above.

$\therefore \quad \Delta P' = \Delta P/2 = 668394/2 = 334197$ N/m²

New velocity and thus the new flow rate is calculated as follows :

$$\frac{\Delta P'}{\rho} = \frac{4 f L u^2}{2 D}$$

$$f u^2 = \frac{\Delta P' \times D}{\rho \times 2L} = \frac{334197 \times 0.025}{9000 \times 60 \times 2} = 0.0696$$

$$\therefore \quad u = 0.264/(f)^{0.5} \quad \text{... (1)}$$

$$N_{Re} = \frac{Du\rho}{\mu} = \frac{0.025 \times 1000 \times u}{0.0008} = 31250\, u \quad \text{... (2)}$$

We have

$$f = \frac{0.078}{(N_{Re})^{0.25}} = \frac{0.078}{(31250\, u)^{0.25}}$$

$$\therefore \quad f = 0.00587 / (u)^{0.25} \quad \text{... (3)}$$

From equations (1) and (3), we get

$$u = \frac{0.264\, (u)^{\frac{1}{4} \times \frac{1}{2}}}{[0.00587]^{0.5}} = 3.446\, u^{1/8}$$

$$\therefore \quad u^{1-1/8} = 3.446$$

$$u^{7/8} = 3.446$$

$$\therefore \quad u = 4.11 \text{ m/s}$$

Putting value of u in equation (3), we get

$$f = 0.00587 / (u)^{0.25} = 0.00587/(4.11)^{0.25} = 0.00412$$

$$\therefore \quad u = 4.11 \text{ m/s and } f = 0.00412$$

OR : assume a certain value of 'f ', calculate u with the help of Equation (1), then N_{Re} by Equation (2) and finally calculate 'f' by using Equation (3) till 'f' assumed approximately matches 'f' calculated and take the corresponding value of 'u' for the calculation of the new flow rate.

f, assumed	u , m/s	N_{Re}	f, calculated
0.003	4.82	150625	0.00396
0.004	4.17	1303125	0.0041
0.0041	4.12	128750	0.00412

Take u = 4.11 m/s.

New water velocity through the pipe = 4.11 m/s

$$\text{New mass flow rate of water} = \dot{m} = \rho u A$$
$$= 1000 \times 4.11 \times 4.909 \times 10^{-4}$$
$$= \mathbf{2.02\ kg/s}$$ ... **Ans.**

$$\text{New } \Delta P = \frac{1000 \times 4 \times 0.00412 \times 60 \times (4.11)^2}{2 \times 0.025} = 334058\ N/m^2$$

which is approximately equal to ΔP' of 334197 N/m².

Example 7.59 : *Sulphuric acid is pumped at a rate of 3 kg/s through a pipeline of 25 mm diameter. Calculate the drop in pressure over a length of 60 m. If the pressure drop falls by one half, what will be the new flow rate ?*

Data : *Density of acid = 1840 kg/m³*

Viscosity of acid = 25 (mN.s)/m²

Solution : Mass flow rate of acid = $\dot{m}$ = 3 kg/s, D = 25 mm = 0.025 m, L = 60 m

$$\text{Cross-sectional area of pipe} = \pi/4\ (0.025)^2 = 4.909 \times 10^{-4}\ m^2$$
$$\text{Velocity of acid} = 3 / 1840 \times 4.909 \times 10^{-4} = 3.32\ m/s$$
$$N_{Re} = D\,u\,\rho/\mu$$
$$= 0.025 \times 3.32 \times 1840 / (25 \times 10^{-3}) = 6109$$
$$f = 0.078 / (N_{Re})^{0.25}$$
$$= 0.078 / (6109)^{0.25} = 0.0088$$

The pressure drop due to friction is given by

$$\Delta P = \rho . \frac{4\,f\,L\,u^2}{2D}$$
$$= 1840 \times 4 \times 0.0088 \times 60 \times (3.32)^2 / 2 \times 0.025 = 856678\ N/m^2$$

The new pressure drop is 50% of the original.

$\therefore$ ΔP' = 0.5 (ΔP) = 0.5 (856678) = 428339 N/m²

The new velocity will be calculated as follows :

$$\Delta P' = \frac{\rho\,4\,f\,L\,u^2}{2D}$$

and $$N_{Re} = Du\rho/\mu$$

So $$u = N_{Re}\,\mu/D\rho$$

Substituting this value of u in the equation for ΔP' gives

$$\Delta P' = \frac{(\rho\,.\,4\,f\,L\,(N_{Re}\,\mu/D\rho)^2)}{2\,D}$$

Rearranging the above equation, we get

$$N_{Re}^2 f = \Delta P'\,D^3\,\rho/2\,L\,\mu^2$$
$$= 428339 \times (0.025)^3 \times 1840 / 2 \times 60 \times (25 \times 10^{-3})^2$$
$$= 164197$$
$$N_{Re}\sqrt{f} = 405.2$$

The relation between f and N_{Re} for turbulent flow is

$$1/\sqrt{f} = 4 \log (N_{Re} \sqrt{f}) - 0.4$$

$$1/\sqrt{f} = 4 \log (405.2) - 0.4$$

$$\therefore \quad f = 0.00994$$

$$\Delta P' = \frac{\rho \, 4 \, f \, L \, u^2}{2 \, D}$$

$$u^2 = \Delta P' \, 2 \, D / \rho \, 4 \, f \, L$$

$$= 428339 \times 2 \times 0.025 / 1840 \times 4 \times 0.00994 \times 60$$

$$= 4.88$$

$$\therefore \quad u = 2.21 \text{ m/s}$$

The new velocity is 2.21 m/s.

$$\text{New mass flow rate} = \rho \, u \, A$$

$$= 1840 \times 2.21 \times 4.909 \times 10^{-4} = 1.996 \simeq \mathbf{2.0 \text{ kg/s}}$$ **... Ans.**

Example 7.60 : *Crude oil of density 840 kg/m³ is pumped at a rate of 3 l/s through a 52 mm i.d. steel pipe under a pressure drop of 550 kPa over a length of 600 m. Calculate the Fanning friction factor using the Hagen-Poiseuille equation.*

Solution : The Hagen-Poiseuille equation is

$$\frac{\Delta P}{\Delta L} = \frac{32 \, \mu \, u}{D^2}$$

$$\Delta P = 550 \text{ kPa} = 550 \times 10^3 \text{ Pa}$$

$$\Delta L = 600 \text{ m}, \; D = 52 \text{ mm} = 0.052 \text{ m}$$

Cross-sectional area of the pipe= $A = \pi/4 \,.\, (0.052)^2 = 2.124 \times 10^{-3} \text{ m}^2$

Volumetric flow rate $= Q = 3 \; l/s = 3 \times 10^{-3} \text{ m}^3/\text{s}$

Velocity of oil through the pipe $= \dfrac{Q}{A} = \dfrac{3 \times 10^{-3}}{2.124 \times 10^{-3}} = 1.41 \text{ m/s}$

$$\rho = 840 \text{ kg/m}^3$$

$$N_{Re} = Du \, \rho/\mu$$

$$\text{and } f = 16 / N_{Re} \text{ (for laminar flow)}$$

$$f = 16 / (Du\rho/\mu)$$

$$\therefore \quad \mu = f \, Du \, \rho/16$$

$$\frac{\Delta P}{\Delta L} = \frac{32 \, u}{D^2} \left(\frac{f \, Du\rho}{16} \right)$$

$$\frac{\Delta P}{\Delta L} = \frac{2 \, f \, D \, u^2 \, \rho}{D^2}$$

$$f = \frac{\Delta P \, D}{\Delta L \,.\, 2 \, u^2 \, \rho} = \frac{550 \times 10^3 \times 0.052}{600 \times 2 \times (1.41)^2 \times 840}$$

$$f = 0.0143$$

For laminar flow, we have

$$N_{Re} = 16/f = 16/0.0143 = 1119$$

Fanning friction factor = f = **0.0143** **... Ans.**

Example 7.61 : *Water at 303 K (30°C) flows through a horizontal pipe 25 mm in diameter, in which the pressure drop per metre length is to be limited to 2.35 Pa/m. Calculate the volumetric flow rate.*

Data : *ρ of water = 996 kg/m³, μ of water = 0.8 mPa.s*

Solution : D = 25 mm = 0.025 m

$\Delta P/L = 2.35$ Pa/m (N/m² per m)

The pressure drop due to friction is given by

$$\frac{\Delta P}{\rho} = \frac{4\,f\,L\,u^2}{2\,D}$$

$$\frac{\Delta P}{L} = \frac{4\,f\,u^2\,\rho}{2D}$$

$$N_{Re} = \frac{Du\rho}{\mu} \quad \therefore \quad u = N_{Re}\,\mu / D\rho$$

$$\frac{\Delta P}{L} = \frac{2\,f\,N_{Re}^2\,\mu^2\,.\,\rho}{D^3\,\rho^2}$$

$$\therefore \quad N_{Re}^2\,f = \frac{\Delta P}{L}\cdot\frac{D^3\,.\,\rho}{2\,\mu^2}$$

where D = 0.025 m, ρ = 996 kg/m³, μ = 0.8 mPa.s = 0.8×10^{-3} Pa.s

$$\Delta P / L = 2.35 \text{ Pa/m}$$

$$\therefore \quad N_{Re}^2\,f = \frac{2.35 \times (0.025)^3 \times 996}{2 \times (0.8 \times 10^{-3})^2} = 28572$$

$$\therefore \quad N_{Re}\sqrt{f} = 169$$

For turbulent flow, the relation between f and N_{Re} is

$$1/\sqrt{f} = 4 \log (N_{Re}\sqrt{f}) - 0.4 = 4 \log (169) - 0.4 = 8.51$$

$$\therefore \quad f = 0.0138$$

$$\frac{\Delta P}{\rho} = \frac{4\,f\,L\,u^2}{2\,D}$$

$$\therefore \quad u^2 = \frac{\Delta P}{L} \times \frac{2\,D}{4\,f} \times \frac{1}{\rho} = \frac{2.35 \times 2 \times 0.025}{4 \times 0.0138 \times 996} = 2.13 \times 10^{-3}$$

$$\therefore \quad u = 0.046 \text{ m/s}$$

Cross-sectional area of the pipe= $\pi/4\,.\,(0.025)^2 = 4.909 \times 10^{-4}$ m²

Volumetric flow rate of water = u . A

= $0.046 \times 4.909 \times 10^{-4}$ = **2.26×10^{-5} m³/s = 22.6 cm³/s** ... **Ans.**

Example 7.62 : *Calculate the flow rate of natural gas flowing through a 300 mm i.d. pipe with a frictional pressure drop of 21.5 kPa/km. For the calculation purpose consider the gas to be an incompressible fluid having a density of 3.48 kg/m³ and a viscosity of 0.012 mPa.s at the average temperature and pressure at its transportation.*

Solution : D = 300 mm = 0.3 m, ρ = 3.48 kg/m³

μ = 0.012 mPa.s = 0.012×10^{-3} Pa.s ((N.s)/m²)

$\Delta P/L$ = 21.5 kPa/km = 21.5 Pa/m (i.e., N/m² per m)

The pressure drop due to friction is given by the relation

$$\frac{\Delta P}{\rho} = \frac{4\,f\,L\,u^2}{2\,D}$$

$$\therefore \quad f\,u^2 = \frac{\Delta P}{L}\cdot\frac{D}{2\,\rho} = \frac{21.5 \times 0.30}{2 \times 3.48} = 0.927$$

$$\therefore \quad u = (0.927/f)^{0.5} = 0.963/\sqrt{f}$$

We have,

$$N_{Re} = \frac{Du\rho}{\mu} = 0.30 \times \left(\frac{0.963}{\sqrt{f}}\right) \times \frac{3.48}{0.012 \times 10^{-3}}$$

$$\therefore \quad N_{Re}\sqrt{f} = 83781$$

For turbulent flow, f is related to N_{Re} by

$$1/\sqrt{f} = 4 \log (N_{Re}\sqrt{f}) - 0.4$$

Substituting the value of $N_{Re}\sqrt{f}$ gives

$$1/\sqrt{f} = 4 \log (83781) - 0.4 = 19.3$$

$$\therefore \quad f = 0.0027$$

$$u = 0.963 / \sqrt{f} = 0.963/\sqrt{0.0027} = 18.6 \text{ m/s}$$

Velocity of gas = 18.6 m/s

Cross-sectional area of the pipe= $\pi/4 . (0.3)^2 = 0.0707$ m²

Volumetric flow rate = u.A = 18.6 × 0.0707

= **1.315 m³/s** ... **Ans.**

OR : Friction factor f for turbulent flow may be calculated by

$$f = 0.078 / (N_{Re})^{0.25} \quad \text{... (1)}$$

We have :

$$u = 0.963 / (f)^{0.5} \quad \text{... (2)}$$

$$N_{Re} = \frac{Du\rho}{\mu}$$

$$N_{Re} = \frac{0.3 \times 3.48}{0.012 \times 10^{-3}} u = 87000 u \quad \text{... (3)}$$

Here we have to adopt a trial and error procedure.

Assume a suitable value of f and calculate u with the help of Equation (1) and then calculate N_{Re} with the help of Equation (2) and then finally calculate value of 'f' with the help of Equation (3). This value should be close to the assumed value.

f, assumed	u, m/s	N_{Re}	f, calculated
0.004	15.23	1325010	0.0023
0.003	17.58	1529460	0.0021
0.0025	19.26	1675620	0.00217
0.0020	21.53	1873110	0.0021
0.0021	21.0	1827000	0.0021

The velocity of gas is therefore,

u = 21 m/s

Volumetric flow rate = 21 × 0.0707

= **1.485 m³/s** ... **Ans.**

By this method the deviation in volumetric flow rate as compared to the previous method is :

$$\frac{1.485 - 1.315}{1.315} \times 100 = 12.92\%$$

Example 7.63 : *Water at 294 K (21°C) is flowing at a velocity of 3 m/s through the annulus between a pipe with an outside diameter of 25 mm and the other pipe with an internal diameter of 50 mm in a double pipe heat exchanger. Estimate the pressure drop due to friction per 1 m length of the annulus.*

Data : *ρ of water = 1000 kg/m³, μ of water = 1.0 mPa.s*

Solution : Let us determine D_e.

$$D_e = \text{equivalent diameter of the annulus.}$$

$$D_e = 4.r_H$$

$$= 4\frac{\text{(Cross-sectional area for flow)}}{\text{Wetted perimeter for fluid flow}}$$

$$= \frac{4\left[\pi/4 \,.\, D_2^2 - \pi/4 \,.\, D_1^2\right]}{\left[\pi D_1 + \pi D_2\right]}$$

$$D_e = \frac{D_2^2 - D_1^2}{(D_1 + D_2)} = D_2 - D_1$$

We have : $D_2 = 50$ mm and $D_1 = 25$ mm.

$D_e = 50 - 25 = 25 \text{ mm} = 0.025 \text{ m}$

D_1 is the outside diameter of the inside pipe and

D_2 is the inside diameter of the outside pipe.

$\rho = 1000 \text{ kg/m}^3$, $\mu = 1.0 \text{ mPa.s} = 1 \times 10^{-3} \text{ Pa.s}$

N_{Re} for flow in the annulus is given by

$$N_{Re} = \frac{D_e\, u\, \rho}{\mu} = \frac{0.025 \times 3 \times 1000}{1 \times 10^{-3}} = 75000$$

Since $N_{Re} > 4000$, 'f' may be calculated by

$$f = 0.078 / (N_{Re})^{0.25}$$

$$= 0.078 / (75000)^{0.25} = 0.0047$$

The pressure drop due to friction is given by

$$\frac{\Delta P}{\rho} = \frac{4\, f\, L\, u^2}{2\, D_e}$$

$$\frac{\Delta P}{L} = \frac{4 f \rho u^2}{2 D_e}$$

$$\therefore \quad \frac{\Delta P}{L} = \frac{4 \times 0.0047 \times 1000 \times (3)^2}{2 \times 0.025} = 3384 \text{ Pa/m or N/m}^2 \text{ per m}$$

$\approx$ **3.4 kPa/m or kN/m² per m** ... **Ans.**

Example 7.64 : *A kerosene storage tank drains by gravity to a tank truck. The pipeline between the tank and the truck is of 60 m length and of 25 mm internal diameter. Both tank and truck are open to the atmosphere. The flow rate of kerosene through the pipeline is 800 cm³/s. Calculate the difference between the level in the tank and that in the truck.*

Data : *Density of kerosene = 800 kg/m³, Viscosity of kerosene = 5 × 10⁻⁴ Pa.s*

For turbulent flow use the relation : $f = 0.078 / (N_{Re})^{0.25}$

Solution : Volumetric flow rate of kerosene = $Q = 800 \text{ cm}^3\text{/s} = 8 \times 10^{-4} \text{ m}^3\text{/s}$

$D = 25 \text{ mm} = 0.025 \text{ m}$

Cross-sectional area of the pipe = $A = \pi/4 \,.\, (0.025)^2 = 4.909 \times 10^{-4} \text{ m}^2$

Velocity of kerosene in the pipeline = $\frac{Q}{A} = 8 \times 10^{-4} / 4.909 \times 10^{-4} = 1.63 \text{ m/s}$

$\rho = 800 \text{ kg/m}^3$, $\mu = 5 \times 10^{-4} \text{ Pa.s}$

$N_{Re} = Du\rho/\mu = 0.025 \times 1.63 \times 800 / 5 \times 10^{-4} = 65200$

Since $N_{Re} > 4000$, the flow is turbulent and f and N_{Re} are thus related by

$f = 0.078/(N_{Re})^{0.25}$

$f = 0.078 / (65200)^{0.25} = 0.0049$

The head loss due to friction is given by

$$h_f = \frac{4 f L u^2}{2 g D}$$

$$= 4 \times 0.0049 \times 60 \times (1.63)^2 / (2 \times 9.81 \times 0.025)$$

$$= 6.37 \text{ m}$$

The Bernoulli equation in energy units is

$$\frac{P_1}{\rho g} + \frac{u_1^2}{2g} + Z_1 = \frac{P_2}{\rho g} + \frac{u_2^2}{2g} + Z_2 + h_f$$

$P_1 = P_2$, since tanks open to atmosphere and $u_1 = u_2$

Setting $Z_2 = 0$, we get

$Z_1 = h_f = 6.37$ m

∴ Difference between the levels in the tank and that in the truck = **6.37 m.** **... Ans.**

Example 7.65 : *Calculate the pressure drop due to friction in a 300 m long pipe of 150 mm i.d. through which water is flowing at a rate of 0.05 m³/s.*

Data : *ρ of water = 1000 kg/m³, μ of water = 1.0 × 10⁻³(N.s)/m²*

Solution : Volumetric flow rate = Q = 0.05 m³/s

D = 150 mm = 0.15 m, L = 300 m

$\rho = 1000$ kg/m³, $\mu = 1.0 \times 10^{-3}$ (N.s)/m² ≡ kg/(m.s)

Cross-sectional area of pipe = A = π/4 . (0.15)² = 0.01767 m²

Velocity of water through the pipe = Q/A = 0.05/0.01767 = 2.83 m/s

Reynolds number, $N_{Re} = Du\rho/\mu$

$$= 0.15 \times 2.83 \times 1000 / 1.0 \times 10^{-3} = 424500$$

Since $N_{Re} > 4000$, the flow is turbulent and therefore, f may be calculated by

$$f = 0.078 / (N_{Re})^{0.25}$$

$$= 0.078 / (424500)^{0.25} = 0.00305$$

The pressure drop due to friction is

$$\Delta P = \rho . \frac{4 f L u^2}{2 D}$$

$$= 1000 \times 4 \times 0.00305 \times 300 \times (2.83)^2 / (2 \times 0.15)$$

$$= 97708 \text{ N/m}^2$$

= 97.708 kN/m² = **97.71 kN/m² (kPa)** **... Ans.**

Example 7.66 : *Acetic acid flows through a 75 mm internal diameter pipe at a rate of 0.015 m³/s. Calculate the pressure drop in the horizontal pipe of length 70 m.*

Data : *Viscosity of acid = 2.5 (mN.s)/m²*
Density of acid = 1060 kg/m³

Solution : Volumetric flow rate = 0.015 m³/s

D = 75 m = 0.075 m, L = 70 m

$\rho = 1060$ kg/m³, $\mu = 2.5$ (mN.s)/m² = 2.5×10^{-3} (N.s)/m² ≡ kg/(m.s)

Cross-sectional area of the pipe = π/4 . (0.075)² = 4.418 × 10⁻³ m²

Velocity of acid in the pipe = 0.015 / 4.418 × 10⁻³ = 3.4 m/s

Reynolds number, $N_{Re} = Du\rho/\mu$

$$= 0.075 \times 3.4 \times 1060/2.5 \times 10^{-3} = 108120$$

For turbulent flow, the Fanning friction factor, f may be calculated by

$$f = 0.078 / (N_{Re})^{0.25}$$

$$= 0.078 / (108120)^{0.25} = 0.0043$$

The pressure drop through a pipe is given by

$$\Delta P = \rho\left[\frac{4\,f\,L\,u^2}{2\,D}\right]$$

$$= 1060\left[\frac{4 \times 0.0043 \times 70 \times (3.4)^2}{2 \times 0.0075}\right]$$

$$= 98355\ N/m^2$$

$$= \mathbf{98.35\ kN/m^2\ (kPa)}$$ **... Ans.**

Example 7.67 : *A petroleum product having a viscosity of 0.5 (mN.s)/m² and specific gravity of 0.7 is pumped through a pipe of 150 mm inside diameter to a storage tank placed 100 m away. The pressure drop along the pipe is 70 kN/m². This pipe line has to be repaired and therefore it is necessary to pump it using an alternative route which consists of 70 m long 200 mm i.d. pipe followed by 50 m long 100 mm i.d. pipe. The existing pump is capable of developing a pressure of 300 kN/m². Show by calculations whether it is suitable for use for the period required for the repairs or not ?*

Solution : $\mu = 0.5\ (mN.s)/m^2 = 0.5 \times 10^{-3}\ (N.s)/m^2$, $\rho = 700\ kg/m^3$

D of existing pipe = 150 mm = 0.15 m, L = 100 m

$\Delta P = 70\ kN/m^2 = 70 \times 10^3\ N/m^2$

$$\frac{\Delta P}{\rho} = \frac{4\,f\,L\,u^2}{2\,D}$$

$$N_{Re} = Du\rho/\mu \quad \therefore\ u^2 = N_{Re}^2\,\mu^2/D^2\,\rho^2$$

$$\frac{\Delta P}{\rho} = \frac{4\,f\,L\,N_{Re}^2\,\mu^2}{2\,D\,.\,D^2\,\rho^2}$$

$$N_{Re}^2\,f = \frac{\Delta P\,D^3\,\rho}{2\,L\,\mu^2} = \frac{70 \times 10^3 \times (0.15)^3\,(700)}{2 \times 100 \times\ (0.5 \times 10^{-3})^2} = 2.205 \times 10^{10}$$

$$N_{Re}\sqrt{f} = (2.205 \times 10^{10})^{1/2} = 14849$$

$$\frac{1}{\sqrt{f}} = 4\log\,(N_{Re}\sqrt{f}\,) - 0.40$$

$$\frac{1}{\sqrt{f}} = 4\log\,(14849) - 0.4$$

$$\therefore \quad f = 0.00243$$

$$\frac{\Delta P}{\rho} = \frac{4\,f\,L\,u^2}{2\,D}$$

$$u^2 = \Delta P\,.\,2D/\rho\,4\,f\,L = 70 \times 10^3 \times 2 \times 0.15/700 \times 4 \times 0.00243 \times 100$$

$$\therefore \quad u = 5.55\ m/s$$

Cross-sectional area of the existing pipe = $\pi/4\,.\,(0.15)^2 = 0.0177\ m^2$

Volumetric flow rate = $u.A = 5.55 \times 0.0177 = 0.098\ m^3/s$

u_1 = velocity in 200 mm pipe = $0.098 / \pi/4\,.\,(0.2)^2 = 3.12\ m/s$

u_2 = velocity in 100 mm pipe = $0.098/ \pi/4\,.\,(0.1)^2 = 1.56\ m/s$

N_{Re} in 200 mm diameter pipe :

$N_{Re_1} = 0.2 \times 3.12 \times 700 / 0.5 \times 10^{-3} = 873600$

$\therefore \quad f_1 = 0.078 / (N_{Re})^{0.25} = 0.078 / (873600)^{0.25} = 0.00255$

N_{Re} in 100 mm diameter pipe :

$N_{Re_2} = 0.1 \times 1.56 \times 700 / 0.5 \times 10^{-3} = 218400$

$f_2 = 0.078 / (N_{Re})^{0.25} = 0.078 / (218400)^{0.25} = 0.00361$

Pressure drop in the alternate route :

$$\Delta P = \rho\left[\frac{4 f_1 L_1 u_1^2}{2 D_1} + \frac{4 f_2 L_2 u_2^2}{2D_2}\right]$$

$L_1 = 70$ m , $L_2 = 50$ m

$D_1 = 0.2$ m, $D_2 = 0.10$ m

$$\Delta P = 700\left[\frac{4 \times 0.00255 \times 70 \times (3.12)^2}{2 \times 0.20} + \frac{4 \times 0.00361 \times 50 \times (1.56)^2}{2 \times 0.10}\right] = 181313 \text{ N/m}^2 = 181.313 \text{ kN/m}^2$$

Hence, the existing pump is satisfactory for this duty. **... Ans.**

Example 7.68 : *98% sulphuric acid of viscosity 0.025 (N.s)/m² and density 1840 kg/m³ is pumped at a rate of 685 cm³/s through a 25 mm pipe line. Estimate the value of Reynolds number.*

Solution : Flow rate of sulphuric acid = Q = 685 cm³/s = 685×10^{-6} m³/s

Density of acid = ρ = 1840 kg/m³, Viscosity of acid = 0.025 (N.s)/m²

Diameter of pipe = D = 25 mm = 0.025 m

Average velocity of acid through pipe = u = Q/A

A = Cross-sectional area of the pipe = $(\pi/4)\ D^2 = (\pi/4)\ (0.025)^2 = 4.909 \times 10^{-4}$ m²

$u = (685 \times 10^{-6}) / 4.909 \times 10^{-4} = 1.398$ m/s

$$\text{Reynolds number, } N_{Re} = \frac{Du\rho}{\mu} = \frac{0.025 \times 1.398 \times 1840}{0.025} = \mathbf{2572}$$ **... Ans.**

As the Reynolds number is in between 2100 and 4000, the flow is in a transition region.

Example 7.69 : *A crude oil of viscosity 0.9 poise and relative density of 0.90 is flowing through a horizontal pipe of 120 mm and length 12 m. Calculate the difference in pressure at the two ends of pipe, if 785 N weight of the oil is collected in a tank in 25 s.*

Solution : Weight of oil = 785 N

The force becomes weight when the body acts under gravitational acceleration (g).

$$\text{Weight} = \frac{m \cdot g}{g_c}$$

where,

$$g = 9.81 \text{ m/s}^2$$

$$g_c = 1 \text{ (kg·m)/(N·s}^2)$$

$$\text{Weight} = 785 \text{ N}$$

$$785 = \frac{m \times 9.81}{1}$$

$$m = 80.02 \text{ kg}$$

$$\text{Units of weight} \Rightarrow \frac{(\text{kg}) \times (\text{m/s}^2)}{(\text{kg·m})/(\text{N·s}^2)} \Rightarrow \text{N}$$

80.02 kg oil is collected in 25 s. Therefore,

$$\text{Mass flow rate of oil} = \dot{m} = \frac{80.02}{25} = 3.2 \text{ kg/s}$$

$$\text{Diameter of pipe} = D = 120 \text{ mm} = 12 \text{ cm} = 0.12 \text{ m}$$

$$\text{Viscosity of oil} = \mu = 0.9 \text{ P} = 0.09 \text{ kg/(m.s)}$$

$$\text{Density of oil} = \rho = 0.90 \text{ g/cm}^3 = 900 \text{ kg/m}^3$$

$$\text{Area of pipe} = A = \pi/4\, D^2 = \pi/4\,(0.12)^2 = 0.0113 \text{ m}^2$$

We have :

$$\dot{m} = u\rho A$$

$$u = \frac{\dot{m}}{\rho A} = \frac{3.2}{900 \times 0.0113} = 0.3146 \approx 0.315 \text{ m/s}$$

$$\text{Reynold's number} = N_{Re} = \frac{Du\rho}{\mu} = \frac{0.12 \times 0.315 \times 900}{0.09} = 378$$

Since N_{Re} is less than 2100, the flow is laminar.

The Hagen-Poiseuille equation is

$$\Delta P = \frac{32\,\mu u L}{D^2}$$

We have,

$$\mu = 0.09 \text{ kg/(m.s)}$$

and

$$1 \text{ kg/(m.s)} = 1 \text{ (N.s)/m}^2$$

$$\text{Units of } \Delta P \Rightarrow \frac{\text{(N.s)}}{\text{m}^2} \times \frac{\text{m}}{\text{s}} \times \frac{\text{m}}{\text{m}^2} \Rightarrow \text{N/m}^2 = \text{Pa (pascal)}$$

$\therefore$

$$\mu = 0.09 \text{ (N.s)/m}^2$$

$$\Delta P = \frac{32 \times 0.09 \times 0.315 \times 12}{(0.12)^2}$$

$$= 756 \text{ N/m}^2 = \mathbf{0.756\ kN/m^2 = 0.756\ kPa}$$

... **Ans.**

Example 7.70 : *Fuel oil is pumped up in a 300 mm diameter and 16 km long pipeline at the rate of 15 kg/s. The pipe line is laid at a slope of 1 in 300. The density of fuel oil is 950 kg/m³ and its kinematic viscosity is 21.4 stokes. Find the power required to pump the oil.*

Solution :

$$\text{Diameter of pipe} = 300 \text{ mm} = 0.30 \text{ m}$$

$$\text{Length of pipe} = L = 16 \text{ km} = 16000 \text{ m}$$

$$\text{Slope of the pipe} = i = 1/300$$

$$\text{Density of oil} = \rho = 950 \text{ kg/m}^3$$

$$\text{Mass flow rate of oil} = \dot{m} = 15 \text{ kg/s}$$

$$\text{Kinematic viscosity} = \upsilon = \frac{\mu}{\rho} = 21.4 \text{ stokes} = 21.4 \times 10^{-4} \text{ m}^2/\text{s}$$

$$A = \frac{\pi}{4} D^2$$

$$= \frac{\pi}{4}(0.3)^2 = 0.07068 \text{ m}^2$$

We have : $\dot{m} = \rho u A$

$$u = \frac{\dot{m}}{\rho A} = \frac{15}{950 \times 0.07068} = 0.223 \text{ m/s}$$

$$\text{Reynolds number} = N_{Re} = \frac{Du\rho}{\mu} = \frac{Du}{\upsilon} = \frac{0.3 \times 0.223}{21.4 \times 10^{-4}} = 31.26$$

Since the Reynolds number is less than 2100, the flow is laminar. Therefore,

$$f = \frac{16}{N_{Re}} = \frac{16}{31.26} = 0.5118 \approx 0.512$$

$$\text{Head loss due to friction} = h_f = \frac{4fLu^2}{2gD}$$

$$h_f = \frac{4 \times 0.512 \times 16000 \times (0.223)^2}{2 \times 9.81 \times 0.30} = 276.85$$

The total head against which the pump has to work is

$$H = h_f + L \cdot i$$

$$= 276.85 + 16000 \times \frac{1}{300} = 330.18 \text{ m}$$

The power required to pump the oil is

$$P = \frac{\rho Q H}{75}, \text{ hp}$$

We have : $Q = \dfrac{\dot{m}}{\rho}$

$\therefore$ $$P = \frac{\dot{m}H}{75} = \frac{15 \times 330.18}{75} = \mathbf{66 \text{ h.p.}}$$ ... **Ans.**

OR : $$P = QH\rho g$$

$$= \dot{m} \cdot H \cdot g$$

$$= 15 \times 9.81 \times 330.18$$

$$= 48585.987 \text{ J/s} = 48585.987 \text{ W}$$

$$= \mathbf{48586 \text{ W} = 48.586 \text{ kW}}$$ ... **Ans.**

Example 7.71 : *Water is flowing at a rate of 500 cm³/s through an orifice of 25 mm diameter installed in a 75 mm diameter pipe. What will be the difference in the level on a mercury manometer connected across the meter ? The coefficient of orifice meter is 0.65.*

Solution : Volumetric flow rate of water = 500 cm³/s

The flow equation for an orifice meter is

$$Q = \frac{C_o \cdot A_o}{\sqrt{1 - \beta^4}} \cdot \sqrt{2 g \Delta H} \quad \text{... (1)}$$

where Q = Volumetric flow rate = 500 cm³/s = 5×10^{-4} m³/s

C_o = Coefficient of meter = 0.65

A_o = Orifice flow area

$= \dfrac{\pi}{4} D_o^2$ where D_o = 25 mm = 0.025 m

$$g = 981 \text{ m/s}^2$$

$$\beta = \frac{D_o}{d} = \frac{0.025}{0.075} = 0.333$$

ΔH = Pressure drop across the meter in terms of m of H_2O

$$A_o = \frac{\pi D_o^2}{4} = \frac{\pi}{4} \times (0.025)^2 = 4.909 \times 10^{-4} \text{ m}^2$$

Substituting the values, Equation (1) gives

$$5 \times 10^{-4} = \frac{0.65 \times 4.909 \times 10^{-4}}{\sqrt{1-(0.333)^4}} \sqrt{2 \times 9.81 \times \Delta H}$$

Solving, we get

$$\Delta H = 0.1236 \text{ m of } H_2O$$

Δh = difference in levels in the mercury manometer.

$$\Delta H = \Delta h \frac{(\rho_{Hg} - \rho_{H_2O})}{\rho_{H_2O}}$$

$$0.1236 = \Delta H = \Delta h \frac{(\rho_{Hg} - \rho_{H_2O})}{\rho_{H_2O}}$$

where ρ_{Hg} = 13600 kg/m³

ρ_{H_2O} = 1000 kg/m³

$$0.1236 = \Delta h \frac{(13600 - 1000)}{1000}$$

$\Delta h = 9.81 \times 10^{-3}$ m = **9.81 mm** **... Ans.**

Example 7.72 : *A venturimeter is installed in a pipe line for the measurement of flow rate of water. The pressure drop across the throat and upstream of the meter is ten centimeters of mercury. Calculate the volumetric flow rate of water in m³/s.*

Data : *Diameter of throat = 15 mm*

Diameter of pipe = 25 mm

Coefficient of meter = 0.98

Density of water = 1000 kg/m³

Density of mercury = 13600 kg/m³

Solution : The flow equation for a venturi meter is

$$Q = \frac{C_v \cdot A_T}{\sqrt{1-\beta^4}} \sqrt{2g\,\Delta H}$$

where Q = volumetric flow rate in m³/s

C_v = 0.98

A_T = cross-sectional area of throat

$$\beta = \frac{\text{diameter of throat}}{\text{diameter of pipe}} = \frac{D_T}{D}$$

ΔH = pressure drop in meters of flowing fluid

$$A_T = \frac{\pi}{4} D_T^2, \quad \text{where } D_T = 15 \text{ mm} = 0.015 \text{ m}$$

$$= \frac{\pi}{4}(0.015)^2$$

$$= 1.767 \times 10^{-4} \text{ m}^2$$

$$\beta = \frac{D_T}{D} = \frac{15}{25} = 0.6$$

$$\Delta H = \Delta h \frac{(\rho_{Hg} - \rho_{H_2O})}{\rho_{H_2O}}$$

where Δh = pressure drop in meters of mercury

Δh = 10 cm = 0.10 m

ρ_{Hg} = 13600 kg/m³, ρ_{H_2O} = 1000 kg/m³

$$\therefore \quad \Delta H = 0.1\left(\frac{13600 - 1000}{1000}\right) = 1.26 \text{ m of } H_2O \text{ (flowing fluid)}$$

Substituting the values of various terms involved in the above equation of Q, we get

$$Q = \frac{0.98 \times 1.767 \times 10^{-4}}{\sqrt{1-(0.6)^4}} \times \sqrt{2 \times 9.81 \times 1.26} = \mathbf{9.23 \times 10^{-4}\ m^3/s} \quad \text{... Ans.}$$

Example 7.73 : *Orifice meter is installed in a pipe line for measurement of flow rate of water. The pressure drop across the orifice meter is 10 centimetres of mercury. Estimate the volumetric flow rate in m³/s.*

Data : Diameter of orifice = 25 mm

Diameter of pipe = 50 mm

Coefficient of orifice = 0.62

Density of water = 1000 kg/m³

Density of mercury = 13,600 kg/m³

Solution : The flow equation of an orifice meter is

$$Q = \frac{C_o A_o}{\sqrt{1-\beta^4}} \times \sqrt{2g\,\Delta H} \quad \text{... (1)}$$

where Q = Volume flow rate in m³/s

C_o = Coefficient of orifice = 0.62

g = 9.81 m/s²

A_o = Area of orifice, m²

ΔH = Pressure drop in metres of flowing fluid.

$$\beta = \frac{\text{Diameter of throat}}{\text{Diameter of pipe}} = \frac{25}{50} = 0.5$$

$$A_o = \frac{\pi}{4} D_o^2$$

where D_o = 25 mm = 0.025 m

$$A_o = \frac{\pi}{4}(0.025)^2 = 4.909 \times 10^{-4} \text{ m}^2$$

$$\Delta H = \Delta h \left[\frac{\rho_{Hg} - \rho_{H_2O}}{\rho_{H_2O}}\right]$$

where $\Delta h = 10 \text{ cm Hg (given)} = 0.10 \text{ m Hg}$

$$\Delta H = 0.1\left[\frac{13600 - 1000}{1000}\right]$$

$$= 1.26 \text{ m of } H_2O$$

Substituting the values of the terms involved in Equation (1), we get

$$Q = \frac{0.62 \times (4.909 \times 10^{-4})}{\sqrt{1-(0.5)^4}} \times \sqrt{2 \times 9.81 \times 1.26}$$

Volumetric flow rate $=$ **1.56×10^{-3} m³/s** **... Ans.**

Example 7.74 : *A venturimeter is installed in a 25 mm internal diameter pipe line. The pressure drop across the upstream side and throat of the venturimeter is two meters of water. Calculate the volumetric flow rate of water in m^3/s through the pipe line.*

Data : *Diameter of throat of venturimeter = 15 mm*

Density of water = 1000 kg/m³

Coefficient of venturi = 0.98

Solution : D = I.D. of pipe = 25 mm = 0.025 m

D_T = diameter of throat = 15 mm = 0.015 m

C_V = venturi coefficient = 0.98

Density of water = ρ = 1000 kg/m³

ΔH = pressure drop at the upstream side

= 2 m of H_2O

For a venturi meter, the flow equation is

$$Q = \frac{C_V \cdot A_T}{\sqrt{1-\beta^4}} \times \sqrt{2g\,\Delta H}$$

$$g = 9.81 \text{ m/s}^2,\ C_V = 0.98$$

$$\beta = \frac{\text{Diameter of throat}}{\text{Diameter of pipe}} = \frac{15}{20} = 0.6$$

$$\Delta H = 2 \text{ m of water}$$

$$A_T = \text{area of throat} = \frac{\pi}{4} D_T^2 = \frac{\pi}{4}(0.015)^2 = 1.767 \times 10^{-4} \text{ m}^2$$

Q = volumetric flow rate, m³/s

$$Q = \frac{0.98 \times (1.767 \times 10^{-4})}{\sqrt{1-(0.60)^4}} \times \sqrt{2 \times 9.81 \times 2}$$

$$= \mathbf{1.163 \times 10^{-3} \text{ m}^3/\text{s}}$$

Example 7.75 : *An orifice meter is used to measure the flow rate of water flowing in a pipe line of 78 mm I.D. The orifice diameter is 15 mm. Mercury manometer reads 18 cm. The volumetric flow rate in this case is 719 cm³/s.*

(i) Calculate the coefficient of discharge of the meter.

(ii) If the pressure drop is decreased to 9 cm of Hg, what will be the flow rate ?

Solution : Pipe diameter = 78 mm = 0.078 m

Orifice diameter = 15 mm = 0.015 m

Δh = 18 cm = 0.18 m

Q = 719 cm³/s = 7.19×10^{-4} m³/s

Density (mercury) = 13,600 kg/m³

Density (water) = 1000 kg/m³

$g = 9.81 \text{ m/s}^2$

The discharge through an orifice meter is given by

$$Q = \frac{C_o A A_o \sqrt{2g\,\Delta H}}{\sqrt{A^2 - A_o^2}} \qquad \text{... (1)}$$

where Q = volumetric flow rate in m³/s

C_o = coefficient of discharge

A = area of pipe in m²

A_o = area of orifice in m²

g = acceleration due to gravity in m/s²

Δh = pressure drop in metres of flowing fluid

(i) We have to find C_o. Thus rearranging Equation (1), we get

$$C_o = \frac{Q\sqrt{A^2 - A_o^2}}{A A_o \sqrt{2g\,\Delta H}} \qquad \text{... (2)}$$

$$Q = 719 \text{ cm}^3/\text{s} = 0.719 \times 10^{-3} \text{ m}^3/\text{s}$$

$$A = \frac{\pi D^2}{4} = \frac{\pi \times (0.078)^2}{4} = 4.778 \times 10^{-3} \text{ m}^2$$

$$A_o = \frac{\pi D_o^2}{4} = \frac{\pi \times (0.015)^2}{4} = 1.767 \times 10^{-4} \text{ m}^2$$

$$\Delta H = \Delta h \left[\frac{\rho_{Hg} - \rho_{H_2O}}{\rho_{H_2O}}\right] = 0.18 \left[\frac{13600 - 1000}{1000}\right] = 2.268 \text{ m of } H_2O$$

Substituting these values in Equation (2), we get

$$C_o = \frac{0.719 \times 10^{-3}\sqrt{(4.778 \times 10^{-3})^2 - (1.767 \times 10^{-4})^2}}{4.778 \times 10^{-3} \times 1.767 \times 10^{-4} \times \sqrt{2 \times 9.81 \times 2.268}}$$

∴ C_o = **0.6096 ≃ 0.61** ... **Ans.**

(ii) Given : Pressure drop is decreased to 9 cm of Hg and for this we have to find Q by making use of C_o obtained.

Δh = 9 cm

C_o = 0.61

$A = 4.778 \times 10^{-3}$ m²

$A_o = 1.767 \times 10^{-4}$ m²

$g = 9.81$ m/s²

$$\Delta H = 0.09\left[\frac{13600 - 1000}{1000}\right] = 1.134 \text{ m}$$

Substituting these values in Equation (1) gives

$$Q = \frac{0.61 \times 4.778 \times 10^{-3} \times 1.767 \times 10^{-4}\sqrt{2 \times 9.81 \times 1.134}}{\sqrt{(4.778 \times 10^{-3})^2 - (1.767 \times 10^{-4})^2}}$$

∴ Q = **5.1×10^{-4} m³/s** ... **Ans.**

Example 7.76 : *A 250 mm diameter pipe carries oil at a rate of 120 l/s and pressure at a point A (station 1) is 19.62 kN/m². g. If the point A is 3.5 m above the datum line, calculate the total energy at point A in m of oil. The specific gravity of oil is 0.8 and the density of water is 1000 kg/m³.*

Solution : D = 250 mm = 0.25 m

$$\text{Specific gravity of oil} = \frac{\text{density of oil}}{\text{density of water}}$$

∴ Density of oil = $0.8 \times 1000 = 800$ kg/m³

Pressure at station A = 19.62 kN/m².g

= 19.62×10^3 N/m².g

= 120945 N/m² (Pa)

= $19.62 \times 10^3 + 101325$

Volumetric flow rate of the oil = 120 l/s = 120×10^{-3} m³/s

Cross-sectional area of the pipe = $\pi/4\ (0.25)^2 = 0.04909$ m²

Velocity of oil through the pipe = $120 \times 10^{-3} / 0.04909 = 2.44$ m/s

The total energy at point A in terms of m of oil is given by

$$\frac{P}{\rho g} + \frac{u^2}{2g} + Z = \frac{120945}{800 \times 9.81} + \frac{(2.44)^2}{2 \times 9.81} + 3.5$$

$$= 15.41 + 0.30 + 3.5 = 19.21 \text{ m of oil.}$$

$$\text{Total energy in J/kg} = \frac{P}{\rho} + \frac{u^2}{2} + gZ$$

$$= \frac{120945}{800} + \frac{(2.44)^2}{2} + 9.81 \times 3.5 = 188.5 \text{ J/kg}$$

Total energy in J/kg = energy in m of oil × g

= $19.21 \times g = 19.21 \times 9.81$

= **188.45 J/kg.** ... **Ans.**

Example 7.77 : *A 300 mm pipe carries water at a velocity of 24 m/s. At stations A and B measurements of pressure and elevation were 361 kN/m² and 288 kN/m² and 30.5 m and 33.5 m, respectively. For steady flow, find the loss of head between stations A and B.*

Solution : Total energy in terms of metres of water is given by

$$\frac{P}{\rho g} + \frac{u^2}{2g} + Z$$

$$\text{Total energy at point A} = \frac{361 \times 10^3}{1000 \times 9.81} + \frac{(2u)^2}{2 \times 9.81} + 30.5$$

= 96.66 m of H_2O

$$\text{Total energy at point B} = \frac{288 \times 10^3}{1000 \times 9.81} + \frac{(24)^2}{2 \times 9.81} + 33.5$$

= 92.21 m of H_2O

Loss of head = $96.66 - 92.21$ = **4.45 m** ... **Ans.**

Example 7.78 : *A pipe 300 m long has a slope of 1 in 100 and tapers from 1.2 m diameter at high end to 0.6 m diameter at the low end. Water is flowing at a rate of 90 l/s. If the pressure at the high end is 68.67 kPa, find the pressure at the lower end. Neglect the losses.*

Solution : Volumetric flow rate of water = 90 a/s = 90×10^{-3} m³/s

Higher end : station 1

Lower end : station 2

Pipe diameter at the higher end = 1.2 m

Cross-sectional area of pipe at the higher end = $\pi/4\ (1.2)^2 = 1.131\ m^2$

Velocity at station 1 = $u_1 = 90 \times 10^{-3} / 1.131 = 0.08$ m/s

Pipe diameter at the lower end = 0.6 m

Cross-sectional area of pipe at the lower end = $\pi/4\ (0.6)^2$

$$= 0.283\ m^2$$

Velocity at station 2 = $u_2 = 90 \times 10^{-3} / 0.0283 = 0.318$ m/s

The Bernoulli equation between the higher and lower ends of the pipe is

$$\frac{P_1}{\rho g} + \frac{u_1^2}{2g} + Z_1 = \frac{P_2}{\rho g} + \frac{u_2^2}{2g} + Z_2 \quad \text{... (1)}$$

Assuming a datum line/level passing through the lower end of the pipe, we have

$$Z_2 = 0$$

Length of the pipe = 300 m, slope of the pipe line = 1 in 100

$$\therefore \quad Z_1 = \frac{1}{100} \times 300 = 3\ m.$$

$P_1 = 68.67$ kPa $= 68.67 \times 10^3$ Pa (N/m^2)

$\rho = 1000$ kg/m^3 and $g = 9.81$ m/s^2

Substituting the values in Equation (1), we get

$$\frac{68.67 \times 10^3}{1000 \times 9.81} + \frac{(0.08)^2}{2 \times 9.81} + 3.0 = \frac{P_2}{1000 \times 9.81} + \frac{(0.318)^2}{2 \times 9.81} + 0$$

$$P_2 = 98053\ N/m^2\ \text{(or Pa)}$$

$$= \mathbf{98.05\ kN/m^2\ (or\ kPa)} \quad \text{... Ans.}$$

Example 7.79 : *A 15 kW pump with 80% efficiency is discharging oil of specific gravity 0.85 to a overhead tank from a storage tank. The surface of the oil in the storage tank from a datum line is 5 m and that in the overhead tank from the datum line is 25 m. Both the tanks are open to atmosphere. If the losses in the piping system are 1.75 m of flowing fluid, calculate the volumetric flow rate of oil.*

Solution :

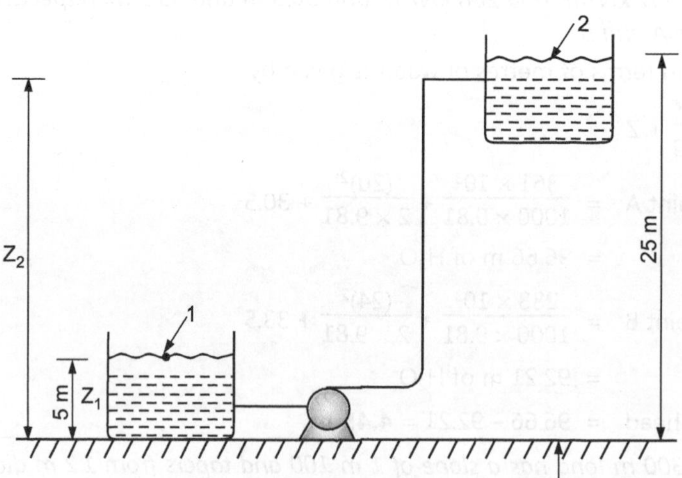

Fig. Ex. 7.79

Frictional losses in the system = 1.75 m

Density of oil = $0.85 \times 1000 = 850$ kg/m^3

$P_1 = P_2$ = 101.325 kPa

The velocity at both 1 and 2 stations being small is neglected.

$$Z_1 = 5 \text{ m} \quad , \quad Z_2 = 25 \text{ m}$$

$$\therefore \quad (Z_2 - Z_1) = 25 - 5 = 20 \text{ m}$$

$$\text{Total head} = 20 + 1.75 = 21.75 \text{ m}$$

$$\text{Efficiency of the pump} = 80\ \% = 0.80$$

$$\text{Power requirement of the pump} = 15 \text{ kW} = 15000 \text{ W}$$

We have : $\text{Power} = (\text{mass flow rate} \times \text{head} \times g)/\eta$

$$\therefore \quad \text{Mass flow rate} = \frac{\text{Power} \times \eta}{\text{Head} \times g}$$

$$= \frac{15000 \times 0.80}{21.75 \times 9.81}$$

$$\dot{m} = 56.24 \text{ kg/s}$$

$$\text{Volumetric flow rate of the oil} = \dot{m} / \rho$$

$$= \frac{56.24}{850}$$

$$= 0.0662 \text{ m}^3\text{/s or}$$

= **66.2 *l*/s** **... Ans.**

The Bernoulli equation between stations 1 and 2 is

$$\frac{P_1}{\rho g} + \frac{u_1^2}{2g} + Z_1 + \eta W_p = \frac{P_2}{\rho g} + \frac{u_2^2}{2g} + Z_2 + h_f$$

We have : $P_1 = P_2$, $u_1 = u_2$, $Z_1 = 5$ m , $Z_2 = 25$ m and $h_f = 1.75$ m

$$\eta W_p = Z_2 - Z_1 + h_f$$

$$\therefore \quad \eta W_p = 25 - 5 + 1.75 = 21.75 \text{ m}$$

$$W_p = 21.75/0.8 = 27.2 \text{ m of oil}$$

$$\text{Power requirement} = 15000 = \dot{m} \times 27.2 \times 9.81$$

$\dot{m}$ = **56.21 kg/s ... almost same as previously calculated.** **... Ans.**

Example 7.80 : *A pump draws a solution of sp. gr. 1.84 from a storage tank through 75 mm pipe. The velocity in the suction line is 0.914 m/s. The pump discharges through a pipe of 50 mm diameter to an overhead tank. The end of the discharge pipe is 15.2 m above the level of the solution in the tank. Frictional losses in the entire piping system are 29.9 J/kg. What pressure must the pump develop and what is the power requirement of the pump if efficiency of the pump is 60% ?*

Solution :

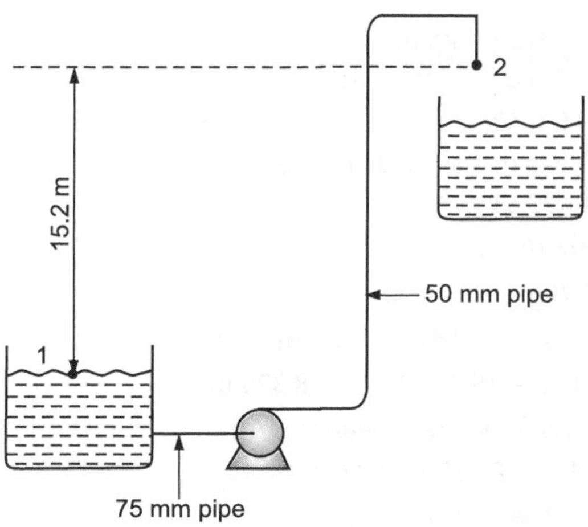

Fig. Ex. 7.80

Take station 1 at the surface of the liquid in the tank and station 2 at the discharge end of the 50 mm pipe. As the pressure at both the ends is atmospheric, $P_1 = P_2$.

Take datum level for elevation through station 1, so $Z_1 = 0$ and $Z_2 = 15.2$ m.

The velocity at station 1 is negligible because the diameter of the tank is large in comparison with that of pipe ($u_1 \approx 0$).

ρ of the solution = 1.84 × 1000 = 1840 kg/m³

Diameter of the suction line = 75 mm = 0.075 m

Velocity in the suction line = 0.914 m/s

Diameter of the discharge line = 50 mm = 0.05 m

$$\dot{m} = \rho_1 u_1 A_1 = \rho_2 u_2 A_2 \text{ but } \rho_1 = \rho_2$$

$$\therefore \quad u_2 A_2 = u_1 A_1$$

u_2 = velocity through the discharge line

$$u_2 = u_1 A_1 / A_2$$

$$= 0.914 \times \frac{\pi/4\,(0.075)^2}{\pi/4\,(0.05)^2}$$

$$= 2.06 \text{ m/s}$$

h_f = frictional losses in the entire piping = 29.6 J/kg

$$\therefore \quad \text{Work done by the pump} = \frac{u_2^2}{2} + g\,Z_2 + h_f$$

$$= \frac{(2.06)^2}{2} + 9.81 \times 15.2 + 29.6$$

$$= 181 \text{ J/kg.}$$

$$\text{Mass flow rate of the solution} = \dot{m} = \rho\, u_2 A_2$$

$$= 1840 \times 2.06 \times \pi/4\,(0.05)^2$$

$$= 7.44 \text{ kg/s}$$

Efficiency of the pump is 60%.

$$\text{Power required by the pump} = 7.44 \times 181/0.60$$

$$= 2244 \text{ W}$$

$$= \mathbf{2.24\ kW}$$ **... Ans.**

$$\left[\begin{array}{l}\text{Frictional head loss in}\\ \text{terms of m of solution}\end{array}\right] = 29.6 \text{ J/kg/g}$$

$$= 29.6 \frac{\text{N.m}}{\text{kg}} = \frac{\text{kg.m}}{\text{s}^2} \cdot \frac{\text{m}}{\text{kg}}$$

$$= 29.6 \text{ m}^2/\text{s}^2$$

$$= 29.6/g\ (\text{m}^2/\text{s}^2) \times (1/(\text{m/s}^2))$$

$$= 29.6/9.81 \text{ m}$$

$$= 3.02 \text{ m}$$

$$\text{Velocity head} = u_2^2/2g$$

$$= (2.06)^2/2 \times 9.81 = 0.105 \text{ m}$$

$$\text{Total head} = 0.105 + 15.2 + 3.02 = 18.325 \text{ m}$$

$$\text{Power required} = \text{Mass flow rate} \times \text{head} \times g / \eta$$

$$= 7.44 \times 18.325 \times 9.81/0.60$$

$$= 2229 \text{ W}$$

$$= \mathbf{2.23\ kW}$$ **... Ans.**

Pressure developed by the pump can be calculated by applying the Bernoulli's equation over the pump itself.

Let P_a and P_b be the pressure at the pump suction and discharge. Neglect the level difference between pump suction and discharge. So $Z_a = Z_b$.

$$\frac{P_a}{\rho} + \frac{u_a^2}{2} + g\,Z_a + \eta W_P = \frac{P_b}{\rho} + \frac{u_b^2}{2} + g\,Z_b$$

$$\text{Pressure developed by the pump} = P_b - P_a$$

$$P_b - P_a = \rho\left[\eta W_P + \frac{u_a^2 - u_b^2}{2}\right]$$

$$= 1840\left[181 + \frac{(0.914)^2 - (2.06)^2}{2}\right]$$

$$= 329904 \text{ N/m}^2 \text{ or Pa}$$

$$= \mathbf{329.9\ kN/m^2\ (or\ kPa)}$$ **... Ans.**

Units of ρ : kg/m³

Units of ηW_p or $u^2/2$: J/kg or m² / s²

$$\text{Units of } (P_b - P_a) \;:\; \frac{kg}{m^3}\cdot\frac{m^2}{s^2}$$

$$= \frac{kg}{m.s^2} = \left(\frac{kg.m}{s^2}\right)\left(\frac{1}{m^2}\right) = N/m^2 \text{ [as } N = (kg.m)/s^2]$$

Example 7.81 : *Water flows at the rate of 0.147 m³/s through a 150 mm diameter orifice inserted in a 300 mm diameter pipe. If pressure gauges fitted at the upstream and the downstream of the orifice plate give readings of 176.58 kN/m² and 88.29 kN/m² respectively, find the coefficient of discharge of the orifice meter.*

Data : *Density of water = 1000 kg/m³*

Solution : The flow equation of an orifice meter is

$$Q = \frac{C_o A_o}{\sqrt{1-\beta^4}} \times \frac{\sqrt{2\,(P_a - P_b)}}{\rho}$$

D_o = 150 mm = 0.15 m, D = 300 mm = 0.30 m

$$\beta = \frac{D_o}{D} = \frac{150}{300} = 0.5$$

Volumetric flow rate = 0.147 m³/s

ρ = 1000 kg/m³

P_a = 176.58 kN/m² = 176580 N/m²

P_b = 88.29 kN/m² = 88290 N/m²

$$A_o = \pi/4\, D_o^2 = \pi/4\,(0.15)^2 = 0.01767 \text{ m}^2$$

Rearranging the flow equation gives

$$C_o = \frac{Q\sqrt{1-\beta^4}}{A_o \times \sqrt{2\,(P_a - P_b)/\rho}} = \frac{0.147\sqrt{1-(0.5)^4}}{0.01767\sqrt{\dfrac{2\,(176580-88290)}{1000}}}$$

$$= 0.606 \simeq 0.61$$

Coefficient of discharge of the orifice meter = **0.61** **... Ans.**

Example 7.82 : *A venturimeter is to be fitted in a pipe of 250 mm diameter where pressure head is 7.6 m of flowing fluid and the flow rate is 8.1 m^3/min. Determine the diameter of the throat. Take the coefficient of venturimeter as 0.96.*

Solution : Q = Volumetric flow rate = 8.1 m^3/min = 8.1/60 = 0.135 m^3/s

D = 250 mm = 0.250 m, D_T = diameter of throat.

The flow equation for a venturi meter is

$$Q = \frac{C_v A_T \sqrt{2\,g\,\Delta H}}{\sqrt{(1-\beta^4)}}$$

where $\Delta H = 7.6$ m , $g = 9.81$ m/s²

$$\beta = D_T/D = D_T/0.250 = 4\,D_T$$

$$A_T = \text{area of throat} = \pi/4\,D_T^2$$

$$\therefore \quad 0.135 = \frac{0.96 \times \pi/4 \times D_T^2 \times \sqrt{2 \times 9.81 \times 7.6}}{\sqrt{1-(4\,D_T)^4}}$$

Solving, we get $\sqrt{1-(4\,D_T)^4} = 68.2\,D_T^2$

$$1-(4\,D_T)^4 = (68.2\,D_T^2)^2$$

$$1-256\,D_T^4 = 4651.24\,D_T^4$$

$$\therefore \quad D_T^4 = (1/4907.24)$$

$$D_T = (1/4907.24)^{0.25} = 0.1195 \text{ m}$$

Diameter of the throat = D_T = 0.1195 m = **119.5 mm** ... **Ans.**

Example 7.83 : *Find the size of an orifice that would give a pressure difference of 0.333 m of petroleum product for the flow of a petroleum product at a rate of 0.05 m^3/s in a 150 mm diameter pipe.*

Data : *Coefficient of orifice meter = 0.62.*

Solution : Volumetric flow rate of the petroleum product = 0.05 m^3/s

D = 150 mm = 0.15 m and C_o = 0.62.

The flow equation of an orifice meter is

$$Q = \frac{C_o\,A_o}{\sqrt{1-\beta^4}} \cdot \sqrt{2g\,\Delta H}$$

where Q = 0.05 m^3/s, C_o = 0.62 and ΔH = 0.333 m

Let D_o be the diameter of orifice in m.

$$\therefore \quad A_o = \pi/4\,D_o^2$$

$$\beta = \frac{D_o}{D} = \frac{D_o}{0.15} = 6.67\,D_o$$

$$0.05 = \frac{0.62 \times \pi/4 \times D_o^2}{\sqrt{1-(6.67\,D_o)^4}} \sqrt{2 \times 9.81 \times 0.333}$$

$$0.04017\sqrt{1-(6.67\,D_o)^4} = D_o^2$$

$$1.614 \times 10^{-3}\,[1-1979.3\,D_o^4] = D_o^4$$

$$1.614 \times 10^{-3} - 3.1946\,D_o^4 = D_o^4$$

$$D_o^4 = 3.8478 \times 10^{-4}$$

$$D_o = [3.8478 \times 10^{-4}]^{1/4} = 0.140 \text{ m}$$

Size of the orifice = 0.140 m = **140 mm** ... **Ans.**

Example 7.84 : *A sharp edged circular orifice is to be used to measure the flow rate of water at 293 K (20°C) in a pipeline of internal diameter of 250 mm. The orifice diameter is 50 mm. The reading of a mercury manometer is 242 mm.*

Calculate the flow rate in l/s.

Data : *Density of water = 1000 kg/m³, Viscosity of water = 1.0 mPa.s*

Density of mercury = 13600 kg/m³

For turbulent flow, $N_{Re,O} > 30000$, Orifice coefficient = 0.61.

Solution : D = 250 mm = 0.25 m, D_o = 50 mm = 0.05 m

$\beta = D_o/D = 50/250 = 0.20$

ρ of water = 1000 kg/m³, ρ of Hg = 13600 kg/m³

A_o = area of orifice = π/4 (0.05)² = 0.00196 m²

$$\Delta h = 242 \text{ mm Hg} = 0.242 \text{ m Hg}$$

$$\Delta H = \Delta h \left(\frac{\rho_{Hg} - \rho_{H_2O}}{\rho_{H_2O}}\right) = 0.242 \left(\frac{13600 - 1000}{1000}\right) = 3.05 \text{ m of } H_2O$$

The volumetric flow rate using this meter is calculated by

$$Q = \frac{C_o A_o}{\sqrt{1-\beta^4}} \sqrt{2g\,\Delta H}$$

Taking C_o = 0.61,

$$Q = \frac{0.61 \times 0.00196}{\sqrt{1-(0.2)^4}} \sqrt{2 \times 9.81 \times 3.05} = 0.00926 \text{ m}^3/\text{s} = \mathbf{9.26\ l/s}$$

Volumetric flow rate of water = **9.26 *l*/s** ... **Ans.**

$$N_{Re,o} = \frac{D_o u_o \rho}{\mu}$$

$$u_o = 0.00926/0.00196 = 4.72 \text{ m/s}$$

$$\mu = 1 \text{ mPa.s} = 1 \times 10^{-3} \text{ Pa.s}$$

$$N_{Re,o} = \frac{0.05 \times 4.72 \times 1000}{1 \times 10^{-3}} = 236000$$

As $N_{Re,o} > 30000$, C_o = 0.61 taken is correct.

Example 7.85 : *The flow rate of water in a 150 mm pipeline is measured with the help of a venturimeter having a throat diameter of 50 mm. When the pressure drop over the converging section is 120 mm of water, the water flow rate is 3 kg/s. Find out the coefficient of discharge for the converging cone and the Reynolds number.*

Data : *ρ of water = 1000 kg/m³, μ of water = 1.0 mPa.s*

Solution : Mass flow rate of water = 3 kg/s

D = 150 mm = 0.15 m

D_T = 50 mm = 0.05 m

ρ of water = 1000 kg/m³, μ of water = 1×10^{-3} Pa.s

Volumetric flow rate of water = Q = 3/1000 = 0.003 m³/s

ΔH = 120 mm of H_2O = 0.12 m of water

A_T = Area of throat = π/4 (0.05)² = 0.00196 m²

The flow equation of a venturi meter is

$$Q = \frac{C_v A_T}{\sqrt{1-\beta^4}} \sqrt{2g\,\Delta H}$$

$\beta = D_T/D = 0.05/0.15 = 0.333,\ C_v = ?$

$$0.003 = \frac{C_v \times 0.00196\sqrt{2 \times 9.81 \times 0.12}}{\sqrt{1-(0.333)^4}}$$

$$\therefore \quad C_v = 0.99$$

Coefficient of discharge of the venturi meter = **0.99** **... Ans.**

Velocity of water through the pipe = Q/A = 0.003 / π/4 (0.15)² = 0.17 m/s

$$N_{Re} = \frac{Du\rho}{\mu} = \frac{0.15 \times 0.17 \times 1000}{1 \times 10^{-3}} = \mathbf{25500}$$ **... Ans.**

Example 7.86 : *Water flowing at a rate of 1.5 l/s through a 50 mm diameter pipe is metered by means of a simple orifice of diameter 25 mm. The coefficient of discharge of the orifice meter is 0.62. What will be the reading on a mercury – under – water manometer connected across the meter ?*

What is the Reynolds number for the flow in the pipe ?

Data : *ρ of water = 1000 kg/m³, μ of water = 1.0 (mN.s)/m²*

ρ of mercury = 13600 kg/m³

Solution : Volumetric flow rate of water = 1.5 *l*/s = 1.5×10^{-3} m³/s

$D = 50$ mm = 0.05 m, $D_o = 25$ mm = 0.025 m, $C_o = 0.62$

A_o = area of orifice = $\pi/4\,(0.025)^2 = 4.909 \times 10^{-4}$ m²

β = 25/50 = 0.5

The flow equation of an orifice meter is

$$Q = \frac{C_o A_o}{\sqrt{1-\beta^4}}\sqrt{2g\,\Delta H}$$

$$1.5 \times 10^{-3} = \frac{0.62 \times 4.909 \times 10^{-4}}{\sqrt{1-(0.5)^4}}\sqrt{2 \times 9.81 \times \Delta H}$$

$$\therefore \quad \Delta H = 1.16 \text{ m of } H_2O$$

The reading on the water manometer is 1.16 m

Let us convert ΔH to Δh.

$$\begin{bmatrix}\text{Reading on the mercury}\\ \text{–under–water manometer}\end{bmatrix} = \Delta h = \frac{\Delta H \times \rho}{(\rho_{Hg} - \rho)}, \text{ since } \Delta H = \Delta h\left(\frac{\rho_{Hg} - \rho_{H_2O}}{\rho_{H_2O}}\right)$$

$$= \frac{1.16 \times 1000}{(13600 - 1000)} = 0.0921 \text{ m Hg} = \mathbf{92.1\ mm\ Hg}$$ **... Ans.**

Example 7.87 : *Water flows through an orifice of 25 mm diameter installed in a 75 mm i.d. pipe line at the rate of 0.3 l/s. What will be the difference in level on a water manometer connected across the meter. The coefficient of discharge of the meter is 0.61 ?*

Solution : Volumetric flow rate of water = Q = 0.3 *l*/s = 0.3×10^{-3} m³/s

$D_o = 25$ mm = 0.25 m, D = 75 mm = 0.075 m, $\beta = D_o/D = 25/75 = 0.333$, $C_o = 0.61$

A_o = area of orifice = $\pi/4\,(0.025)^2 = 4.909 \times 10^{-4}$ m²

The flow equation of an orifice meter is

$$Q = \frac{C_o A_o}{\sqrt{1-\beta^4}}\sqrt{2g\,\Delta H}$$

$$0.3 \times 10^{-3} = \frac{0.61 \times 4.909 \times 10^{-4}}{\sqrt{1-(0.333)^4}}\sqrt{2 \times 9.81 \times \Delta H}$$

$$\therefore \quad \Delta H = 0.0505 \text{ m of } H_2O$$

∴ Difference in the levels on the water manometer = 0.0505 m = **50.5 mm** **... Ans.**

Example 7.88 : *Sulphuric acid is flowing through a pipe of 50 mm internal diameter. A thin orifice of 10 mm diameter is fitted in the pipe and the differential pressure shown by a mercury manometer of the orifice is 10 cm. Calculate :*

(a) the mass flow rate of acid in kg/s and

(b) the approximate loss of pressure in kN/m² caused by the orifice.

Data : *Sp.gr. of acid = 1.3, ρ of water = 1000 kg/m³*

ρ of mercury = 13600 kg/m³ and coefficient of orifice meter = 0.61

Solution : $D = 50$ mm = 0.05 m, D_o = diameter of orifice = 10 mm = 0.01 m

sp.gr. of acid = 1.3, density (ρ) of acid = 1.3 × 1000 = 1300 kg/m³

ρ of mercury = 13600 kg/m³

Differential pressure on the mercury manometer shows 10 cm difference in the limbs of manometer.

$\Delta h = 10 \text{ cm} = 0.10 \text{ m of Hg}$

Let us convert Δh to ΔH.

$$\Delta H = \Delta h \left[\frac{\rho_{Hg} - \rho_{acid}}{\rho_{acid}}\right] = 0.10\left[\frac{13600 - 1300}{1300}\right] = 0.946 \text{ m of acid}$$

$C_o = 0.61, \beta = 10/50 = 0.20$

$$\text{Area of orifice} = A_o = \pi/4\,(0.01)^2 = 7.854 \times 10^{-5} \text{ m}^2$$

The flow equation of an orifice meter is

$$Q = \frac{C_o A_o}{\sqrt{1-\beta^4}}\sqrt{2\,g\,\Delta H} = \frac{0.61 \times 7.854 \times 10^{-5}}{\sqrt{1-(0.2)^4}}\sqrt{2 \times 9.81 \times 0.946} = 2.06 \times 10^{-4} \text{ m}^3\text{/s.}$$

Mass flow rate of acid $= Q \times \rho$ of acid $= 2.06 \times 10^{-4} \times 1300 =$ **0.27 kg/s** ... **Ans. (a)**

Approximate loss of pressure or the drop in pressure is

$$\Delta P = h\,\rho\,g$$

$= 0.946 \times 1300 \times 9.81 = 12064$ N/m² = **12.064 kN/m²** ... **Ans. (b)**

Pressure balance at a – a,

$$P_1 + Z_m\,\rho\,g = P_2 + (Z_m - h_m)\,\rho g + h_m\,\rho_m\,g$$

$$(P_1 - P_2)\,\Delta P = h_m\,(\rho_m - \rho)\,g$$

$= 0.10\,(13600 - 1300) \times 9.81 = 12066$ N/m² = **12.066 kN/m²** ... **Ans. (b)**

Example 7.89 : *A rotameter tube with an internal diameter of 25 mm at the top and 20 mm at the bottom is 0.3 m long. The diameter of the float is 20 mm. Water flows through the rotameter. Calculate the flow rate of water when the float is halfway up the tube.*

Data : *sp. gr. of float material = 4.80, density of water = 1000 kg/m³*

volume of float = 6.0 cm³, coefficient of discharge = 0.70

Solution : The mass flow rate through a rotameter is given by

$$\dot{m} = C_D\,A_2\sqrt{\frac{2\,g\,V_f\,(\rho_f - \rho)\,\rho}{A_f\,[1-(A_2/A_1)^2]}}$$

V_f = Volume of float = 6.0 cm³ = 6×10^{-6} m³

ρ_f = Density of float material = 4.80 × 1000 = 4800 kg/m³

ρ = Density of water = 1000 kg/m³

Diameter of float = 20 mm = 0.02 m

A_f = Area of float = $\pi/4\,(0.02)^2 = 3.141 \times 10^{-4}$ m²

A_1 – Cross-sectional area of the tube at float position (i.e., at the highway up).

A_2 – Area of annulus between tube and float at the float position (i.e., at the halfway up).

Diameter at 300 mm is 25 mm

Diameter at 0 mm is 20 mm

Diameter at halfway up i.e. at 150 mm is D_1

$$D_1 = 20 + \frac{(25-20)}{300} \times 150 = 22.5 \text{ mm} = 0.0225 \text{ m}$$

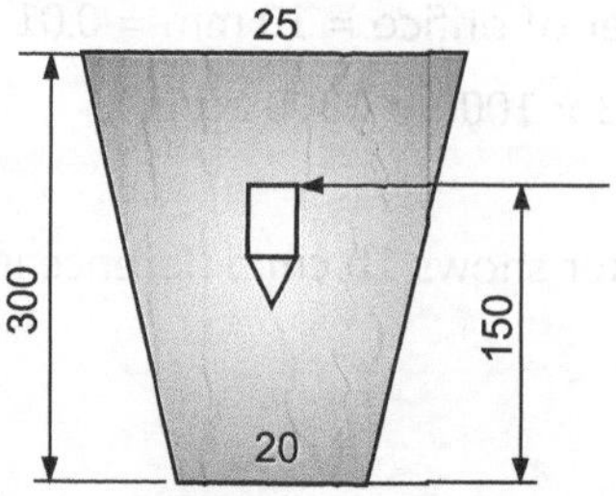

Fig. Ex. 7.89

$$A_1 = \pi/4 \,.\, (0.0225)^2 = 3.976 \times 10^{-4} \text{ m}^2$$

$$A_2 = A_1 - A_f = 3.976 \times 10^{-4} - 3.141 \times 10^{-4} = 0.835 \times 10^{-4} \text{ m}^2$$

$$A_2/A_1 = 0.835 \times 10^{-4} / 3.976 \times 10^{-4} = 0.21$$

$$\dot{m} = 0.7 \times 0.835 \times 10^{-4} \sqrt{\frac{2 \times 9.81 \times 6 \times 10^{-6}\,(4800-1000) \times 1000}{3.141 \times 10^{-4}\,[1-(0.21)^2]}} = 0.071 \text{ kg/s}$$

Mass flow rate of water when the float is halfway = **0.071 kg/s.** ... **Ans.**

Example 7.90 : *A rotameter with a stainless steel float has a maximum capacity of 1.2 l/s of water at 301 K (28°C). Find out the maximum capacity for kerosene in l/s for the same rotameter and the same float.*

sp. gr. of stainless steel = 7.92

sp. gr. of kerosene = 0.82

Solution : For the same rotameter and for the same float position, V_f, A_f A_2 and A_1 are constant. Generally, floats are of such a shape that they set up eddy currents, and C_D is nearly constant.

$$\therefore \quad Q \propto \left(\frac{\rho_f - \rho}{\rho}\right)^{1/2}$$

For water : $$Q_1 = k\left(\frac{\rho_f - \rho_1}{\rho_1}\right)^{1/2} \quad \text{... (1)}$$

For kerosene : $$Q_2 = k\left(\frac{\rho_f - \rho_2}{\rho_2}\right)^{1/2} \quad \text{... (2)}$$

$\rho_f = 7920$ kg/m³, $\rho_1 = 1000$ kg/m³, $\rho_2 = 820$ kg/m³, $Q_1 = 1.2$ l/s $= 1.2 \times 10^{-3}$ m³/s

Dividing equation (2) by equation (1),

$$\frac{Q_2}{Q_1} = \frac{[(\rho_f - \rho_2)/\rho_2]^{1/2}}{[(\rho_f - \rho_1)/\rho_1]^{1/2}}$$

$$\frac{Q_2}{1.2 \times 10^{-3}} = \frac{[(7920-820)/820]^{1/2}}{[(7920-1000)/1000]^{1/2}}$$

$$\therefore \quad Q_2 = 1.34 \times 10^{-3} \text{ m}^3\text{/s} = 1.34 \text{ l/s}$$

Maximum capacity for kerosene = **1.34 l/s** ... **Ans.**

Example 7.91 : *A rotameter tube is 0.3 m long with an internal diameter of 25 mm at the top and 20 mm at the bottom. The diameter of float is 20 mm. Find the height/position of the float when metering water at the rate 100 cm^3/s.*

Data : Sp. gr. of float = 4.80, volume of float = 6.0 cm^3

Coefficient of discharge of the meter = 0.72, density of water = 1000 kg/m^3.

Solution : Volumetric flow rate of water = 100 cm^3/s = 100×10^{-6} m^3/s

The volumetric flow rate through a rotameter is given by

$$Q = C_D A_2 \sqrt{\frac{2 g V_f (\rho_f - \rho)}{\rho A_f [1 - (A_2/A_1)^2]}}$$

where, $C_D = 0.72$, $V_f = 6$ $cm^3 = 6 \times 10^{-6}$ m^3,

$\rho = 1000$ kg/m^3, $\rho_f = 4.80 \times 1000 = 4800$ kg/m^3,

Diameter of float = 20 mm = 0.02 m,

A_f = area of float = $\pi/4\ (0.02)^2 = 3.141 \times 10^{-4}$ m^2,

A_1 = Cross-sectional area of the tube at float position (top of the float)

A_2 = Area of the annulus between tube and float at float position (top of the float)

$A_2 = A_1 - A_f = A_1 - 3.141 \times 10^{-4}$ m^2

Substituting the values gives

$$100 \times 10^{-6} = 0.72 A_2 \sqrt{\frac{2 \times 9.81 \times 6 \times 10^{-6} (4800 - 1000)}{1000 \times 3.141 \times 10^{-4} [1 - (A_2/A_1)^2]}}$$

$$1.164 \times 10^{-4} = A_2 / \sqrt{1 - (A_2/A_1)^2} \quad \ldots (1)$$

Trial and error procedure is to be adopted.

$$A_2 = A_1 - A_f = A_1 - 3.141 \times 10^{-4} \ m^2$$

$$A_1 = \pi/4\ D_1^2$$

D_1 is the diameter of tube at the float position.

D_1, m	A_1, m^2	A_2, m^2	RHS of Eq. (1)
0.022	3.80×10^{-4}	0.659×10^{-4}	6.69×10^{-5}
0.025	4.909×10^{-4}	1.768×10^{-4}	5.26×10^{-4}
0.023	4.155×10^{-4}	1.014×10^{-4}	1.046×10^{-4}
0.024	4.524×10^{-4}	1.383×10^{-4}	1.45×10^{-4}
0.0235	4.337×10^{-4}	1.196×10^{-4}	1.244×10^{-4}
0.0232	4.227×10^{-4}	1.086×10^{-4}	1.124×10^{-4}
0.0234	4.3×10^{-4}	1.159×10^{-4}	1.20×10^{-4}
0.0233	4.264×10^{-4}	1.123×10^{-4}	1.164×10^{-4}

For D_1 = 0.0233 m, LHS of Eqn. (1) = RHS of Eqn. (1)

$\therefore \quad D_1 = 0.0233 \text{ m} = 23.3 \text{ mm}$

Diameter of tube at the top (i.e., at 300 mm) = 25 mm

Diameter of tube at the bottom (i.e., at 0 mm) = 20 mm

D_1 is the diameter of tube at a distance x mm from the bottom.

$$\therefore \quad D_1 = 20 + \left(\frac{25-20}{300}\right) x$$

$$23.3 = 20 + \frac{5}{300} \cdot x$$

Solving, we get $x = 198$ mm

Height of the float when metering 100 cm³/s of water = **198 mm** ... **Ans.**

Example 7.92 : *A venturimeter is to be installed in a 100 mm diameter line to measure the flow of water. The maximum flow is expected to be 73.8 m³/h. The mercury is to be used as a manometric fluid and the corresponding manometer reading is 1.27 m of mercury. Find the throat diameter required for the venturi and the power that would be required to operate the meter at the full load if the permanent loss in pressure is 10% of the venturi differential (i.e., pressure drop across the meter).*

Data : *Density of water = 1000 kg/m³, Density of mercury = 13600 kg/m³,*

Coefficient of venturimeter = 0.98.

Solution : Volumetric flow rate = Q = 73.8 m³/h = 0.0205 m³/s

ρ of water = 1000 kg/m³, ρ of mercury = 13600 kg/m³

Δh = 1.27 m of mercury

We have : $\Delta H = \Delta h \left(\frac{\rho_{Hg} - \rho_{H_2O}}{\rho_{H_2O}}\right)$

Let us convert Δh to ΔH.

$$\Delta H = 1.27 \left(\frac{13600-1000}{1000}\right) = 16 \text{ m of } H_2O$$

D = 100 mm = 0.10 m, D_T = diameter of throat in m

$\beta = D_T/D = D_T/0.1 = 10\,D_T$

A_T = Area of throat = $\pi/4\ D_T^2$, $C_v = 0.98$

The flow equation of a venturi meter is

$$Q = \frac{C_v A_T}{\sqrt{1-\beta^4}} \sqrt{2\,g\,\Delta H}$$

$$0.0205 = \frac{0.98 \times \pi/4 \times D_T^2}{\sqrt{1-(10\,D_T)^4}} \sqrt{2 \times 9.81 \times 16}$$

$$\sqrt{1-(10\,D_T)^4} = 665\,D_T^2$$

$$1 - 10000\,D_T^4 = 442225\,D_T^4$$

$$\therefore \quad D_T = 0.0386 \text{ m}$$

$$D_T = 38.6 \text{ mm}$$

Diameter of venturi = **38.6 mm** ... **Ans.**

$\dot{m}$ = Mass flow rate of water = 0.0205 × 1000 = 20.5 kg/s

Pressure differential across the meter = 1.27 m of Hg = 1270 mm Hg = $\frac{1270}{760} \times 101325$ = 169320 N/m²

The permanent loss in pressure = 10% of the differential = 0.10 (169320) = 16932 N/m²

$$\left[\begin{array}{l}\text{Power required to} \\ \text{operate venturi at full flow}\end{array}\right] = \frac{16932}{1000} \times 20.5 \frac{N}{m^2} \times \frac{1}{\left(\frac{kg}{m^3}\right)} \times \frac{kg}{s}$$

$$= 347 \left(\frac{N.m}{kg}\right) \cdot \left(\frac{kg}{s}\right) = \left(\frac{J}{kg}\right)\left(\frac{kg}{s}\right) = \mathbf{347\ J/s = 347\ W}$$ ... **Ans.**

Example 7.93 : *An orifice meter is to be installed to measure the flow rate of crude oil to a cracking unit. The oil is flowing through a 100 mm i.d. pipe and an adequate run of the straight horizontal pipe is available for the installation of the meter. The expected maximum flow rate is 79.5 m³/h. Mercury is to be used as a manometer fluid, and glycol is to be used in the leads as a sealing liquid. The maximum reading of the meter is 762 mm. Calculate*

(a) The diameter of the orifice and (b) The power loss if 68% of the orifice differential is permanently lost.

Data : *sp. gr. of oil = 0.89, sp. gr. of glycol = 1.11, sp. gr. of mercury = 13.6,*
ρ of water 1000 kg/m³, coefficient of meter = 0.61.

Solution : Volumetric flow rate = 79.5 m³/h = 0.0221 m³/s

Pressure differential across the meter = 762 mm Hg = 0.762 m Hg

$$\Delta P \text{ across the meter } = h_m (\rho_m - \rho)\, g = 0.762\,(13600 - 1100) \times 9.81 = 93440\ N/m^2$$

[ρ of mercury = 13600 kg/m³, ρ of glycol = 1100 kg/m³]

ρ of crude = 890 kg/m³

$\dot{m}$ = mass flow rate of the crude oil = 0.0221 × 890 = 19.67 kg/s

Differential pressure permanently lost is 68% of the orifice differential.

$$\text{Permanent loss in the pressure } = 0.68 \times 93440 = 63539.2\ N/m^2 = \frac{63539.2}{890} \frac{N}{m^2}\left(\frac{1}{\frac{kg}{m^3}}\right) = \frac{N.m}{kg} = 71.39\ J/kg$$

Power consumption in J/s = Loss in pressure (J/kg) × mass flow rate (kg/s)

Maximum power consumption of the meter = 19.67 × 71.39 = 1404 J/s = 1404 W = **1.4 kW** ... **Ans. (b)**

D_o = orifice diameter in meters

D = 100 mm = 0.10 m

$A_o = \pi/4\ D_o^2$

$\beta = D_o/D = D_o/0.10 = 10\ D_o$

$C_o = 0.61$

The flow equation for an orifice meter is

$$Q = \frac{C_o A_o}{\sqrt{1 - \beta^4}} \sqrt{\frac{2\,\Delta P}{\rho}}$$

$$0.0221 = \frac{0.61 \times \pi/4 \times D_o^2}{\sqrt{1 - (10\,D_o)^4}} \sqrt{\frac{2 \times 93440}{890}}$$

$$\sqrt{1 - (10\,D_o)^4} = 314.13\ D_o^2$$

$$1 - 10000\ D_o^4 = 98678\ D_o^4$$

$$\therefore \quad D_o = 0.055\ m$$

$$D_o = 55\ mm$$

Diameter of orifice = **55 mm** ... **Ans. (a)**

Example 7.94 : *A pitot tube is a pipe in which air is flowing, is connected to a manometer containing water. If the difference in water levels in the manometer is 87.5 mm, what is the velocity of flow in the pipe assuming a tube coefficient of 0.99 ?*

Solution : The velocity of flow is given by $u = c\sqrt{2g\Delta H}$

where c = tube coefficient = 0.99

ΔH = manometer reading expressed in terms of m of the flowing fluid (air)

$\Delta H'$ = manometer reading in terms of water

$= 87.5 \text{ mm} = 8.75 \text{ cm} = 0.0875 \text{ m of } H_2O$

$$\Delta H = \Delta H' \left[\frac{\rho_{H_2O} - \rho_{air}}{\rho_{air}}\right]$$

where, $\rho_{H_2O} = 1 \text{ g/cm}^3 = 1000 \text{ kg/m}^3$

$\rho_{air} = 1.29 \times 10^{-3} \text{ g/cm}^3 = 1.29 \text{ kg/m}^3$

$$\Delta H = 0.0875 \left[\frac{1000 - 1.29}{1.29}\right] = 67.74 \text{ m of air}$$

Velocity of flow of air = $u = 0.99 \, [2 \times 9.81 \times 67.74]^{1/2}$ = **36.09 m/s** ... **Ans.**

In this problem, the density of air is taken as 1.29×10^{-3} g/cm³.

Example 7.95 : *A pitot tube is installed in the centre of a pipe 80 mm diameter. Find the velocity of water in the centre of the pipe if the water rises 300 mm in the tube.*

Solution : $h = 300 \text{ mm} = 0.30 \text{ m}$

The velocity of water in the centre of the pipe is

$$u = \sqrt{2gh} = \sqrt{2 \times 9.81 \times 0.30} = 2.426 \approx \mathbf{2.43 \text{ m/s}}$$ ... **Ans.**

Example 7.96 : *A rectangular notch 2.5 m wide has a constant head of 0.40 m. Find the rate of discharge of water through the notch if the coefficient of discharge for the notch is 0.60.*

Solution : We have, B = Width of the notch = 2.5 m

H = Head of water = 0.40 m

$g = 9.8 \text{ m/s}^2$

C_d = Coefficient of discharge = 0.60

The relation to be used is

$$Q = \frac{2}{3} C_d B \sqrt{2g} (H)^{1.5} = \frac{2}{3} \times 0.60 \times 2.5 \, (2 \times 9.81)^{1/2} (0.40)^{1.5}$$

$$= 1.120 \text{ m}^3\text{/s or } 1120 \, l\text{/s}$$

The rate of discharge of water is **1.12 m³/s or 1120 *l*/s** ... **Ans.**

Example 7.97 : *A rectangular notch with $C_d = 0.60$ has a discharge of 360 l/s when the head of water is half the length of the notch. Find the length of the notch.*

Solution : Given : C_d = Coefficient of discharge = 0.60

Q = Rate of discharge = 360 *l*/s = 0.36 m³/s

B = Length/width of the notch

H = Head of water (depth) = Half the length of the notch = $B/2$

For the rectangular notch, Q is given by

$$Q = \frac{2}{3} C_d B \sqrt{2g} (H)^{1.5}, \text{ but } H = B/2$$

$$Q = \frac{2}{3} C_d B \sqrt{2g} (B/2)^{1.5}$$

$$0.36 = \frac{2}{3} \times 0.6 \times \sqrt{2 \times 9.81} \frac{B^{2.5}}{(2)^{1.5}} \quad \therefore (B)^{2.5} = 0.5747$$

$\therefore$ $B = 0.80 \text{ m} = 800 \text{ m}$

Length of the rectangular notch = **800 mm.** ... **Ans.**

Example 7.98 : *A right-angled V-notch was used to measure the discharge of a centrifugal pump. Find the discharge through the notch, if the depth of water at the notch is 200 mm. Take C_d = 0.60.*

Solution : θ = Angle of the notch = 90°

H = Depth of water at the notch = 200 mm = 0.20 m

C_d = Coefficient of discharge = 0.60

Q = Discharge through the notch

$$Q = \frac{8}{15} C_d \sqrt{2g} \tan\frac{\theta}{2} (H)^{2.5} = \frac{8}{15} \times 0.60 \times \sqrt{2 \times 9.81} \tan\frac{90}{2} (0.20)^{2.5}$$

$= \mathbf{0.02535\ m^3/s}$ **or** $\mathbf{25.35}$ ***l*/s** ... **Ans.**

Or : For a 90° V-notch for C_d = 0.60, we have

$$Q = 1.417 H^{5/2} \text{ ... in SI units} = 1.417 (0.20)^{5/2}$$

$= \mathbf{0.02535\ m^3/s}$ or $\mathbf{25.35}$ ***l*/s** ... **Ans.**

Example 7.99 : *In an experiment in a laboratory, 50 l of water flowing over a right-angled V-notch was collected in one minute. If the head of the sill is 50 mm, calculate the coefficient of discharge of the notch.*

Solution : Water collected in one minute = 50 *l*

$$\therefore \quad Q = \text{rate of flow of water} = \frac{50}{1} = 50\ l/\text{min} = \frac{50 \times 10^{-3}}{60} = 8.333 \times 10^{-4} \text{ m}^3/\text{s}$$

Head of the sill = H = 50 mm = 0.05 m

θ = 90° ... given

We have :

$$Q = \frac{8}{15} C_d \sqrt{2g} \tan\frac{\theta}{1} (H)^{2.5}$$

$$C_d = \frac{15Q}{8\sqrt{2g} \cdot \tan\frac{\theta}{2} \cdot (H)^{2.5}} = \frac{15 \times 8.333 \times 10^{-4}}{8 \times \sqrt{2 \times 9.81} \times \tan\frac{90}{2} \times (0.05)^{2.5}}$$

$C_d = 0.6309 \approx \mathbf{0.63}$ ... **Ans.**

Example 7.100 : *A crude oil of viscosity 0.9 poise and relative density of 0.90 is flowing through a horizontal pipe of 120 mm and length 12 m. Calculate the difference in pressure at the two ends of pipe, if 785 N weight of the oil is collected in a tank in 25 s.*

Solution : Weight of oil = 785 N

The force becomes weight when the body acts under gravitational acceleration (g).

$$\text{Weight} = \frac{m \cdot g}{g_c}$$

where,
$$g = 9.81 \text{ m/s}^2$$
$$g_c = 1 \text{ (kg·m)/(N·s}^2)$$
$$\text{Weight} = 785 \text{ N}$$
$$785 = \frac{m \times 9.81}{1}$$
$$m = 80.02 \text{ kg}$$
$$\text{Units of weight} \Rightarrow \frac{(\text{kg}) \times (\text{m/s}^2)}{(\text{kg·m})/(\text{N·s}^2)} \Rightarrow \text{N}$$

80.02 kg oil is collected in 25 s. Therefore,
$$\text{Mass flow rate of oil} = \dot{m} = \frac{80.02}{25} = 3.2 \text{ kg/s}$$
$$\text{Diameter of pipe} = D = 120 \text{ mm} = 12 \text{ cm} = 0.12 \text{ m}$$
$$\text{Viscosity of oil} = \mu = 0.9 \text{ P} = 0.09 \text{ kg/(m.s)}$$
$$\text{Density of oil} = \rho = 0.90 \text{ g/cm}^3 = 900 \text{ kg/m}^3$$
$$\text{Area of pipe} = A = \pi/4\, D^2 = \pi/4\,(0.12)^2 = 0.0113 \text{ m}^2$$

We have :
$$\dot{m} = u\rho A$$
$$u = \frac{\dot{m}}{\rho A} = \frac{3.2}{900 \times 0.0113} = 0.3146 \approx 0.315 \text{ m/s}$$
$$\text{Reynold's number} = N_{Re} = \frac{Du\rho}{\mu} = \frac{0.12 \times 0.315 \times 900}{0.09} = 378$$

Since N_{Re} is less than 2100, the flow is laminar.

The Hagen-Poiseuille equation is
$$\Delta P = \frac{32\,\mu u L}{D^2}$$

We have,
$$\mu = 0.09 \text{ kg/(m.s)}$$

and
$$1 \text{ kg/(m.s)} = 1 \text{ (N.s)/m}^2$$
$$\text{Units of } \Delta P \Rightarrow \frac{(\text{N.s})}{\text{m}^2} \times \frac{\text{m}}{\text{s}} \times \frac{\text{m}}{\text{m}^2} \Rightarrow \text{N/m}^2 = \text{Pa (pascal)}$$

$\therefore$
$$\mu = 0.09 \text{ (N.s)/m}^2$$
$$\Delta P = \frac{32 \times 0.09 \times 0.315 \times 12}{(0.12)^2}$$
$$= 756 \text{ N/m}^2 = \mathbf{0.756 \text{ kN/m}^2 = 0.756 \text{ kPa}}$$ ... **Ans.**

Example 7.101 : *Fuel oil is pumped up in a 300 mm diameter and 16 km long pipeline at the rate of 15 kg/s. The pipe line is laid at a slope of 1 in 300. The density of fuel oil is 950 kg/m³ and its kinematic viscosity is 21.4 stokes. Find the power required to pump the oil.*

Solution :
$$\text{Diameter of pipe} = 300 \text{ mm} = 0.30 \text{ m}$$
$$\text{Length of pipe} = L = 16 \text{ km} = 16000 \text{ m}$$
$$\text{Slope of the pipe} = i = 1/300$$
$$\text{Density of oil} = \rho = 950 \text{ kg/m}^3$$
$$\text{Mass flow rate of oil} = \dot{m} = 15 \text{ kg/s}$$

$$\text{Kinematic viscosity} = \upsilon = \frac{\mu}{\rho} = 21.4 \text{ stokes} = 21.4 \times 10^{-4} \text{ m}^2/\text{s}$$

$$A = \frac{\pi}{4} D^2 = \frac{\pi}{4} (0.3)^2 = 0.07068 \text{ m}^2$$

We have :
$$\dot{m} = \rho u A$$

$$u = \frac{\dot{m}}{\rho A} = \frac{15}{950 \times 0.07068} = 0.223 \text{ m/s}$$

$$\text{Reynolds number} = N_{Re} = \frac{Du\rho}{\mu} = \frac{Du}{\upsilon} = \frac{0.3 \times 0.223}{21.4 \times 10^{-4}} = 31.26$$

Since the Reynolds number is less than 2100, the flow is laminar. Therefore,

$$f = \frac{16}{N_{Re}} = \frac{16}{31.26} = 0.5118 \approx 0.512$$

$$\text{Head loss due to friction} = h_f = \frac{4fLu^2}{2gD}$$

$$h_f = \frac{4 \times 0.512 \times 16000 \times (0.223)^2}{2 \times 9.81 \times 0.30} = 276.85$$

The total head against which the pump has to work is

$$H = h_f + L \cdot i = 276.85 + 16000 \times \frac{1}{300} = 330.18 \text{ m}$$

The power required to pump the oil is

$$P = \frac{\rho Q H}{75}, \text{ hp}$$

We have :
$$Q = \frac{\dot{m}}{\rho}$$

$$\therefore \quad P = \frac{\dot{m}H}{75} = \frac{15 \times 330.18}{75} = \mathbf{66 \text{ h.p.}}$$ ... **Ans.**

OR :
$$P = QH\rho g = \dot{m} \cdot H \cdot g = 15 \times 9.81 \times 330.18$$
$$= 48585.987 \text{ J/s} = 48585.987 \text{ W} = \mathbf{48586 \text{ W} = 48.586 \text{ kW}}$$ ... **Ans.**

Example 7.102 : *Water in a tank comes out through a sharp edged orifice of 10 cm² area. Head of water above the centre of orifice is 3 m. If $C_c = 0.6$ and $C_v = 0.98$, find the discharge and area of jet at vena contracta.*

Solution : Given :

$$C_v = \text{coefficient of velocity} = 0.98$$

$$C_c = \text{coefficient of contraction} = 0.60$$

$$\text{Area of orifice} = 10 \text{ cm}^2 = 10 \times 10^{-4} \text{ m}^2 = 10^{-3} \text{ m}^2$$

$$C_c = \frac{\text{area of jet at vena contracta}}{\text{area of orifice}} = \frac{a_c}{a}$$

$$\therefore \quad a_c = a \cdot C_c = 10 \times 0.60 = 6 \text{ cm}^2$$

Area of jet at vena contracta = 6 cm² ... **Ans.**

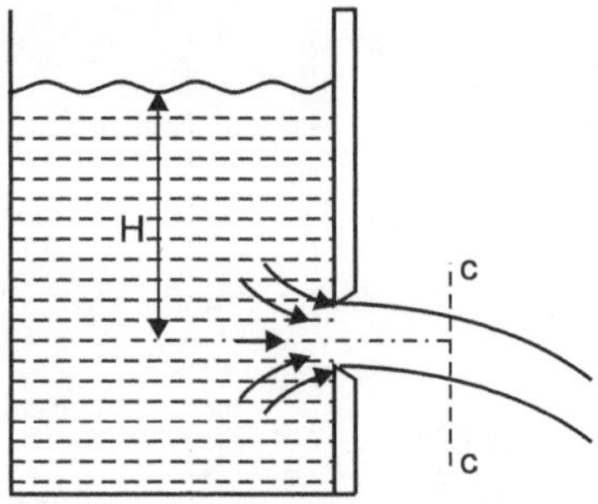

Fig. Ex. 7.102

Section c-c is known as vena contracta. Where jet is more or less horizontal where maximum contraction takes place - which is slightly on the downstream side of the orifice.

C_d = coefficient of discharge = $C_v \cdot C_c = 0.98 \times 0.60 = 0.588$

H = Head of water = 3 m

Actual discharge = $C_d \cdot a \cdot \sqrt{2gH} = 0.588 \times 10^{-3} (2 \times 9.81 \times 3)^{1/2}$

= $\mathbf{4.511 \times 10^{-3}\ m^3/s}$ ... **Ans.**

= **4.511 *l*/s** ... **Ans.**

Example 7.103 : *A crude oil of specific gravity 0.92 and viscosity of 1.2 poise is pumped through a pipe of 100 mm i.d. laid at a slope of 1 in 100. The oil is pumped at the rate of 15 l/s. Find the difference between the pressure gauge readings if they are fitted 300 m apart.*

Solution : Sp. gr. of oil = 0.92

Density of oil = 0.92 g/cm³

Viscosity of oil = 1.2 poise = 1.2 g/(cm.s)

Diameter of pipe = 100 mm = 10 cm

Volumetric flow rate = Q = 15 *l*/s = 15000 cm³/s

$$\text{Slope of pipe } i = 1 \text{ in } 100 = \frac{1}{100} = 0.01$$

$$\text{Velocity of oil} = u = \frac{Q}{\pi/4\ D^2} = \frac{15000}{\pi/4\ (10)^2} = 190.98 \approx 191 \text{ cm/s}$$

$$\text{Reynolds' number} = N_{Re} = \frac{Du\rho}{\mu} = \frac{10 \times 191 \times 0.92}{1.2} = 1464$$

Since N_{Re} is less than 2100,

$$f = \frac{16}{N_{Re}} = \frac{16}{1464} = 0.01093$$

Length = L = 300 m = 30000 cm

$$\text{Head loss due to friction} = h_f = \frac{4fLu^2}{2gD} = \frac{4 \times 0.01093 \times 30000 \times (191)^2}{2 \times 9.81 \times 10} = 2438.76 \text{ cm}$$

Taking the lower point of pipe as station-1 and the upper point as station-2 apply Bernoulli's equation.

$$z_1 + \frac{P_1}{\rho} + \frac{u_1^2}{2g} = z_2 + \frac{P_2}{\rho} + \frac{u_2^2}{2g} + h_f$$

With $u_1 = u_2$ and $z_1 = 0$, it becomes

$$0 + \frac{P_1}{\rho} = 30000 \times \frac{1}{100} + \frac{P_2}{\rho} + 2438.76$$

$$\frac{P_1}{\rho} - \frac{P_2}{\rho} = 300 + 2438.76 = 2738.76 \text{ cm}$$

$$P_1 - P_2 = 2738.76 \times \rho = 2738.76 \times 0.92 = 2519.66 \text{ gm/cm}^2$$

= $\mathbf{2.51966 \approx 2.52\ kg/cm^2}$ ... **Ans.**

Important Points

- A fluid is a substance which is capable of flowing if allowed to do so.
- A fluid is a substance which undergoes continuous deformation when subjected to a shear force.
- If the density of a fluid is affected appreciably by changes in temperature and pressure, then the fluid is said to be compressible. e.g., air.
- If the density of a fluid is not appreciably affected by moderate changes in temperature and pressure, the fluid is said to be incompressible. e.g., water.
- The viscosity of a fluid at a given temperature is a measure of its resistance to flow. The SI unit of viscosity is $P_a{\cdot}s$.
- Fluids which follow Newton's law of viscosity are called Newtonian fluids and fluids which do not follow Newton's law of viscosity are called non-Newtonian fluids.
- The vertical height or the free surface above any point in a liquid at rest is called the pressure head.
- The equation of continuity is : $\dot{m} = \rho uA$ = constant.
- The average velocity of a fluid through a pipe is defined as the ratio of the volumetric flow rate of the fluid to the cross-sectional area of the pipe.
- The mass flow rate of a fluid through a pipe is related to the volumetric flow rate by $\dot{m} = Q\rho$, where ρ is the density of the fluid.
- $P/\rho g + z + u^2/2g$ = constant – it is the Bernoulli equation in which each term represents energy per unit mass of the fluid and has the dimension of length.
- Various manometers that are used for pressure measurement or differential pressure measurement include U-tube manometer, inclined tube manometer and differential manometer.
- The flow in which the fluid flows in parallel, straight lines without lateral mixing is called laminar flow.
- The flow in which the fluid flows eratically in the form of cross currents and eddies and there is a complete mixing in the radial direction is called turbulent flow.
- The relation used to find the frictional loss through a straight pipe is $h_{fs} = 4fLu^2/2D$.
- The friction fraction chart is a logarithmic plot of friction factor v/s Reynolds number.
- If N_{Re} is less than 2100, the flow is laminar and if $N_{Re} > 4000$, the flow is turbulent.
- For turbulent flow, the friction factor (f) is related to the Reynolds number (N_{Re}) by the relation : $f = 0.0078/ (N_{Re})^{0.25}$.
- For laminar flow, the friction factor (f) is related to the Reynolds number (N_{Re}) by $f = 16/N_{Re}$
- The equivalent length of a fitting is that length of a straight pipe of the same nominal size as that of the fitting, which would cause the same frictional loss as that caused by the fitting.
- Orifice and venturi meters are variable head meters used for flow measurement.
- Advantages of orifice meter are: simple in construction, easy to install, cheap and greater flexibility.
- Advantage of venturi meter are: high pressure recovery, lower power loss, high accuracy, used for flow measurements of compressible as well as incompressible fluids and higher reproducibility.
- Rotameters provide a direct reading of the flow rate.

Practice Questions

1. Give the classification of fluids.
2. State Newton's law of viscosity.
3. Define viscosity and give its SI unit.
4. Define kinematic viscosity and pressure head.
5. Define velocity gradient and pressure gradient.
6. Define hydrostatic equilibrium.

7. Give the relationship between absolute pressure, gauge pressure and atmospheric pressure and define each of them.
8. Derive Newton's viscosity law relation.
9. State applications of hydrostatic equilibrium.
10. Explain briefly fluid statics with its applications.
11. Explain briefly fluid dynamics with its applications.
12. Define the terms – Newtonian and non-Newtonian fluids.
13. Darw a neat sketch of U-tube manometer.
14. Define critical velocity.
15. Define laminar flow and turbulent flow.
16. Give the relation between mass flow rate and volumetric flow rate.
17. Define the terms point velocity and average velocity.
18. Explain briefly Reynolds experiment with a neat sketch.
19. Give the relation between Fanning friction factor and Reynolds number for laminar flow and turbulent flow.
20. Give the relation between Darcy friction factor and Reynolds number for laminar flow and turbulent flow.
21. Give the relation between point velocity and maximum velocity for laminar flow through a pipe.
22. Derive the equation of continuity.
23. Derive the relation between point velocity and maximum velocity for laminar flow through a pipe.
24. Define steady flow.
25. Define mass velocity and give its relation with volumetric flow rate.
26. Water flows through a circular pipe for which Reynolds number is 1700. Find the Fanning friction factor and Darcy friction factor.
27. Find the mass velocity of water flow through a pipe of internal diameter 25 mm, velocity = 3 m/s and density = 1000 kg/m^3.
28. Explain briefly Reynolds experiment with its sketch.
29. Explain in brief different flow regimes.
30. Derive the flow equation for a venturi meter.
31. Derive the flow equation for an orifice meter.
32. Draw a neat sketch of orifice meter and explain its construction.
33. Draw a neat sketch of venturi meter and explain its construction.
34. Draw a neat sketch of rotameter and explain its construction.
35. Differentiate between orifice meter and venturi meter.
36. Differentiate between variable head meter and area meter.
37. Derive Bernoulli's equation for incompressible fluid without fluid friction.
38. State Bernoulli's equation and give the meaning of each term in it.
39. Explain briefly the procedure for calibrating a given rotameter.
40. Explain brief the procedure for calibrating a given orifice meter.
41. Explain in brief with a sketch the construction of rotameter.
42. Define the term critical velocity and give the physical significance of Reynolds' number.
43. Give assumptions made in deriving Bernoulli's equation.
44. State Bernoulli's theorem.
45. Give limitations of Bernoulli's equation.

❑❑❑

CHAPTER EIGHT

PIPE FITTINGS AND VALVES

8.1 PIPE, TUBING AND THEIR COMPARISON

Piping systems in chemical plants consist of pipes (various sizes and materials), pipe fittings such as socket, elbow, etc. and valves for transferring materials from one equipment to another equipment within a processing area or for transferring materials from tank farm to processing area and vice versa. So in this chapter, we will discuss pipes, tubes, alongwith their sizes and materials of construction, pipe fittings and valves.

In the chemical industry, fluids are usually transferred from one point to another through pipe or tubing. Pipe or tubing is circular in cross-section and may be made in different diameters and in different wall thickness, from any available material of construction depending upon the service conditions. The materials of construction include iron, steel, brass, copper, glass, plastics, etc.

- Pipe is heavy walled, whereas tubing is thin walled. Metal pipe is available in standard lengths of about 6 m, whereas tubing is available in coils several meters long.
- Tubes from 6 to 50 mm are frequently made from non-ferrous metals, such as brass, copper or aluminium and are widely used in heat exchangers. Pipe is relatively larger in diameter than tubing and are made of cast iron steel, etc.
- Pipe surface is slightly rough, whereas the tube surface is very smooth.
- Pipe sections may be joined by screwing, flanging or welding, whereas tube pieces may often be joined by brazing, soldering or by flared fittings.
- In case of pipes, the schedule number specifies the thickness while in case of tubing, BWG specifies the wall thickness.
- The size of a pipe is specified in terms of nominal pipe size, whereas the size of a tube is specified by its outside diameter.

8.2.1 Safe Working

- It is a practice of performing a task with minimum risk to people, equipments, materials, environment and processes.

8.2.2 Pressure and Allowable Stress

- Pressure is defined as the force per unit area (F/A). That is, the force applied/acting perpendicular to the surface of an object per unit area over which the force is distributed. SI units : N/m^2, Pascal (Pa).
- **Maximum Allowable Working Pressure (MAWP) :** It is the maximum or highest pressure at which the vessel or equipment may be operated at its design pressure.
- **Operating pressure (Working pressure) :** It is the pressure at which the equipment operates during its life.
- **Design pressure :** It is the pressure in excess of the operating pressure for the sake of safety.
  Design pressure is 10 to 25% more than the operating pressure.
  $$\text{Design pressure} = 1.1 \text{ to } 1.25 \text{ (operating pressure)}$$
- **Allowable stress :** It is the maximum stress (tensile, compressive or bending) which is allowed to be applied on an object. OR It is the maximum stress that an object is expected to support. The allowable stress is calculated using the following relation :
  $$\text{Allowable stress} = \frac{\text{Yield strength}}{\text{Factor of safety}}$$
  For example, for A36 steel, yield strength is 36000 psi and factor of safety is 4. With this, the allowable stress of A36 steel is (= 36000/4) 9000 psi.
  [Yield strength or stress : It is the stress at which an object will be permanently damaged.]

8.2.3 Standard Sizes of Pipes

Nominal Pipe Size (NPS) :

Nominal pipe size is a North American set of standards used to designate diameter and thickness of a pipe. . Pipe size is specified in terms of a nominal pipe size for diameter based on inches and a schedule/schedule number for wall thickness. The term NB (nominal bore) is frequently used interchangeably with NPS. Based on the NPS and schedule of a pipe, the outside diameter and wall thickness of the pipe can be obtained from reference tables. The nominal pipe size of steel pipe ranges from 1/8 to 30 inches. For NPS 1/8 to 12 inches, the NPS and OD values are different (the nominal pipe size is slightly less than the outside diameter). For example, the OD of an NPS 12 pipe is actually 12.75 inches. For NPS 14 inches and larger, the NPS and OD values are equal (the nominal pipe size is the outside diameter). For example, the OD of an NPS 14 pipe is 14 inches. For a given NPS, the OD remains fixed and the wall thickness increases with schedule/schedule number. For a given schedule number, the OD increases with NPS whose wall thickness remains constant or increases. Pipes of other materials are also made with the same outside diameter as steel pipes for the same NPS.

8.2.4 Schedule Number / Pipe Schedule Number

Pipe schedule is the term used to indicate the thickness of a pipe. Schedule number is a dimensionless number that specifies the thickness of a pipe. The Schedule number is given by the expression

$$\text{Schedule number} = 1000\left(\frac{P}{S}\right)$$

where P is the internal working pressure in the pipe in psig and S is the allowable stress (ultimate tensile strength) for the pipe material at the conditions of use in psi.

The higher the schedule number is, the thicker the pipe is (in other words, the schedule number increases with the thickness of the pipe). Normally for pipes, the outer diameter is fixed for each nominal pipe size. Thus a particular nominal pipe size will have different inside pipe diameter (i.e., different wall thickness) depending on the schedule number specified.

There are eleven schedule numbers that are commonly used : 5, 20, 30, 40, 60, 80, 100, 120, 140 and 160. The most commonly used schedule numbers today are 40, 80 and 160. The most popular schedule number is 40.

8.2.5 Birmingham Wire Gauge [BWG]

Birmingham Wire Gauge (BWG) is a dimensionless number that specifies the thickness of a tube. BWG number increases with increase in thickness of the tube.

The tube wall thickness specified by the BWG number which range from 24 (very light) to 7 (very heavy).

8.2.6 Materials of Construction for Pipes and Tubes

Different materials that are used for construction of pipes and tubes include ferrous materials, non-ferrous materials, plastics, glass and lined metal.

Ferrous materials account for a large part of all the piping in process industries. In this category we have carbon steel, iron and stainless steel. Carbon steel pipe is strong, ductile and cheaper than the pipe made from other materials. Carbon steel pipe is natural choice, if it meets the requirements of pressure, temperature, corrosion resistance and hygiene. Iron pipe is made from cast iron and ductile iron and are used for underground water, gas and sewage lines. Galvanised iron pipes (GI pipes) are used for conveying raw water and distribution of treated water. Various stainless steels (SS-304, SS-316, etc.) are used for pipes and tubes because of their particular corrosion resistance to process chemicals.

Non-ferrous pipes and tubes are made from copper and copper alloys-brass, bronze and cupronickel, nickel and nickel alloys – monel and inconel, lead and aluminium. These materials (including stainless steels) are relatively expensive but are selected for pipes and tubes because of their particular resistance to process chemicals, good heat transfer and tensile strength at higher temperatures.

Non-metallic pipes and tubes are made of materials such as plastics, glass, rubber and ceramics. Plastic pipes are used for transporting corrosive fluids and especially useful for handling corrosive or hazardous gases and dilute mineral acids. They are used as all plastic pipe, as filled plastic materials (glass-fibre reinforced) and as lining or coating materials. Commonly used plastics are polypropylene, rigid polyvinyl chloride, acrylonitrile butadiene styrene and polyvinylidene chloride (Saran).

Lining carbon steel pipe with a material able to withstand chemical attack permits its use to carry corrosive fluids. Lining materials include rubber, plastics, glass and metals and thus we have rubber lined pipe, lead lined pipe, etc.

8.3 VARIOUS TYPES OF FITTINGS AND APPLICATIONS

Fittings :

Fittings refer to the pieces that can be employed in pipelines for one of the following purposes :

(a) Joining two pipe pieces., e.g., coupling/socket, union, nipple, hex nipple and barrel nipple.

(b) Changing the pipeline diameter, e.g., reducer.

(c) Termination of the pipeline, e.g., plug.

(d) Changing the direction of flow, e.g., elbow, bend.

(e) Branching of the pipeline, e.g., tee, cross.

(f) Controlling the flow through a pipeline, e.g., valves.

Fittings may be threaded or flanged. Common fittings are shown in Fig. 8.1.

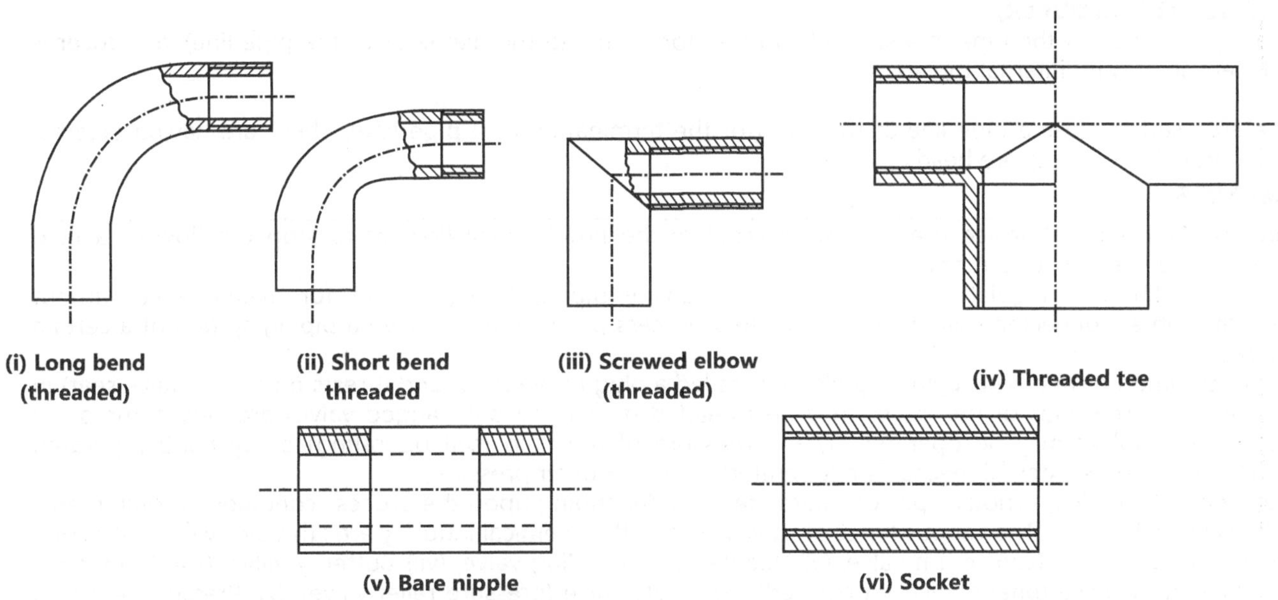

(a) Threaded pipe fittings (two lines that are very close to each other indicate threading)

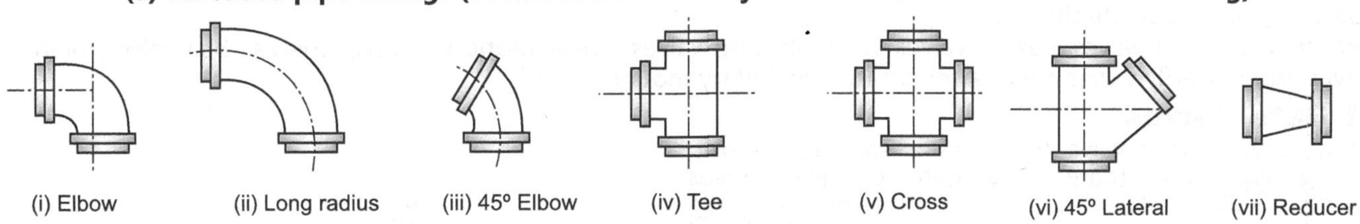

(b) Flanged pipe fitting

Fig. 8.1

For using screwed fittings, the ends of the pipe are threaded externally with a threading tool. Polytetrafluoroethylene tape is wrapped around the threaded end to ensure a good seal before making a threaded joint. Screwed fittings are used with pipe smaller than 3 in. Lengths of pipe larger than 2 in are usually connected by flanges or by welding. Flanges are metallic circular discs bolted together and compressing a gasket interposed between their faces. The flanges are attached to the pipe by screwing or by welding or brazing. Fittings are available in the same materials as piping.

BEND :

Bends are used to change the direction of fluid flow. They are having external threads. The frictional losses in bends are small compared to those in elbows due to smooth change in the direction of flow.

ELBOW

Elbows are used to change the direction of fluid flow. There is sudden change in the direction of flow and therefore, pressure drop is more in elbows than in bends.

TEE

Pipes are branched off at right angles by means of fittings such as tees. Tees are having internal threads.

NIPPLE

It is a small piece of pipe with external threads throughout (i.e., threaded on the outside). It is screwed into the internally threaded ends of the pipes to be connected. It is used to join two pieces of pipe. In case of bare nipple there are threads only at the ends of small pipe piece (a bare nipple is a piece of pipe of length 75 mm, 100 mm, 150 mm, etc. having threads at the ends).

SOCKET

A socket also called as a coupler is used to join two pieces of pipe with their axes in alignment. It is a small piece of pipe with internal threads throughout. The socket is screwed on half way on the threaded end of one pipe and then the end of the other pipe is screwed into the socket in the remaining half.

REDUCING SOCKET (REDUCER)

To change the size of the pipe in a straight run (i.e., for changing the diameter of the pipe line), a reducer is used. Reducers are having internal threads.

PLUG

These are used to close a pipe line at the end (for the termination of a pipe line). These are having external threads and provided with square heads.

8.4 VALVES

- A valve is a mechanical device used either to regulate the flow/control flow or to stop the flow of a fluid through a pipe line or in or out of a vessel.
- Valves, in addition to regulate the flow, serve to isolate piping or equipment for maintenance without interrupting other connected equipments, regulate a process pressure and relieve a piping system of a certain pressure.
- Valves in common have a flow regulating element called a plug (valve plug) and it rests on a seat (valve seat) in the valve body. The ends of the valves may be threaded or flanged but flanged valves are very common in industries. The valves may be operated by the pressure of the fluid (self operating) or by hand (operated manually) or by some external mechanism (operated by means of air pressure).
- In the chemical industry, various types of valves are used depending upon the process conditions/requirements and fluids to be handled. The various types of valves used in the chemical industry are : (i) Gave valve, (ii) Globe valve, (iii) Ball valve, (iv) Non-return valve, (v) Needle valve, (vi) Plug valve, (vii) Butterfly valve, (viii) Diaphrams valve, (ix) Control valve (pneumatically operated), (x) Safety valve (pressure relief valve), (xi) Pressure reducing valve and (xii) Solenoid valve. Gate, plug and ball valves are shut-off valves whose purpose is to close-off the flow and not to regulate the flow.
- Gate valves, globe valves, needle valves are multi-turn valves (linear motion valves), whereas ball valves, plug valves and butterfly valves are quarter-turn valves (rotary motion valves).

8.4.1 Gate Valves

- A gate valve is a valve that uses a solid to tapered wedge-type disk (gate). The gate or disk moves perpendicular to flow to start or stop the flow in piping. In service, this valve is in either fully open or fully closed position.
- Gate valves are used primarily for on-off applications. These are suited for high pressure and high temperature use with wide variety of fluids but are not suited for slurries and viscous fluids.
- A gate valve is a valve that provides a straight through passage for the flow of a fluid. A gate is moved up and down within a body by a stem whose axis is perpendicular to the body ends.
- In gate valves, the diameter of the opening through which the fluid passes is nearly the same as that of the pipe and also the direction of flow does not change and as a result the pressure drop through them is very low in fully open condition. Therefore, gate valves are used to minimise pressure drop in the open position and to stop the flow instead of to regulate the flow of fluid.

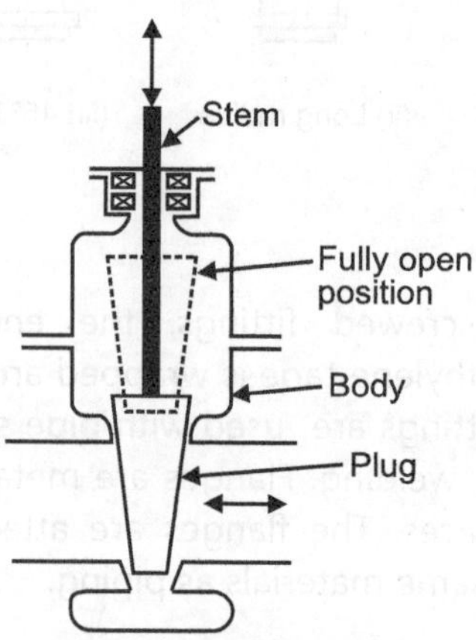

Fig. 8.2 : Gate valve

8.4.2 Globe Valves

- A globe valve is used to control/regulate the flow of fluid. It is the most commonly used valve for efficient regulation of a critical service. It is extensively used for automatic process control and for high temperature applications. These valves are not normally used for on-off service and are commonly used in steam lines.
- A globe valve is a valve having generally a spherical body in which the axis of a stem is perpendicular to the body ends. The pressure drop through globe valves is much greater than that for gate valves due to change in the direction of flow.

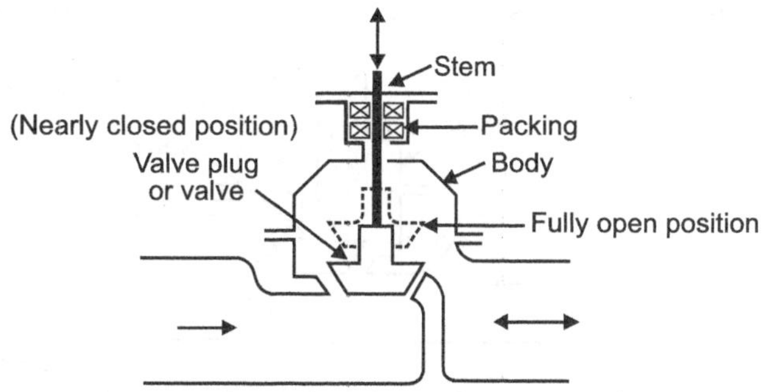

Fig. 8.3 : Globe valve

8.4.3 Ball Valves

- Ball valves are quarter turn operated valves and use a ball to control the flow.
- It is used for flow control and on-off service/applications. It is widely used in industry where conditions of corrosion plus temperature/pressure exist. Ball valves are simple, compact and quick opening and they are operated through 90°. They are good for conditions which require to be fire safe. They are constructed in a very wide size and temperature-pressure ranges. These valves are used for corrosive liquids, cryogenic liquids and gases, water, oil, slurries, etc.
- It is a valve in which a spherical closure element (ball) having a port (passage) through it is turned substantially through 90° from close to open position.
- Easy to maintain, tight shut-off, quarter turn operation, quick operating and low pressure drop.
- When the valve is in open position, the port (through hole) is in line with fluid flow and hence the fluid passes through it. In a closed position the port is perpendicular to the flow.

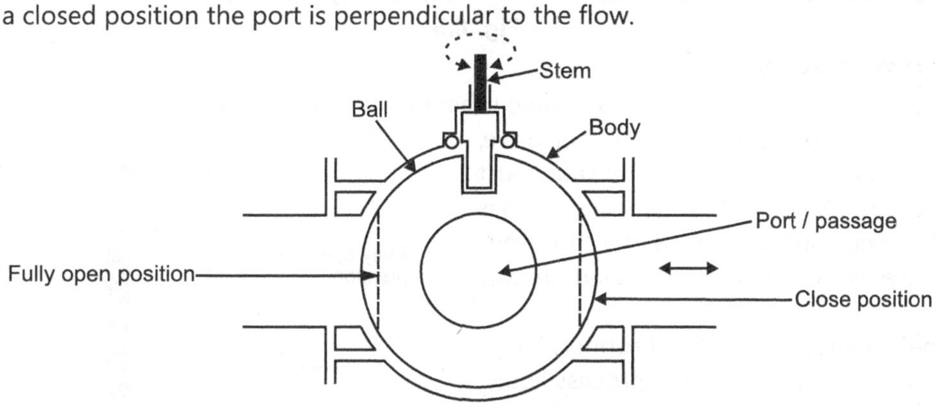

Fig. 8.4 (A) : Ball valve (fully closed)

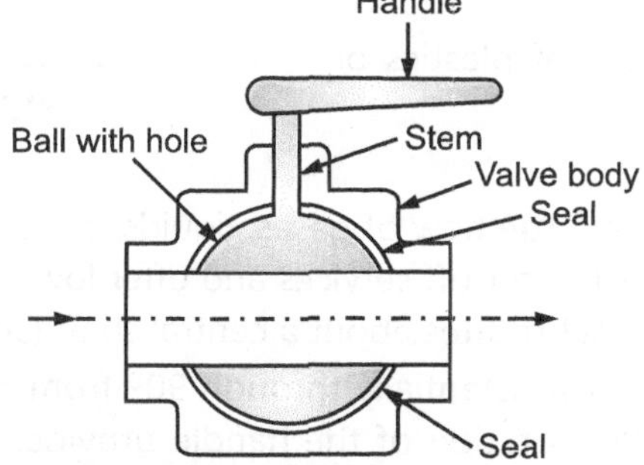

Fig. 8.4 (B) : Ball valve (valve fully open-ON)

8.4.4 Check Valves

- A check valve is a valve which allows fluid to pass/flow through it in only one direction. Therefore, this valve is used when unidirectional flow is desired.

- This valve can be mounted in a horizontal or vertical line and operates automatically. The valve is opened by the pressure of the fluid passing through it and closed by any reversal of flow/weight of a check mechanism when the flow ceases. Check valves (also called non-return valves) may be swing type or lift type.

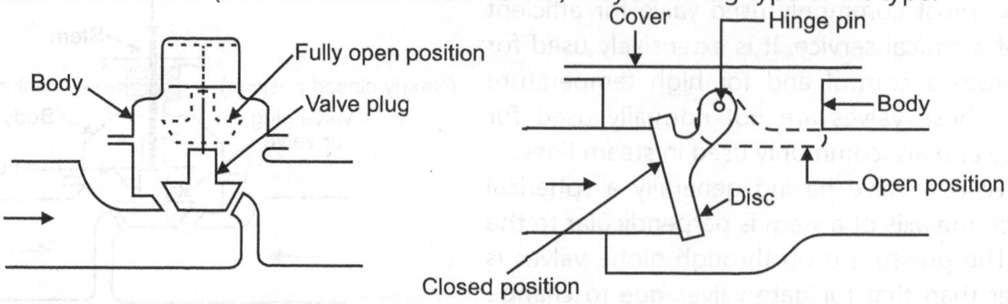

(a) Lift check valve (globe), horizontal line **(b) Swing check valve**

Fig. 8.5

- Vertical lift check valves are used in vertical lines where the flow is normally upward. Globe check valves are used in horizontal lines.

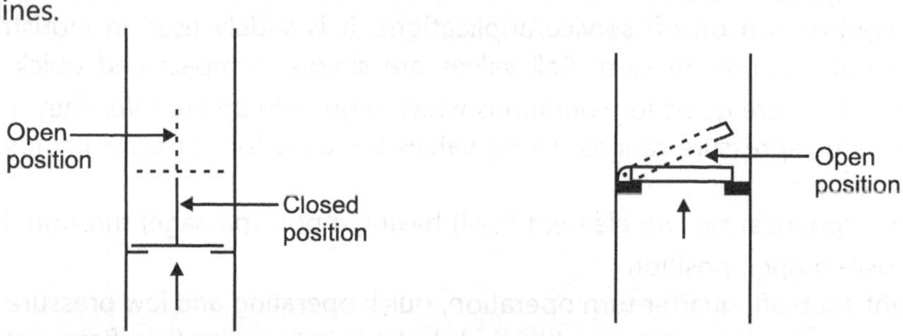

(c) Lift check valve, vertical **(d) Swing type (foot valve)**

Fig. 8.5

8.4.5 Diaphragm Valves

- A diaphragm valve is a valve that contains a flexible rubber (natural or synthetic) diaphragm as a plug.
- It is a glandless type of valve used for corrosive, volatile and toxic fluids especially where leakage must be avoided. They are also suited for handling slurries and are used for regulating the flow and also for on-off service. These valves may be installed in any position.
- In this valve, the diaphragm (usually of rubber) keeps the working parts in isolation from the process liquid. The lined bodies act as an in-built gasket for the connecting flanges.
- Diaphragm valves are constructed out of plastics or metals (rubber lined).

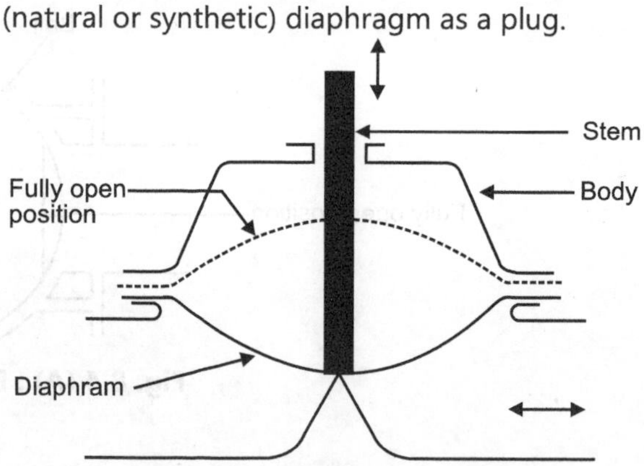

Fig. 8.6 : Diaphragm valve

8.4.6 Butterfly Valves

- Butterfly valves are suited for handling large flow of gases, liquids, and slurries at relatively low pressures. These valves are suitable for throttling as well as on-off services and offer low pressure drop.
- It has a disk-shaped closure element that rotates about a central shaft (stem) and has a short circular body.
- It is a valve in which the disk is turned substantially through 90º from close to open position, on an axis right angles to that of the valve port. A 90° rotation of the handle provides a complete closure or opening of the valve.
- Butterfly valves occupy less space in the line than any other valves, simple in construction, and quick opening. These valves are also used for cooling water, air, gases, fire protection lines in an industry.

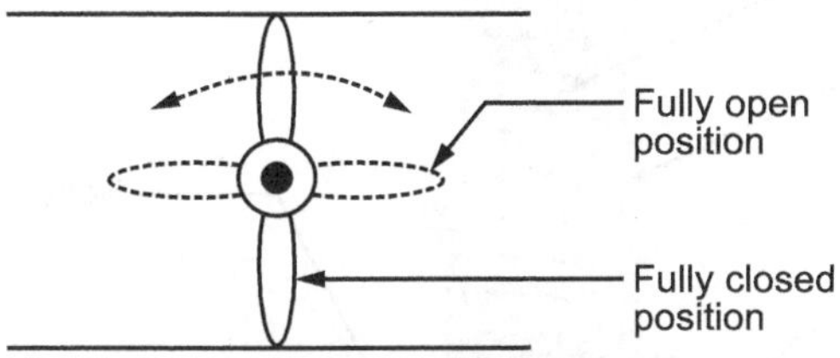

Fig. 8.7 (A) : Butterfly valve

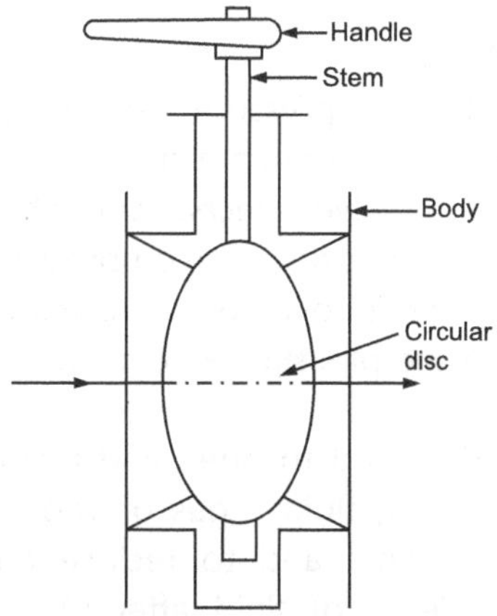

Fig. 8.7 (B) : Butterfly valve

8.4.7 Plug Valves

- A plug valve is a quarter-turn operated valve - that uses a tapered plug to start or stop the flow. The plug has a rectangular port/passage through which a fluid flows/passes when it is in line with the direction of flow. It can be used for high temperature, high pressure services. Plug valves consist of body, bonnet, stem, plug and handle.
- Straight-through flow, positive shut-off, quarter turn operation-open to close, used for multiport operation (i.e., 2, 3, and 4 way), hence simplifying the piping system. Used for general on-off service.
- As it is operated through 90° with the help of a handle, it is easily opened and closed. It is used for water lines (cooling tower water, chilled water) and also for handling fluids with suspended solids, slurries, etc.

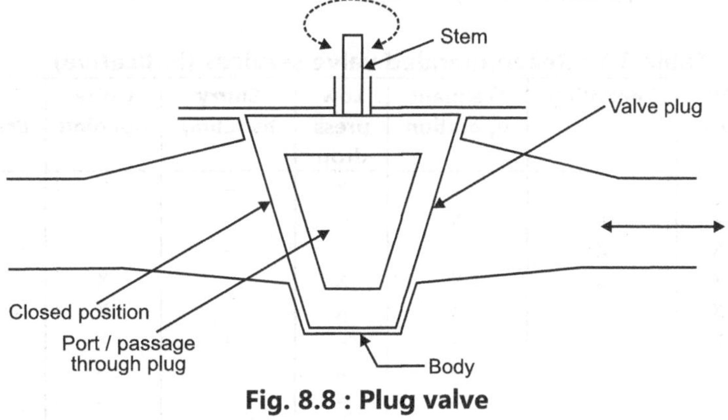

Fig. 8.8 : Plug valve

8.4.8 Control Valves

In automatic process control systems, the most commonly used final control element is a diaphragm motor valve. It consists of a pneumatic diaphragm motor actuator and a process fluid control valve. The actuator positions the valve plug in the orifice in response to a signal from the automatic controller. The valve operated pneumatically may be air to open or air to close type.

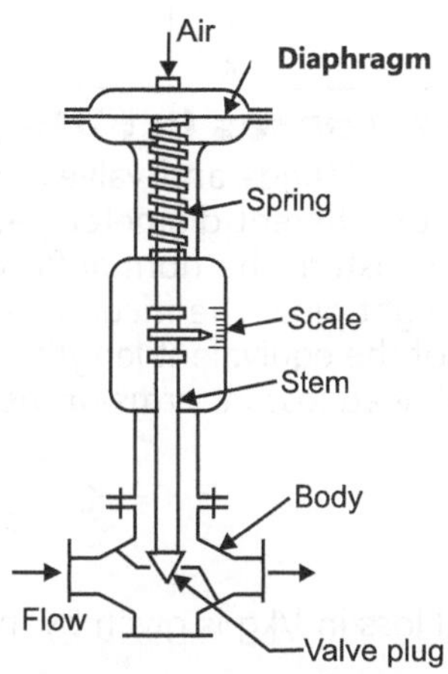

Fig. 8.9 : Control valve

8.4.9 Safety Valve

Chemical plants, boilers, pressure vessels or other system working under pressure may get damaged or destroyed if pressure rises are not controlled.

A safety valve is a valve mechanism which automatically releases a substance (compressible fluid – gas or vapour) from a boiler, pressure vessels or other system, when the pressure exceeds preset limits.

A safety valve is designed to open and relieve excess pressure (or venting fluid – gas or vapour) from vessels or equipments and to reclose and prevent the further release of fluid after normal conditions are restored.

It is a safety device and operates for the purpose of overpressure protection – for protection against excessive pressure only.

It is a mechanical device, mounted vertically on pressure vessels, it closes again when overpressure/excessive pressure ceases, gives protection from overpressure and requires periodic check of calibration.

Two main types of safety valves are : spring loaded safety valves and pilot operated safety valves.

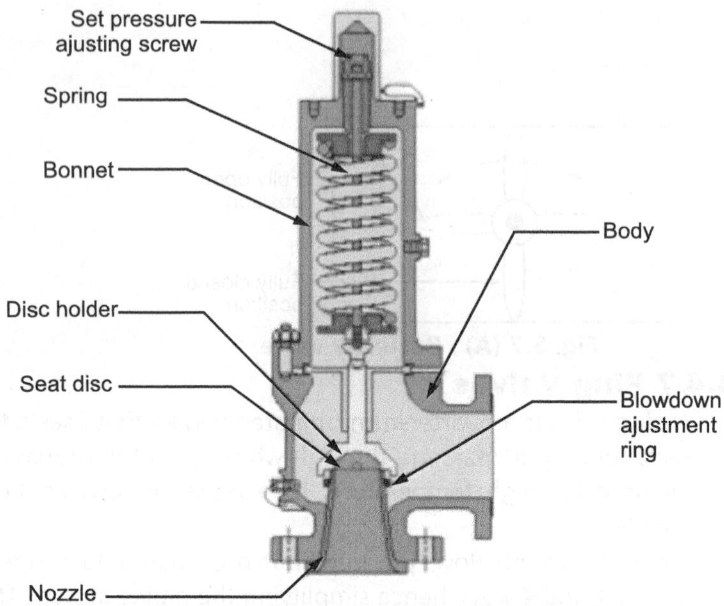

Fig. 8.10 : Spring loaded safety valve

Table 8.1 : Recommended valve services (indicative)

Valve	On-off	Throttling	Frequent operation	Low press drop	Slurry handling	Quick opening	Free draining	Prevent reversal of flow	Prevent over pressure
Gate	X	–	–	X	–	X	X	–	–
Globe	X	X	X	–	–	–	–	–	–
Ball	X	X	X	X	–	X	–	–	–
Plug	X	X	X	X	–	X	X	–	–
Butterfly	X	X	X	X	X	X	X	–	–
Diaphram	X	X	–	–	X	X	X	–	–
Swing check	–	–	–	X	–	–	–	X	–
Lift check	–	–	–	–	–	–	–	X	–
Foot valve	–	–	–	–	–	–	–	X	–
Pressure reducing	–	–	–	–	–	–	–	–	X
Pressure relief/safety valve	X	–	–	–	–	–	–	–	X

8.5.1 Friction Losses in Fittings and Valves

Various types of fittings and valves are used in industrial piping systems to change the direction of flow, for connecting pipes of different diameters, etc. and valves are used to control the flow or to stop the flow of a fluid. Fittings and valves disturb the normal flow-line and cause friction and may lead to greater frictional loss than that caused by the straight pipe. The frictional loss due to fittings and valves may be given either in terms of the velocity head or in terms of the equivalent length.

The frictional head loss in terms of the velocity head is given by the following equation :

$$h_{ff} = K_f \frac{u_1^2}{2g} \qquad \ldots (8.1)$$

The frictional loss in J/kg is given by $h_{ff} = K_f' \frac{u_1^2}{2}$ ... (8.2)

where K_f is the loss factor for the fitting and u_1 is the average velocity of the fluid in the piping leading to the fitting.

8.5.2 Equivalent Length of Pipe Fittings

Another way to express the frictional loss in fittings and valves is through the equivalent length of fittings. The **equivalent length** of a fitting is *that length of straight pipe of the same nominal size as that of the fitting, which would cause the same friction loss as that caused by the fitting.*

The equivalent length of a fitting/valve is usually expressed as a certain number of pipe diameters. Thus, if a bend has an equivalent length of X diameters and if the nominal pipe diameter is 5 cm, then the bend will cause the frictional loss equivalent to that caused by the straight pipe of length 5 X cm.

The equivalent length of a fitting is to be added to the length of a straight pipe to get the total equivalent length of a flow system composed of the straight pipe and the fitting.

Equivalent lengths of some pipe fittings are :

Fitting	Equivalent length, pipe diameters
45° Elbow	15
90° Elbow	32 (upto D = 40 mm)
Tee	60 to 90
Coupler and union joint	Negligible
Gate valve (fully open)	10
Globe valve (fully open)	300

Form-friction losses in the Bernoulli equation

Consider a flow system for the flow of an incompressible fluid as shown in Fig. 8.11, which is composed of two enlarged headers, the connecting pipes and a globe valve.

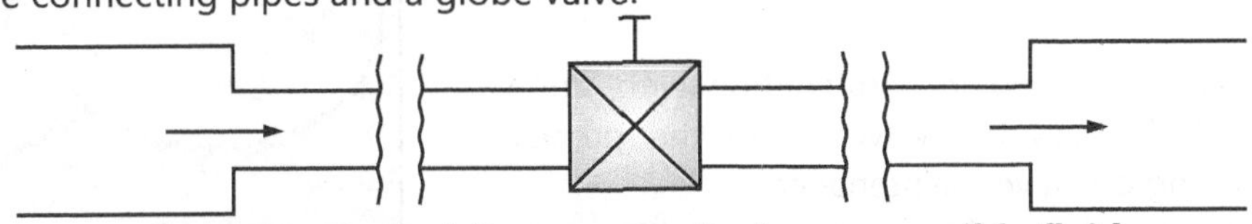

Fig. 8.11 : Typical flow system for incompressible fluid

For obtaining the total frictional loss, the losses in the expansion and contraction headers and the valve must be combined with the skin friction loss of the straight pipe.

$$\text{Skin friction loss in J/kg} = \frac{4\,fL\,u^2}{2\,D} \qquad \text{... (8.3)}$$

where D is the diameter of the straight pipe, u is the average velocity of the fluid through the pipe and L is the length of the straight pipe.

$$h_f = \text{total friction loss in J/kg} = \left[\frac{4fL}{D} + K_c' + K_e' + K_f'\right]\frac{u^2}{2} \qquad \text{... (8.4)}$$

Assume that the piping system under consideration involves a straight pipe of length L, two globe valves and one elbow. Further assume that the equivalent length of the globe valve is given as X pipe diameters (e.g., 190 D, so X = 190) and that of the elbow is given as Y pipe diameters.

In this situation, the total friction loss is obtained as follows :

Equivalent length of elbow = YD

Equivalent length of globe valve = XD

Total equivalent length of the piping system = L' = L + YD + 2 XD

$$h_f = \text{total friction loss in J/kg} = \frac{4\,fL'\,u^2}{2\,D} \qquad \text{... (8.5)}$$

The total head loss due to friction in m of flowing fluid for the system of Fig. 8.11 is given by

$$h_f = \left[\frac{4\,fL}{D} + K_c + K_e + K_f\right]\frac{u^2}{2g} \qquad \text{... (8.6)}$$

where u is the average velocity of the fluid through the pipe of diameter D and f is the Fanning friction factor.

8.6 RUPTURE DISC

- A rupture disc is a non-reclosing pressure relief device that protects the system (e.g., pressure vessel) on which it is installed from over pressurisation.
- A rupture disc is a pressure protection device commonly used in industrial plants.

- Rupture discs are used to protect against overpressure or vacuum within process plants for the protection of man, machine and environment.
- Rupture discs may be round or square and are either flat or domed. They are made of metals (e.g. stainless steel, graphite, inconel and hastelloy) or plastics (PTFE or FEP), from one or more layers.
- A rupture disc may directly be installed between flanges or it may be inserted into a rupture disc holder which is then mounted between flanges. In some cases, the rupture disc is soldered/welded into the holder.
- A rupture disc may be conventional (forward) acting or reverse acting. Conventional discs have a flat or concave surface exposed to the pressure (the dome faces away) and bursting occurs when the pressure overcomes the mechanical resistance of the material. In case of reverse acting rupture discs, a convex surface, i.e., the dome of the rupture disc faces the process (i.e., it is exposed to the pressure).
- The reverse acting rupture disc is used for high pressure applications while the forward acting rupture disc is used for low to moderate pressures.
- A rupture disc provides a leak-tight seal within a vessel until the internal pressure rises to a predetermined level. At that point, the rupture disc bursts preventing damage to the vessel. Rupture discs provide instant response to an increase in the system pressure, but once the disc has ruptured it will not reseal.
- Rupture discs are used where extremely rapid pressure rise may be encountered and where safety valves might not be effective and are used at very high pressure in contact with toxic or expensive fluids when leaks are not allowable.
- Rupture discs are also used in combination with safety relief valves. This isolates the valves from the process resulting into saving on valve maintenance.
- When the normal operating pressure within the vessel/equipment (i.e., the process pressure) reaches the bursting pressure of the disc, the disc ruptures and releases the pressure.

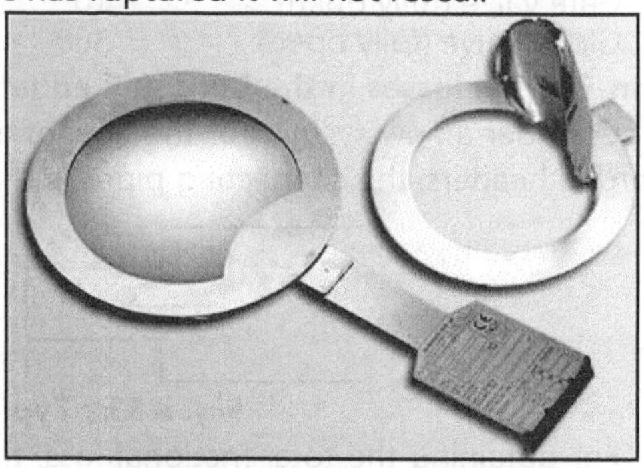

Fig. 8.12 : Rupture disc

- Major advantages of rupture discs include : simple in construction, has no moving parts, provides leak-tight seal (no leak during operation), low cost, reliable, no maintenance, does not require periodic calibration, full opening and faster acting than other pressure relief devices.

Rupture discs (Burst discs) are used in defense, aviation, petrochemical, pharmaceutical food processing, oil and gas and chemical industries.

Practice Questions

1. Distinguish between pipe and tube.
2. Define schedule number and BWS.
3. Define allowable stress, maximum allowable working pressure and operating pressure.
4. Give the function of nipple, plug and elbow.
5. Draw a neat sketch of (i) Reducer, (ii) Elbow, (iii) Socket.
6. Define valve and mention its types.
7. Draw a sketch of globe valve and give its applications.
8. Draw a sketch of diaphragm valve and give its applications.
9. Draw a sketch of swing type check valve (non-return valve).
10. Write in brief on rupture disc.
11. Give the materials of construction of rupture disc.
12. Define equivalent length of pipe fitting.

❑❑❑

CHAPTER NINE

LIQUID PUMPING DEVICES [Transportation of Liquids]

9.1 PUMPING OF FLUIDS : IMPORTANCE OF PUMPS

- In the chemical process industries for the transportation of fluids from one process equipment to another, or through long pipes or ducts, we have to use some form of a mechanical pumping device. Such a mechanical device may be a *pump, fan, blower or compressor, which increases the fluid energy.* For the transportation (handling) of liquids, pumps are used, whereas for the transportation of gases, fans, blowers or compressors are used.
- The power required by a pump, depends on the height through which the liquid is raised, flow rate, length and diameter of the pipe and the density and viscosity of the liquid.
- Pumps are used to keep materials (liquids) moving through the production processes. Pumps are of different types and sizes. Almost all industries need pumps.
- The pumping of crude oil over a long distance, feeding water to a boiler, transfer of raw materials say for example, sulphuric acid, benzene, etc. from a bulk storage to a processing area, industrial cleaning, vehicle washing, spray drying, building services (agriculture), arigation, reverse osmosis, power generation, water supply (gardening), waste water treatment, sewage treatment, transfer of petrol from its underground storage tank to the fuel tank of a car are some examples of use of pumps.
- Pumps are used in almost all industries. The use varies from industry to industry but the common thing is to raise a liquid to a higher pressure. Pumps are used in industries such as chemical, petrochemical, pharmaceutical, food processing, dairy, alcohol, pulp and paper, refineries, oil and gas, etc.
- The liquid to be pumped may have clear or may contain suspended particles. It may be corrosive or non-corrosive and may have a low or high viscosity.

9.2.1 Classification of Pumps

A pump is a machine which converts mechanical energy supplied to it from some external source into pressure energy which is used to lift a liquid from a lower level to a higher level.

A pump is a machine used for moving liquids, solution or slurries by mechanical means.

Pumps are classified as :

(i) Dynamic pumps/kinetic pumps. (ii) Displacement pumps/positive displacement pumps.

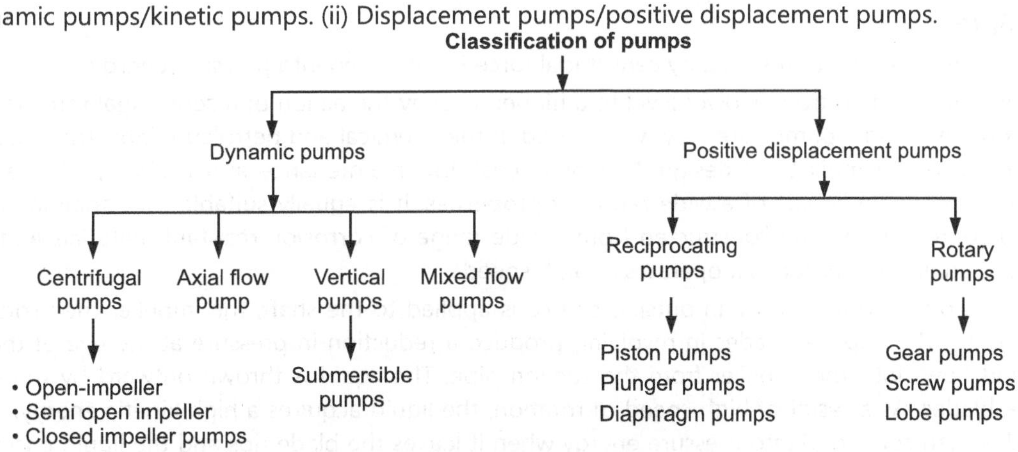

Fig. 9.1 : Classification of pumps

In centrifugal pumps, the kinetic energy of the liquid to be pumped is increased by the action of centrifugal force and is then converted into pressure energy that is needed to pump the liquid.

In positive displacement pumps, a definite or fixed quantity of liquid is pumped in every revolution.

A positive displacement pump is that pump which delivers/discharges a definite/fixed quantity of liquid during the displacement of its piston/plunger/gear/lobe (i.e., during each stroke or each revolution).

Since a fixed quantity of liquid is pumped after each revolution they can develop a very high pressure if the delivery valve or line is closed.

Positive displacement pumps are classified as :

1. Reciprocating pumps, e.g., piston pump, plunger pump, etc.
2. Rotary pumps, e.g., gear pump and mono pump.

In reciprocating pumps, the displacement is by reciprocation of a piston or plunger, while in rotary pumps, the displacement is by rotary action of a gear or vane.

9.2.2 Factors which Influence the Choice of Pump for a Particular Operation

While selecting a pump for a particular duty we have to consider the following factors :

(i) The quantity of liquid to be handled.
(ii) The head against which the liquid to be pumped/raised.
(iii) The nature of liquid to be handled (viscosity, clear liquids, suspensions, corrosiveness).
(iv) The nature of power supply.
(v) The method of operation – continuous or intermittent.
(vi) The flow rate required.
(vii) The pressure on delivery.
(viii) Cost and mechanical efficiency.

9.2.3 Pumps Measurement of Performance

(i) Capacity : It is the rate at which the fluid is pumped by a fluid moving device. It is expressed in various units – based on the type of pumping device. For liquids, it is expressed in litres per minute or gallons per minute and for gases, it is expressed in cubic feet per minute at inlet conditions of gases to the machine.

(ii) Overall efficiency : It is the *ratio of the useful hydraulic work performed to the actual work input, irrespective of the type of drive.*

9.3.1 Centrifugal Pumps

9.3.1.1 Principle

The kinetic energy imparted to the fluid by centrifugal force is converted into pressure energy.

A pump which lifts a liquid from a lower level to a higher level by the action of a centrifugal force is called as a **centrifugal pump**. Centrifugal pumps are very widely used in the chemical and petroleum industries because of its many advantages such as simplicity of design, low initial cost, low maintenance and flexibility of operation. The centrifugal pump can handle liquids of a wide range of properties. It is equally suitable for handling suspensions with a high solid content. It may be constructed from a wide range of corrosion resistant materials and it may be directly coupled to an electric motor as it operates at high speeds.

In a centrifugal pump, power from an outside source is applied to the shaft, the impeller then rotates within the stationary casing. The impeller blades in revolving produce a reduction in pressure at the eye of the impeller. Due to this, liquid flows into the impeller from the suction pipe. The liquid is thrown outward by the centrifugal action along the blades. As a result of high speed of rotation, the liquid acquires a high kinetic energy. The kinetic energy acquired is then converted into pressure energy when it leaves the blade tips and the liquid passes into the volute chamber and finally it is discharged through the outlet (discharge) on the pump. This action of the centrifugal pump is shown in Fig. 9.2.

9.3.1.2 Construction or Component Parts of a Centrifugal Pump

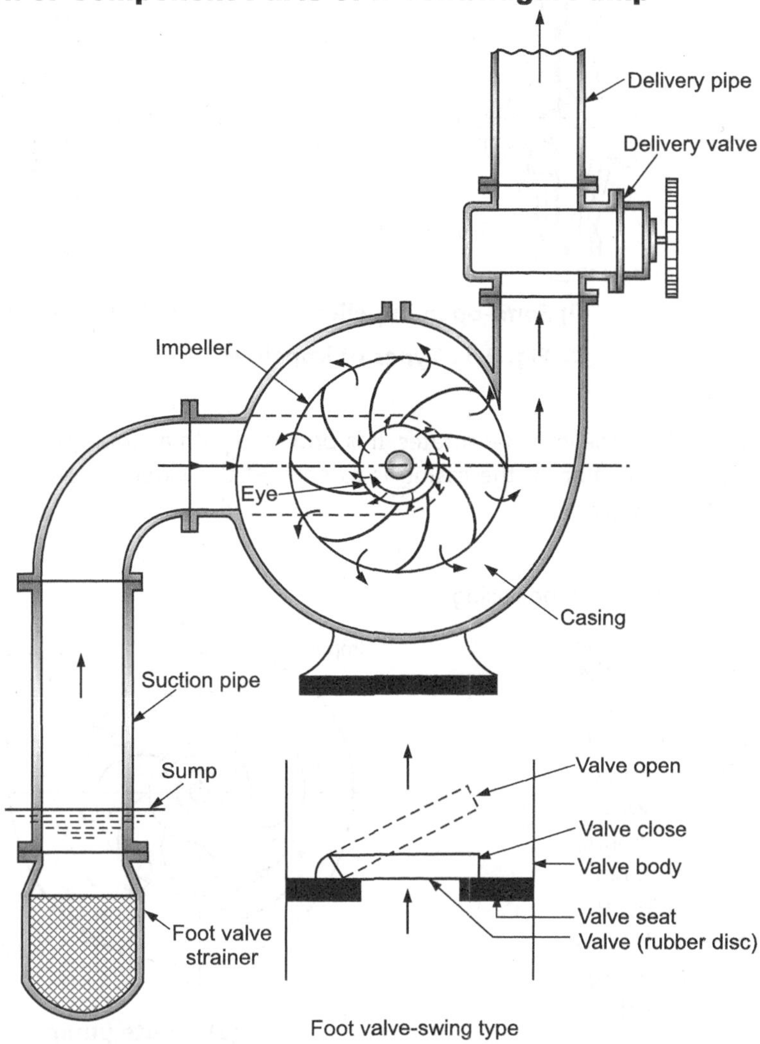

Fig. 9.2 : Component parts of a centrifugal pump

Impeller

It is a wheel or rotor that is provided with a series of curved blades or vanes. It is mounted on a shaft which is coupled to an electric motor (an external source of energy). The blades/vanes are shaped in such a way that the flow within the pump is as smooth as possible. The impeller is the heart of the centrifugal pump.

Impeller types :

(i) Open impeller, (ii) Semi-open impeller, and (iii) Closed or shrouded impeller.

The open impeller has the blades fixed to a central hub. Such impellers are suited for pumping liquids containing suspended solids, e.g., paper pulp, sewage, etc.

The closed impeller has the blades held between two supporting plates / shrouds (crown plate and base plate). This impeller provides better guidance for the liquid and is more efficient. This type of impeller is suited for pumping clear liquids (liquids containing no suspended particles, dirts, etc.).

The semi-open impeller has only one plate (base plate) and no crown plate (i.e., it has a plate on one side of the blades/vanes). Such impellers are suitable for liquids containing some solid particles or dirts.

For viscous liquids or liquids containing solids, open or semi-open type impellers are used. The most efficient impeller is the closed or shrouded type. The impeller may be a single suction type or double suction type. In the former type, the liquid enters the impeller from one side; while in the latter type, the liquid enters from both the sides.

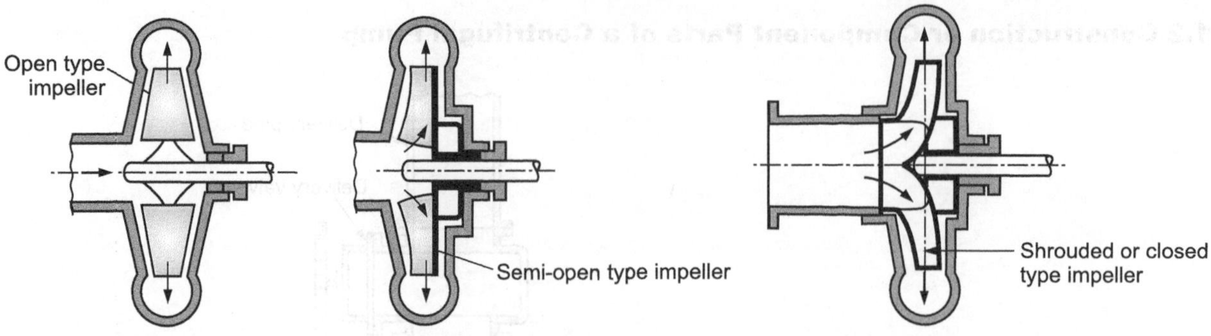

(i) Open impeller **(ii) Semi-open impeller** **(iii) Closed impeller**

Fig. 9.3 : Types of impellers

Casing

It is an airtight chamber in which the impeller rotates. It is provided with an inlet (suction) and outlet (discharge) for the liquid to be pumped. The function of the casing is to convert the kinetic energy imparted to the liquid by the impeller into useful pressure energy.

Types of casing :

(i) volute type casing and (ii) diffuser type casing.

In a volute type casing, the liquid is discharged by the impeller into the volute - a chamber of gradually increasing cross-sectional area towards the outlet [Fig. 9.4 (a)]. In the volute, the fluid velocity decreases gradually thereby increasing fluid pressure, i.e., the volute converts the kinetic energy of the liquid imparted by the impeller into pressure energy. In this design, a considerable loss of energy takes place due to formation of eddies.

A vortex chamber is an improved version of the volute design. In this case, a circular chamber is provided between the impeller and the volute chamber [Fig. 9.4 (b)]. This design reduces eddies to a considerable extent with increase in efficiency.

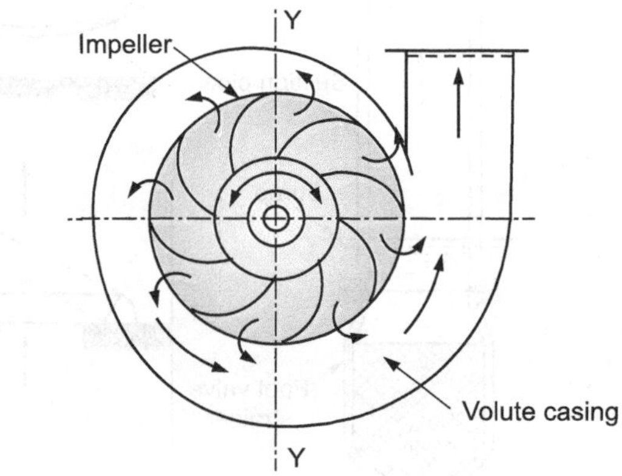

(a) Volute pump

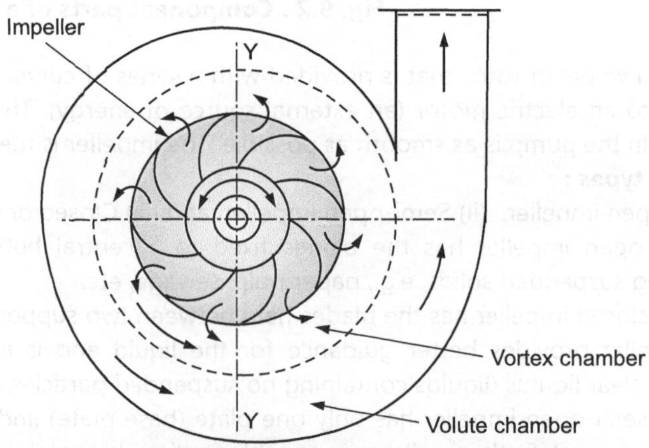

(b) Volute pump with vortex chamber

Fig. 9.4 : (a) Volute type casing
(b) Vortex type casing

In a diffuser type casing (turbine pump), guide vanes or diffusers are interposed between the chamber and the impeller. The impeller is surrounded by a series of guide vanes mounted on a ring called diffuser ring as shown in Fig. 9.5. The conversion of kinetic energy into pressure energy is more efficient with this type compared to the volute type. There is a gradual change in the direction of fluid so that the losses are kept minimum.

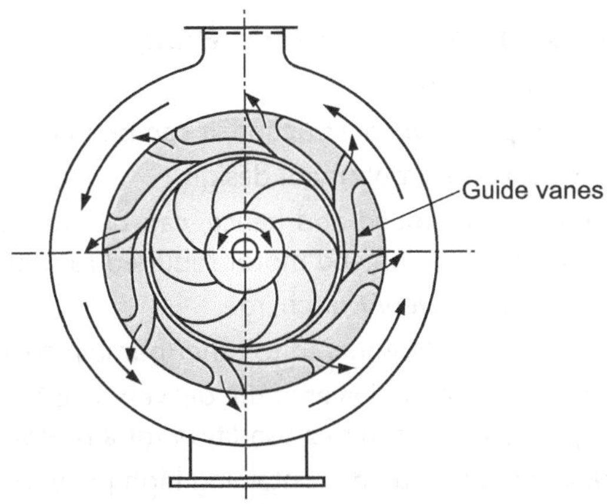

Fig. 9.5 : Diffuser type casing

Other component parts of the centrifugal pump are : shaft, bearing, stuffing box, or mechanical seal, etc. Shaft must be of a corrosion resistant material and must have good mechanical properties. The shaft transmits power from the drive unit to the impeller. Sometimes, a shaft sleeve of corrosion resistant material is provided over the corrosive shaft material from strength criteria.

Shaft is mounted on bearings which handle shaft load without excessive wear. Stuffing box is a means of reducing/avoiding leakage which would otherwise occur at the point of entry of the shaft into the casing. With this the maintenance costs are high. A considerable reduction in expenditure on maintenance can be effected at the price of a small increase in the initial cost by fitting the pump with a mechanical seal. For high pressures and corrosive fluids, mechanical seals are used.

9.3.1.3 Working of a Centrifugal Pump

In the operation of a centrifugal pump before the pump is started, priming of the pump is done. In the priming operation, the suction pipe, pump casing, and portion of the delivery pipe upto a delivery valve are completely filled with the liquid to be pumped so that all the air, gas or vapour from this portion of the pump is expelled out and no air pocket is left. In presence of even very small air pocket in any of these portions, pump will not discharge the liquid. The need to do priming of pump is due to the fact that the pressure generated by a centrifugal pump impeller is directly proportional to the density of fluid that is in contact with it. Therefore, if the impeller is rotated in the presence of air, only negligible pressure would be produced and thus no liquid will be lifted by the pump.

After the pump is primed properly, the delivery valve is kept close and power from an outside source (electric motor) is applied to the shaft. The delivery valve is kept close in order to reduce the starting torque for the motor. The impeller then rotates within the stationary casing. The rotation of the impeller produces a forced vortex which imparts a centrifugal head to the liquid and thus results in an increase of pressure throughout the liquid mass. As long as the delivery valve is closed and impeller is rotated, there will be just churning of the liquid within the casing. When the delivery valve is opened, the liquid is made to flow in an outward radial direction thereby leaving the vanes of the impeller at the outer circumference with high velocity and pressure. Due to centrifugal action, a partial vacuum is created at the eye of the impeller. This causes the liquid from the sump/reservoir (at atmospheric pressure) to flow through the suction pipe to the eye of the impeller thereby replacing the liquid which is being discharged from the entire circumference of the impeller. The high pressure of the liquid leaving the impeller is utilised in lifting the liquid to the required height through the delivery pipe.

During the operation, liquid receives energy from the vanes which results in an increase in both pressure and velocity energy. As such the liquid leaves the impeller with a high absolute velocity. In order that the kinetic energy corresponding to the high velocity of liquid leaving is not wasted in eddies and efficiency of the pump thereby lowered, it is essential that the high velocity of the leaving liquid is gradually reduced to the lower velocity in the delivery pipe, so that a large portion of the kinetic energy is converted into useful pressure energy. This is usually achieved by shaping the casing such that the leaving liquid flows through a passage of gradually expanded area. The gradual increase in the flow area of the casing also helps in maintaining uniform flow velocity throughout.

9.3.1.4 Advantages of Centrifugal Pumps

1. It is simple in construction.
2. Due to its simplicity of construction, it can be made in a wide range of materials.
3. Low initial cost and simplicity of design.
4. It operates at high speed and hence, can be coupled directly to an electric motor. In general, higher the speed, smaller the pump and motor required for a given duty.
5. It gives a steady delivery / discharge.
6. Lower maintenance (compared to other pumps) costs.
7. It does not get damaged even if the delivery line becomes blocked, or the delivery valve is closed, provided the pump does not run in this condition for a prolonged period.
8. It can handle readily liquids containing high proportions of suspended solids.
9. For equal capacity, the centrifugal pump is much smaller than any other type of pump. Therefore, it can be made into a sealed unit with the driving motor and immersed in the suction tank.

9.3.1.5 Disadvantages of Centrifugal Pumps

1. It is not usually self-priming.
2. It operates at low efficiencies (50 - 65%).
3. It cannot handle very viscous liquids efficiently.
4. It does not develop a high pressure. Multistage pumps will develop greater pressure heads but they are much more expensive and cannot be made into corrosion resistant materials because of their greater complex construction.
5. If a non-return valve is not provided in the delivery or suction line, the liquid will run back into the suction tank (reservoir) as soon as the pump stops.

9.3.1.6 Applications of Centrifugal Pumps

These pumps are used in water supply (domestic and industrial), waste water treatment, sewage treatment, irrigation (agriculture), building services (water in buildings), fire protection systems, power plants, fisheries and in industries such as chemical, petrochemical, pharmaceutical, dairy, food processing, oil refineries, etc.

9.3.2 Positive Displacement Pumps

Positive displacements are classified as :

1. Reciprocating pumps, e.g., piston pump, plunger pump and diaphragm pump.
2. Rotary pumps, e.g., gear pump, lobe pump and mono/screw pump.

A positive displacement pump is a pump which provides (discharges) a fixed flow rate irrespective of pressure changes within the system.

Principle of positive displacement pumps :

A positive displacement pump makes a liquid to move by trapping a fixed amount of the liquid and forcing (displacing) that trapped volume into the discharged pipe.

9.3.2.1 Reciprocating Pumps

Construction / Components of a Reciprocating Pump :

A reciprocating pump essentially consists of a piston or plunger which moves to and fro i.e., reciprocates back and forth in a close fitting stationary cylinder. Fig. 9.6 shows a reciprocating pump.

The cylinder is connected to suction and delivery pipes. Each of these pipes are provided with a non-return valve called suction valve and delivery valve respectively. A non-return valve permits unidirectional flow. Thus, the suction valve allows the liquids only to enter the cylinder and the delivery valve allows only its discharge from the cylinder. A piston or plunger is connected to a crank by means of a connecting rod. The crank is rotated by a driving engine or electric motor. When the crank is rotated by the drive, the piston or plunger moves to and fro (forward and backward movement) in the cylinder. Reciprocating pumps are provided with an air vessel at the discharge to obtain a uniform delivery/discharge from them.

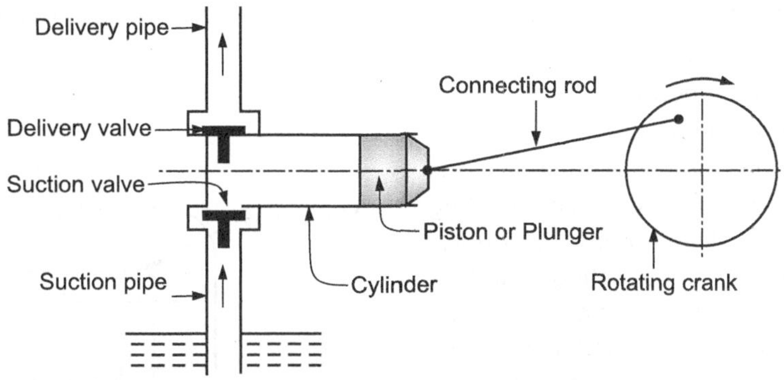

Fig. 9.6 : Single acting reciprocating pump

Working of a Reciprocating Pump :

Assume that the piston or plunger is initially at its extremely left position (i.e., completely inside the cylinder). If the crank rotates through 180°, then the piston or plunger moves to its extreme right position (i.e., moves outwardly from the cylinder). During the outward movement of the piston or plunger, a partial vacuum is created in the cylinder, which enables the atmospheric pressure acting on the liquid surface in the sump below to force the liquid up the suction pipe and fill the cylinder by forcingly opening the *suction valve*. As during this operation of the pump, the liquid is sucked from the sump, it is known as a suction stroke. Hence, at the end of the suction stroke, the piston or plunger is at its extreme right position, crank is at θ = 180°, the cylinder is full of liquid, the suction valve is closed and the delivery valve is just at the point of opening.

When the crank rotates through further 180° (i.e., θ = 180° to 360°), the piston or plunger moves inwardly from its extreme right position towards left. The inward movement of the piston or plunger causes the pressure of the liquid in the cylinder to rise above atmospheric temperature, because of which the suction valve closes and delivery valve opens. The liquid is then forced up the delivery pipe and raised to the required height. As during this operation of the pump, the liquid is actually delivered to the required height, it is known as a *delivery stroke*. At the end of the delivery stroke, the piston or plunger is at the extreme position, crank has one complete revolution, and both the delivery and suction valves are closed. The same cycle is repeated as the crank rotates.

Types of Reciprocating Pumps

Based on the reciprocating member (pressure component) piston, plunger or diaphragm, there are three types of reciprocating pumps :

1. Piston pumps
2. Plunger pumps
3. Diaphragm pumps

According to the liquid being in contact with one side or both the sides of a piston or plunger, the reciprocating pumps may be classified as :

(i) Single acting pump and

(ii) Double acting pump.

(i) Single acting pump

It is the one in which the liquid is in contact with only one side of a piston or plunger (infront of piston or plunger). This pump has one suction pipe and one delivery pipe and in one complete revolution of the crank, there are two strokes - one suction stroke and one delivery stroke.

(ii) Double acting pump

It is the one in which the liquid is in contact with both the sides of a piston or plunger. This pump has two suction pipes and two delivery pipes. During each stroke, the suction takes place on one side of the piston and other side delivers the liquid. So in case of double acting pump in one complete revolution of the crank, there are two suction strokes and two delivery strokes.

Most piston pumps are double acting, while plunger pumps are single acting.

According to the number of cylinders provided, reciprocating pumps may be classified as :

1. Simplex – single cylinder pump.
2. Duplex – double cylinder pump
3. Triplex – triple cylinder pump.

A simplex pump is the one having only one cylinder. The single cylinder pump may be single acting or double acting. A double cylinder pump or duplex pump is the one which has two single acting cylinders, each may be equipped with one suction and one delivery pipe with appropriate valves and a separate piston or plunger for each cylinder. In an alternate arrangement, there may be only one suction pipe, one delivery pipe and one piston for both the cylinders. The working of a duplex pump is similar to that of a double acting single cylinder pump.

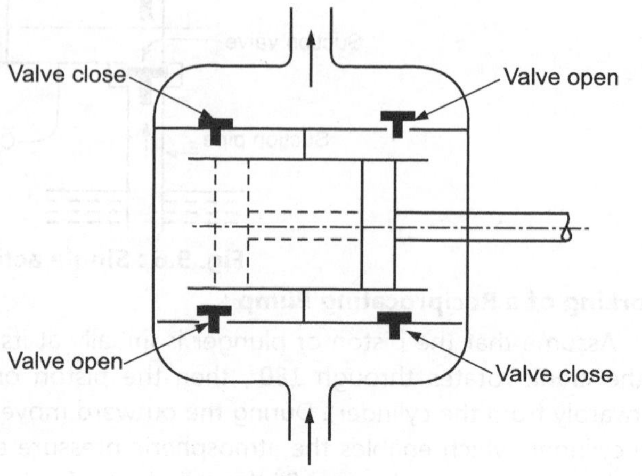

Fig. 9.7 : Duplex pump

A triplex pump has three single acting cylinders, each equipped with one suction and one delivery pipe and separate piston or plunger.

1. Piston pump

It consists of a cylinder with a reciprocating piston. It is equipped with valves at inlet and discharge for the liquid being pumped. The liquid to be pumped enters from the suction line via a suction valve and is discharged at high pressure through a delivery valve. This type of pump may be single acting or double acting and may be direct acting, steam driven or power driven with a crank and a flywheel.

In case of single acting pumps, the liquid is admitted only on one side of the piston (infront of the piston) while in double acting pumps, the liquid is admitted on both sides of the piston.

Both these types are shown in Fig. 9.8.

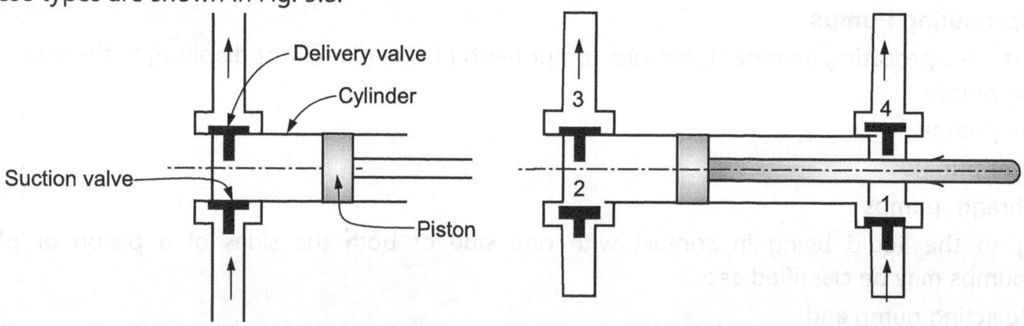

(a) Single acting **(b) Double acting**

Fig. 9.8 : Piston pumps

In case of single acting pumps, as the piston moves backward from the suction port, the liquid is drawn into the chamber through a suction valve which allows the liquid only to flow into the chamber. As the piston moves forward, the liquid is pressed and liquid is pumped out via a delivery valve through the delivery port. In case of double acting pumps, the liquid is drawn into the pump and discharged from the pump during backward as well as forward strokes. In the backward stroke, the liquid is drawn into the pump through the suction port (1) and liquid is discharged through the delivery port (3) and in the forward stroke, the liquid is drawn into the pump through suction port (2) and liquid is discharged through delivery port (4). With single acting pumps, the delivery is zero during return stroke (backward stroke) and with double acting pumps, the delivery is same in the forward and return stroke.

The theoretical delivery of the piston pump is nothing but the total swept volume of the cylinder (number of strokes per second × area of piston × length of stroke). Due to leakages past the piston, valves and inertia of the valves, the actual delivery may be less than the theoretical delivery. The volumetric efficiency of piston pumps which is *the ratio of the actual discharge to the swept volume* is normally greater than 90 per cent.

These type of pumps are provided with an air vessel at the pump discharge to make the flow in the delivery line continuous, i.e., to obtain uniform flow and to reduce energy requirement at the beginning of each stroke. During the forward stroke commencement, the liquid is pumped into the air vessel and the air is compressed and at the end of stroke when discharge decreases, the pressure in the air vessel is sufficiently high to expel some of the liquid in the delivery line.

The discharge of the liquid in the delivery line can be kept approximately constant by incorporation of a large air vessel close to the pump. The air vessel fitted on the delivery line also reduces the frictional losses.

This type of pump is comparatively simple in construction, operates at high mechanical efficiencies over a wide range of operating conditions, does not require priming and can be used to develop pressure upto 6 MPa. The load on a driving mechanism with this type is uneven as the delivery is uneven.

The piston pumps are well suited for pumping small quantities of liquids to high pressures.

Fig. 9.9 shows discharge curves for reciprocating pumps.

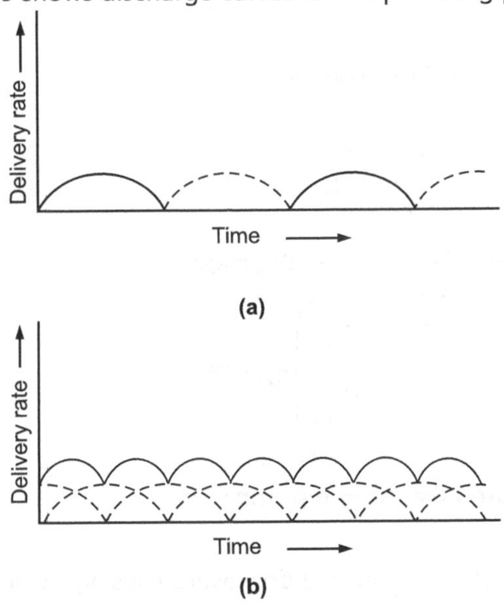

Fig. 9.9 : Delivery from (a) simplex and (b) duplex

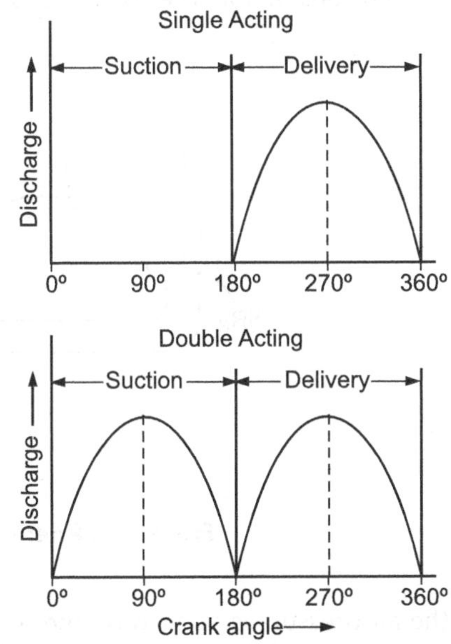

Fig. 9.10 : Delivery from single acting and double acting pump

2. Plunger pumps

The principle of operation of a plunger pump is the same as that of the piston pump. It consists of a heavy-walled cylinder of a small diameter that incorporates a close fitting reciprocating plunger, which is merely an extension of a piston rod. This type of pump is always single acting in the sense that only one end of the plunger is used to pump the liquid and usually are motor driven. These are used for high pressure applications and can discharge against a pressure of 150 MPa or more.

Applications : It may be used for injecting small quantities of inhibitors to high pressure systems and for feeding water to boilers.

3. Metering pumps

Metering pumps are driven by a constant speed electric motor and maintain constant delivery rates irrespective of changes in the pressure against which they operate. For low discharge and high pressure services, they are of the plunger type and for high discharge and low pressure services, they are of the piston type. During the pumping operation also, the delivery rate from the pump can be controlled by controlling the displacement (stroke) of the piston element by manually on the pump or remotely.

Applications Metering pumps are used for constant and accurately controlled delivery rates. They may be used for feeding of reactants and inhibitors to reactors at controlled rates and for the dosing of water supplies.

4. Diaphragm pumps

This pump is divided into two sections by a reciprocating driving member which is nothing but a flexible diaphragm. The diaphragm can be fabricated out of metal, rubber or plastic material (PTFE). In one section of the pump, there is a liquid to be pumped and in the other section there is a piston or plunger working in non-corrosive fluid (e.g., oil) which actuates the diaphragm to and fro. The fluid movement is transmitted by means of the diaphragm (flexible) to the liquid being pumped and hence with this arrangement packing and seals are not exposed to the liquid being pumped. Valves (non-return valves, ball type) at the suction and discharge are the only moving parts which are in contact with the liquid and these valves prevent reverse flow of the liquid.

Pneumatically actuated diaphragm pumps utilize compressed plant air for pumping. Such pumps must have a flooded suction. The pressure they develop is of course, is limited to the available compressed plant air pressure. Fig. 9.11 shows a schematic diagram of pneumatically actuated diaphragm pump.

Applications : Diaphragm pumps are used for handling corrosive liquids, toxic liquids as well as liquids containing suspensions of abrasive solids, i.e., slurries.

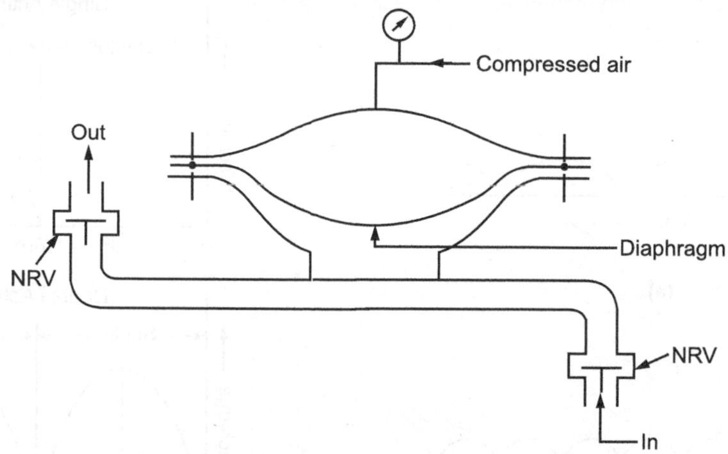

Fig. 9.11 : Pneumatically operated diaphragm pump

Working :

When the air pressure is applied on the diapharm, the diapharm is pushed downward causing the liquid in the pump casing to be pumped into the discharge line. During this discharge stroke, the NRV at the discharge line gets opened and that on the suction line gets closed. When the air pressure is released, the diapharm is returned to its position, pulled upward, and during this suction stroke, the liquid fills the pump casing. During this stroke, the NRV at the suction line gets opened and that at the discharge line gets closed. In this way, the air pressure on the diaphragm is alternately applied and released for the pumping of liquid.

9.3.2.2 Positive Displacement Rotary Pump

In rotary pump as the elements of the pump rotate in a casing, a reduced pressure is created on the inlet side, the liquid is thus forced into the pump, it is then trapped between the rotating elements and the casing and finally is forced out of the discharge side of the pump. The rotary pumps can handle liquids of any viscosity. The discharge rate from this pump is a function of its size and speed of rotation. Rotary pumps include gear pumps and mono pumps.

1. Mono pump/Screw pump

This pump consists of a rotor that rotates in a stator made of rubber or other similar material. The rotor is a true helical metal screw while the stator has a double helical thread pitched opposite to the spiral on the rotor. The liquid to be pumped moves continuously towards a discharge through the voids between the rotor and the stator. This pump is capable of handling highly viscous materials, gritty liquids, etc.

Applications : It is widely used in the chemical industry for feeding slurries containing higher proportion of solids to filtration equipments. Fig. 9.12 shows a schematic diagram of mono pump.

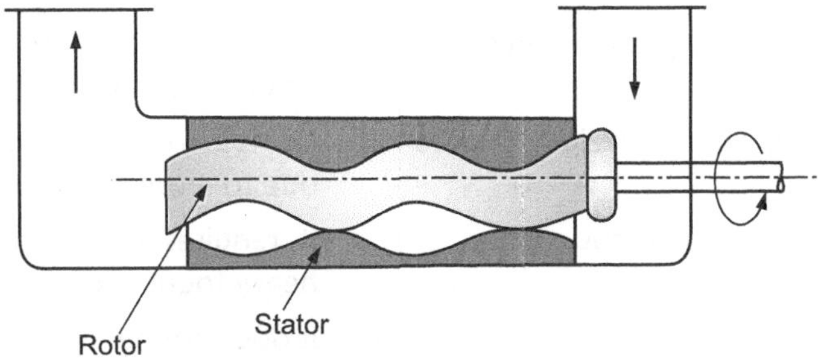

Fig. 9.12 : Mono pump

2. Gear pump

Principle : A gear pump uses the meshing of gears to pump liquid by displacement, i.e., it creates liquid displacement via the meshing of gears.

Construction :

It consists of two toothed gear wheels (spur gears) enclosed in a casing which is provided with inlet and outlet connections for the liquid to be pumped. Of the two gear wheels, one is driven by an electric drive and other rotates in mesh with it. The clearance between the gear wheels as well as between the surface of the gear wheels and the casing is very small. The pump has no valves. For this pump, the delivery/discharge rate of liquid is independent of pressure.

This pump does not need priming. The number of teeth on each gear wheel varies from three to four or to a considerable number. The gear and casing of a gear pump is rigid and this allows gear pumps to pump viscous liquids and operate at high pressures.

Working :

The liquid entering in the pump from the inlet due to creation of a reduced pressure at the inlet is carried round in the space between the gear teeth and the casing during the rotation of gears. Due to further rotation of gears the liquid is pumped out of the discharge as the gear teeth come nto mesh.

A neat sketch of gear pump is shown in Fig. 9.13.

Applications : These pumps are used for handling high viscosity liquids such as molasses, paints, crude oil, lube oils, fuel oils, pure bitumen vegetable fats, vegetable oils, cacao butter, chocolate black liquor resins and adhesives.

Advantages : High speed, high pressure, priming is not required, relatively quiet operation.

Disadvantages : They are not sufficient for liquids carrying suspended solids due to close clearance between gear wheels and teeth.

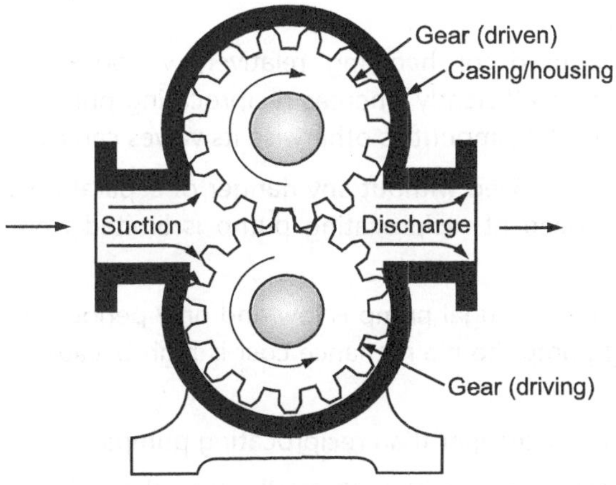

Fig. 9.13 : Gear pump

9.3.3 Comparison of Centrifugal Pump with Reciprocating Pump

Centrifugal Pump	Reciprocating Pump
1. The means of fluid transfer is a centrifugal force.	1. The means of fluid transfer is a displacement of volume by a piston element.
2. Simple in construction and light in weight due to less number of parts.	2. Complex in construction and heavy construction due to more number of parts.
3. It requires less floor space and simple foundation.	3. It requires more floor space and comparatively heavy foundation.
4. It needs priming.	4. It does not need priming.
5. It does not incorporate an air vessel.	5. It incorporates an air vessel.
6. Discharge is steady and even.	6. Discharge is not even, it is pulsating.
7. Used for large capacity and low heads.	7. Used for low capacity and high heads.
8. It is operated at high speeds.	8. It is operated at low speeds.
9. It can handle relatively viscous liquids and liquids containing suspended solids.	9. It cannot handle viscous liquids and liquids containing suspended solids. It can handle clear liquids.
10. They are coupled directly to an electric motor.	10. They are belt driven.
11. They are operated against a closed valve without danger.	11. They are never operated against a closed valve.
12. Efficiencies are low.	12. Efficiencies are high.
13. Maintenance cost is low.	13. Maintenance cost is high.
14. Costwise they are cheaper.	14. Cost of the reciprocating pump is higher than the centrifugal pump of the same power.
15. Designed for high discharge.	15. Designed for high heads.
16. Cannot develop high pressures.	16. Can develop high pressures.
17. Less wear and tear.	17. More wear and tear.

9.3.4 Advantages of Centrifugal Pump over Reciprocating Pump

1. Simple in construction, requires less floor space, relatively simple foundation.
2. Discharge capacity is very much greater than that of the reciprocating pump which can handle relatively small quantities of liquid.
3. Centrifugal pumps can be used for handling relatively viscous liquids and liquids containing large proportion of suspended solids efficiently whereas reciprocating pump is used for low viscosity liquids and liquids free from suspended solids/impurities otherwise its valves can cause frequent trouble.
4. It can be operated at very high speed without any danger of separation and cavitation and can be coupled directly to electric motor. Speed of reciprocating pump is limited from consideration of separation and cavitation.
5. The maintenance cost of the centrifugal pump is low and only periodical checkup is sufficient. On the other hand, for the reciprocating pump the maintenance cost is high because parts such as valves etc. may need frequent replacement.
6. Centrifugal pumps are relatively cheaper than reciprocating pumps.
7. For equal capacity, the centrifugal pump is much smaller than the reciprocating pump.

9.4.1 Power Requirement of a Centrifugal Pump

The work done by a pump for lifting (or delivering) a liquid against gravity through a liquid column of height, ΔH, m is given by

$$W_p = \text{Weight of liquid lifted} \times \text{Height through which liquid is lifted}$$
$$= w \times H$$

We have :

$$\text{Weight} = \text{Mass} \times g$$
$$w = mg$$
$$W_p = mg\,\Delta H \text{ ... in N.m}$$

Units of W_p : $kg \times \frac{m}{s^2} \times m \Rightarrow N.m \Rightarrow J.$

where m is the mass of liquid in kg, ΔH is the total developed head or simply total head in m of the flowing fluid and g is the acceleration due to gravity (9.81 m/s^2).

W_p is the work done by the pump on the flowing fluid.

The power is defined as work done per unit time.

$$\text{Power} = \frac{\text{Work done}}{\text{Time}}$$

Let P_f be the power delivered to the flowing fluid by the pump.

$$P_f = \frac{\text{Work done}}{\text{Time}} = \frac{W_p}{t}$$

where t is in seconds.

$$P_f = \frac{m \cdot g \cdot \Delta H}{t} \text{ ... in J/s = W}$$

But m/t is the mass flow rate of the liquid pumped in kg/s.

Let

$$\frac{m}{t} = \dot{m}$$

$$P_f = \dot{m} g \Delta H \text{ ...... in W}$$

The mass flow rate is related to the volumetric flow rate (Q) by the relation :

$$\dot{m} = Q \cdot \rho$$

where Q is in m^3/s and ρ is the density of the liquid in kg/m^3.

$$\therefore \quad P_f = \rho Q g \Delta H \text{ ...... in W}$$

The power required by the pump (shaft power or power supplied to the pump by an external agency, e.g., an electric motor) is given by

$$P_B = \frac{P_f}{\eta} \text{ ...... in W} \qquad \left[\text{Since } \eta = \frac{P_f}{P_B} = \frac{\text{Power out from pump to liquid}}{\text{Power into pump from motor}}\right]$$

where η is the efficiency of the pump.

9.4.2 Power Requirement of a Reciprocating Pump

Let

D = Diameter of the cylinder/piston

A = Cross-sectional area of the cylinder/piston, m^2

N = rpm of the crank of radius r

L = Length of the stroke, m

$$A = \frac{\pi}{4} D^2$$

Theoretical volume of water delivered in one revolution = AL, m^3

Number of revolutions per second = N/60.

∴ Theoretical discharge = Q_{th} = Discharge in one revolution × Number of revolutions per second

$$= \frac{ALN}{60}, \; m^3/s$$

(1) Theoretical discharge from a single acting reciprocating pump :

$$Q_{th} = \frac{ALN}{60}, \ m^3/s$$

(2) Theoretical discharge from a double acting reciprocating pump :

$$Q_{th} = \frac{2ALN}{60}, \ m^3/s$$

The actual discharge, Q_{actual} is less than the theoretical discharge, Q_{th}, due to leakage through valves, piston, gland and delay in closing the suction and delivery valves.

$$\text{Slip} = Q_{theoretical} - Q_{actual}$$

$$\% \text{ slip} = \frac{Q_{th} - Q_{actual}}{Q_{th}} \times 100$$

Theoretical power of the pump is given by

$$P_{th} = \rho Q_{th} g H$$

where H is the total head in m of the flowing fluid.

$$H = H_s + H_d.$$

For a single acting reciprocating pump :

$$P_{th} = \frac{\rho ALNgH}{60}, \text{ in W}$$

For a double acting reciprocating pump :

$$P_{th} = \frac{2\rho ALNgH}{60}, \text{ in W}$$

If Q_{actual} is the actual flow rate, then the actual power is given as

$$P_{actual} = \rho g \, Q_{actual} H, \text{ in W}$$

The volumetric efficiency of a pump is given by

$$\eta_v = \frac{Q_{actual}}{Q_{theoretical}} \times 100$$

If the mechanical efficiency of a pump is η, then the power requirement/shaft power is given by

$$P_s = \frac{P_{actual}}{\eta}$$

9.5.1 Head of a Centrifugal Pump (Concept and Derivation)

A pump installed in a pipeline which draws a liquid from a reservoir and discharges it at a constant volumetric flow rate at the exit of the pipeline (i.e., at point 2) is shown in Fig. 9.14.

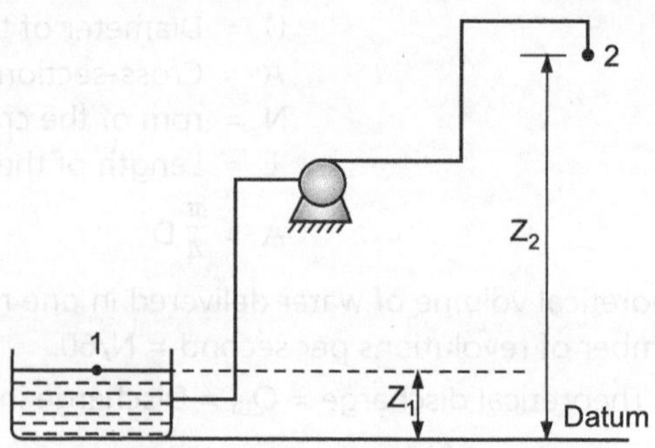

Fig. 9.14 : Pump flow system

The suction line is *that portion of the piping which carries a liquid to a pump* and discharge line is *that portion of the piping which carries the liquid out of the pump.*

Total head developed by a pump is composed of the difference between static pressures and velocity heads plus the frictional losses in the suction and discharge line of the pump.

Let Z_1, Z_2 be the elevations of stations 1 and 2 respectively above a certain datum level. u_1 and u_2 be the velocities and P_1 and P_2 be the pressures at stations 1 and 2 respectively. Let W_p be the pump work per unit mass of fluid and η be the (mechanical) efficiency of the pump.

The Bernoulli's equation between stations 1 and 2 in terms of energy per unit mass of a fluid is

$$\frac{P_1}{\rho} + \frac{u_1^2}{2} + g Z_1 + \eta W_p = \frac{P_2}{\rho} + \frac{u_2^2}{2} + g Z_2 + h_f \quad \text{... (9.1)}$$

$$\eta W_p = g(Z_2 - Z_1) + \frac{1}{2}(u_2^2 - u_1^2) + \frac{1}{\rho}(P_2 - P_1) + h_f \quad \text{... (9.2)}$$

Each term in Equation (9.2) has the units of energy per unit mass.

Let us check the units of the termed involved.

The unit of $g(Z_2 - Z_1)$ is

$$(m/s^2) \times m = m^2/s^2 = \frac{m^2}{s^2} \times \frac{kg}{kg} = \frac{kg.m}{s^2} \cdot \frac{m}{kg} = N.m/kg = J/kg$$

The unit of $\frac{1}{\rho}(P_2 - P_1)$ is :

$$\frac{1}{kg/m^3} \times \frac{N}{m^2} = N.m/kg = J/kg$$

The unit of ηW_p is : J/kg

The unit of h_f is : $h_f = \frac{4 f L u^2}{2D} = \frac{m \times m^2/s^2}{m} = m^2/s^2 = J/kg$

Equation (9.2) can also be written as

$$\eta W_p/g = (Z_2 - Z_1) + \frac{1}{2g}(u_2^2 - u_1^2) + \frac{1}{\rho g}(P_2 - P_1) + (h_f / g) \quad \text{... (9.3)}$$

Each term in the above equation has the units of m of flowing fluid and thus are called as heads. Therefore, Total head =

= difference in static head + difference in velocity head + difference in pressure head + frictional losses in m of fluid – (frictional head)

Let us check the units of each term of Equation (9.3).

The unit of $(Z_2 - Z_1)$ is : m

The unit of $\frac{1}{\rho g}(P_2 - P_1)$ is $= \frac{1}{\left(\frac{kg}{m^3}\right) \times \left(\frac{m}{s^2}\right)} \times N/m^2 = \frac{(kg.m/s^2) \times m^2}{kg/m^3 \times m/s^2} = m$

The unit of $\eta W_p/g$ is : $\frac{J/kg}{m/s^2} = \left(\frac{kg.m}{s^2}\right) \times \frac{1}{kg} \times \frac{1}{m/s^2} = m$

The unit of h_f / g is : $\frac{4 f L u^2}{2 D} \times \frac{1}{g} = \frac{m \times m^2/s^2}{m} \times \left(\frac{1}{m/s^2}\right) = m$

The R.H.S. of Equation (9.3) represents the total head developed by the pump. Thus, the pump work is equal to the total head developed by the pump.

$$\eta W_p/g = \left(Z_2 + \frac{u_2^2}{2g} + \frac{P_2}{\rho g} + h_f/g\right) - \left(Z_1^2 + \frac{u_1^2}{2g} + \frac{P_1}{\rho g}\right) \quad \text{... (9.4)}$$

$$= H_2 - H_1 = \Delta H = \text{total developed head}$$

where H_2 = total discharge head

H_1 = total suction head

$$\eta W_p/g = \Delta H \quad \text{... (9.5)}$$

$$\therefore \quad W_p = \Delta H \cdot g/\eta \quad \text{... (9.6)}$$

The power supplied to the pump or the power input to the pump (P_B) from an external source (e.g., an electric motor) is given by

$$P_B = \dot{m}\, W_P = \frac{\dot{m} \cdot \Delta H \cdot g}{\eta}$$

or $$P_B = \frac{\dot{m} \times \text{total head} \times g}{\eta} \qquad \text{... (9.7)}$$

where $\dot{m}$ is the mass flow rate in kg/s, ΔH is the total head in m of the liquid to be pumped, η is the efficiency of the pump and g is 9.81 m/s^2.

The S.I. unit of P_B is J/s (= W).

Unit of P_B $\Rightarrow$ unit of $\dot{m}$ × unit of ΔH × unit of g, since η is dimensionless

$$\Rightarrow \frac{kg}{s} \times m \times \frac{m}{s^2}$$

$$\Rightarrow \frac{kg}{s} \times m \times \frac{m}{s^2} \Rightarrow \frac{kg{\cdot}m}{s^2} \times \frac{m}{s}$$

$$\Rightarrow \frac{N{\cdot}m}{s} \Rightarrow J/s \Rightarrow W$$

In the S.I. system, power is measured in joules per second, i.e., a unit called the watt (W).

The power delivered/supplied to the fluid by the pump (P_f) or theoretical power is given by

$$P_f = \dot{m} \times \Delta H \times g = \eta \cdot W_p \qquad \text{... (9.8)}$$

Combining Equations (9.7) and (9.8), we get

$$\eta = \frac{P_f}{P_B} \qquad \text{... (9.9)}$$

Sometimes the efficiency of pump is also denoted by η_p.

The mechanical efficiency or simply efficiency of a centrifugal pump is defined as the ratio of the power delivered to the fluid by the pump to the power supplied to the pump from an external source.

$$\eta = \frac{P_f}{P_B} = \frac{\text{Power delivered to fluid}}{\text{Power supplied to the pump}} = \frac{\text{Power delivered to fluid}}{\text{Power input to the pump}}$$

$$= \frac{\text{Theoretical power}}{\text{Actual power input to the pump}} \qquad \text{... (9.9)}$$

If we use electric motor as an external source and if the efficiency of the motor is η_m, then

$$\eta_m = \frac{\text{Power supplied to the pump by the motor}}{\text{Power taken by the motor from the power line}}$$

P_m = Power taken by the motor from the power line

= Power input to the motor from the power line

= Power requirement of the pump-motor set

Power supplied to the pump by the motor = Power output of the motor

$$\eta_m = \frac{P_B}{P_m} \qquad \text{... (9.10)}$$

Combining Equations (9.9) and (9.10), we get

$$\eta_m = \frac{P_f}{\eta P_m}$$

$$P_m = \frac{P_f}{\eta \cdot \eta_m} \qquad \text{... (9.11)}$$

η = Efficiency of the pump, η_m = Efficiency of the motor

If η_o is the overall efficiency, i.e., the efficiency of the pump-motor set, then

$$\eta_o = \eta \cdot \eta_m$$

and
$$P_m = \frac{P_f}{\eta_o} \quad \text{... (9.12)}$$

P_B is also called as shaft power or the power delivered by the motor to the pump and P_f is also called as fluid power.

Case-I : If the stations 1 and 2 are at atmospheric pressure (i.e., are open to the atmosphere), station 1 is arbitrarily chosen as a datum level, and the velocity at station 1 is negligible in comparison with that at station 2, then Equation (9.2) reduces to

$$\eta W_p = g Z_2 + \frac{1}{2} u_2^2 + h_f \text{, J/kg}$$

or
$$\Delta H.g = g Z_2 + \frac{1}{2} u_2^2 + h_f$$

$$\Delta H = Z_2 + \frac{1}{2g} u_2^2 + h_f / g \text{, m} \quad \text{... (9.13)}$$

(as $P_1 = P_2 = P_{atm}$, $Z_1 = 0$, and $u_1 \approx 0$).

9.5.2 Cavitation

The vapour pressure of a liquid is the pressure at which the liquid is converted into vapour at a given temperature. If the pressure in the suction line is less than the vapour pressure of the liquid, then some of the liquid flashes into vapour. This phenomenon is called **cavitation**. When cavitation occurs the liquid will not be pumped.

- Cavitation refers to the formation of bubbles or cavities in liquid, formed in areas of relatively low pressure around an impeller.
- *In case of a centrifugal pump, the phenomenon of flashing of some of the liquid into vapour when the suction pressure falls below the vapour pressure of the liquid* is called **cavitation**.

Effects of cavitations include : a marked increase in unwanted noise and vibrations, mechanical damage of the metallic surface as the vapour bubbles collapse and cavities are formed, and decrease in the efficiency of the pump.

As long as the pressure at the suction is more than the vapour pressure, the liquid is pumped. Therefore, care must be taken so that the pressure in the suction line does not fall below the vapour pressure of the liquid to be pumped. Cavitation will not take place when the sum of the velocity and suction heads at the suction is sufficiently greater than the vapour pressure of the liquid at the temperature of pumping.

The causes of cavitation include : increase in liquid temperature, drop in suction pressure due to frictional losses and sharp elbows in the suction line.

Thus, the cavitation can be eliminated by avoiding sharp bends in the suction line to reduce loss of head, keeping velocity in the suction line low and using low pumping temperatures to keep the vapour pressure low.

The vapour pressure of a liquid is the pressure at a given temperature, at which the liquid will change to a vapour.

9.5.3 Net Positive Suction Head (NPSH)

It is necessary to specify the minimum value of the net positive suction head (NPSH) which must exist at the suction point of the pump by the pump manufacturers.

- In order to avoid cavitation, the pressure at the suction point of the pump must be higher than the vapour pressure of the liquid by a certain value, which is called the net positive suction head.
- The NPSH is *the amount by which the pressure at the suction point of the pump, expressed as a head of the liquid to be pumped (i.e., the sum of the velocity and pressure heads), is in excess of the vapour pressure of the liquid.*

For any installation, NPSH must be calculated by taking into consideration the absolute pressure of the liquid, the level of the pump, the velocity head in the suction line and friction head in the suction line. If the required value of NPSH is not obtained, then partial vaporisation is likely to take place and both suction and delivery heads get reduced. The reduction in suction head may cause the pump to be starved of the liquid.

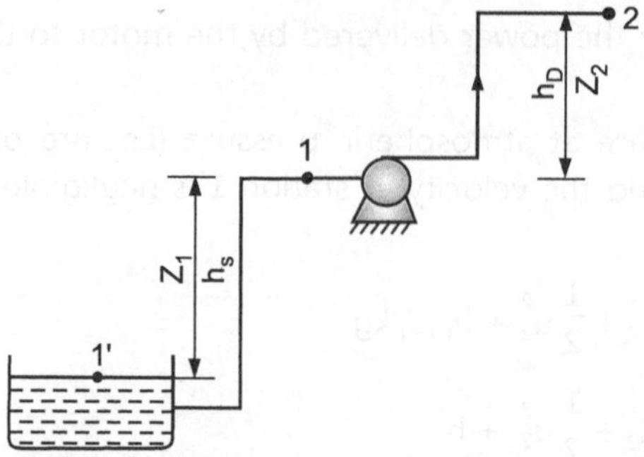

Fig. 9.15 : System with suction lift (Negative suction)

Z_1 is the static suction lift (also denoted by h_s). It is the vertical height of the centre line of the pump shaft above the liquid surface in the pump (reservoir) from which the liquid being raised.

Z_2 is the static delivery lift (also denoted by h_D). It is the vertical height of the liquid surface in the tank to which the liquid is delivered above the centre line of the pump shaft.

For the pumping system shown in Fig. 9.14, the NPSH in m of the flowing liquid is given by

$$\text{NPSH} = \left(\begin{array}{c}\text{absolute pressure head}\\ \text{available at suction point} - 1\end{array}\right) - (\text{vapour pressure head})$$

$$= \frac{u_1^2}{2g} + \frac{P_1}{\rho g} - \frac{P_v}{\rho g} \quad \text{... (9.14)}$$

where P_v = vapour pressure of the liquid at the pumping temperature.

The Bernoulli equation in terms of m of the liquid between stations 1' and 1 is

$$\frac{P_1'}{\rho g} + \frac{u_1'^2}{2g} + Z_1' = \frac{u_1^2}{2g} + Z_1 + \frac{P_1}{\rho g} + h_{fs} \quad \text{... (9.15)}$$

Note that h_{fs} in the above equation is to be expressed in m of the liquid to be pumped.

h_{fs} is the head lost in friction in the suction line.

If $Z_1' = 0$ and $u_1' = 0$, then

$$\frac{P_1'}{\rho g} = \frac{u_1^2}{2g} + Z_1 + \frac{P_1}{\rho g} + h_{fs} \quad \text{... (9.16)}$$

Rearranging Equation (9.16), we get

$$\frac{P_1}{\rho g} + \frac{u_1^2}{2g} = \frac{P_1'}{\rho g} - Z_1 - h_{fs} \quad \text{... (9.17)}$$

From Equations (9.17) and (9.14), we get

$$\text{NPSH} = \frac{P_1'}{\rho g} - \frac{P_v}{\rho g} - Z_1 - h_{fs} \quad \text{... (9.18)}$$

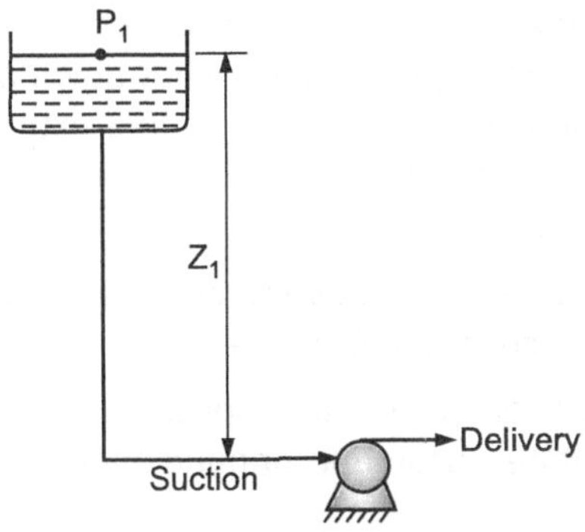

Fig. 9.16 : Suction system of a centrifugal pump (Positive/flooded suction)

For the system shown in Fig. 9.16, in which the pump is taking liquid from a reservoir at an absolute pressure P_1 and in which the liquid level is at a height Z_1 above the suction point of the pump then NPSH is given by the difference between the total head at the suction inlet and the head corresponding to the vapour pressure of the liquid at the pump inlet.

$$\text{NPSH} = \frac{P_1}{\rho g} - \frac{P_v}{\rho g} - h_{fs} + Z_1 \quad \text{... (9.19)}$$

where h_{fs} is the head lost in friction.

The cavitation may sometimes be rectified by closing the valve on the pump delivery or by reducing the pump speed by a small amount. Usually, a small fast-running pump will require a larger NPSH than a larger slow-running pump.

If NPSH = 0, the cavitation will occur as at this value the suction pressure equals the vapour pressure of the liquid.

$$\text{NPSH} = \frac{P_1}{\rho g} - \frac{P_v}{\rho g} - h_{fs} + Z_1 = 0$$

$$\therefore \quad \frac{P_1}{\rho g} + Z_1 - h_{fs} = \frac{P_v}{\rho g}$$

Total head at pump inlet (i.e., at suction) = Head corresponding to the vapour pressure of the liquid at the pump inlet.

Thus, NPSH must be greater than zero. The required value of NPSH is usually 2 or 3 m or more.

NPSH may also be defined as the *head required to make the liquid flow through the suction pipe to the impeller.*

The available NPSH of a pump system (NHSHA) must be greater than the NPSH required of the pump system (NPSHR) with an appropriate design margin to avoid cavitation.

9.5.4 Air Binding and Priming of a Centrifugal Pump

The pressure developed by the impeller of a centrifugal pump is proportional to the density of the fluid in the impeller. If air enters the impeller, the pressure developed is reduced by a factor equal to the ratio of the density of air to the density of liquid and this head in meters of air represents a very small pressure in terms of the liquid to be pumped. Because of this the pump is not capable to force the liquid through the delivery pipe (i.e., if air enters the impeller, it will produce a very small pressure which may not suck liquid from its source through the suction line). This is called as *air binding* of the pump.

If the pump is initially full of air, to avoid air binding it needs priming, i.e., it should be filled up with the liquid to be pumped.

The removal of air from the suction line and pump casing is known as **priming**.

The priming should be done before the pump is started. Priming is done by first filling the suction line and pump casing with the liquid to be pumped. Then pump is started with a delivery valve closed so that rotating impeller pushes the liquid in the delivery pipe. Finally, the delivery valve is opened so that air is displaced by the liquid to be pumped.

Priming may be done by providing a non-return valve in the suction line so that the suction line and casing will be full of liquid to be pumped even though the pump is in a shut down position. Small pumps are usually primed by pouring the liquid through a funnel into the casing from some external source. An air vent provided on the casing is opened to exit of the air. When all the air has been removed from the suction line and the pump casing, the air vent is closed and the pump is said to be primed.

Large pumps are usually primed by evaluating the casing and the suction line with the aid of an air pump or an ejector, the liquid is thus, sucked into the suction line from the sump.

In some pumps, their internal construction is such that special arrangements containing a supply of liquid are provided in the suction line, which facilitate automatic priming of the pump. These pumps are known as self-priming pumps.

The need for priming is eliminated by providing positive suction head.

9.6.1 Centrifugal Pump Performance

Losses in a Centrifugal Pump :

The various losses occurring during the operation of a centrifugal pump are :

(A) Mechanical losses :

These losses occur because of :

(i) disc friction between the impeller and the liquid which fills the clearance spaces between the impeller and the casing,

(ii) friction in bearings and stuffing box packings.

(B) Leakage losses :

These losses are caused by the leakage from the tip of the impeller to the impeller suction. These losses are very high with open impellers.

(C) Recirculation losses :

These losses are caused by the circulation of the liquid within the space between the vanes due to non-uniform velocity in the space between two adjacent vanes of the impeller.

(D) Hydraulic losses :

Hydraulic losses in pump are :

(i) Shock or eddy losses at the entrance to and the exit from the impeller.

(ii) Friction losses in the impeller.

(iii) Friction and eddy losses in the guide vanes (or diffuser) and casing.

Other hydraulic losses consist of friction in the suction and delivery line.

9.6.2 Characteristic Curves

A pump is usually designed for one particular speed, flow rate and head, but in actual practice, it may be operated at some other conditions of head or flow rate. The behaviour of the pump may be quite different for the changed conditions.

In order to predict the behaviour and performance of a pump under varying conditions, tests are performed, and the results of such tests are plotted. The curves thus obtained are known as the *characteristic curves* of the pump.

The pump characteristic is described graphically by the manufacturer as the pump performance curve. In order to find the best suitable pump, we have to go through the characteristic curves.

The characteristic curves showing head, power input and efficiency as a function of discharge rate (capacity) for a particular centrifugal pump at a particular speed are shown in Fig. 9.17.

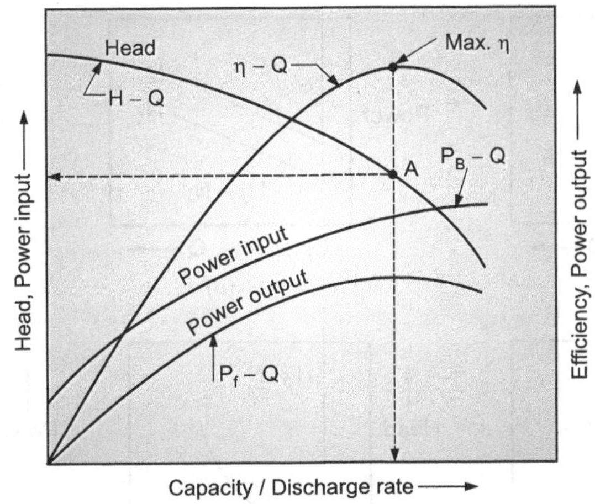

Fig. 9.17 : Characteristic curves of a centrifugal pump for discharge for a given speed

Thus, the characteristic curves of a centrifugal pump are *the graphical relationships between head and discharge rate, efficiency and discharge rate and power input and discharge rate at a given speed.*

The H-Q curve shows the relationship between head and capacity/discharge rate or simply discharge. It is clear from the curve that the head decreases continuously as the discharge rate/capacity is increased. The optimum conditions for operation are those at which the ordinate through the point of maximum efficiency cuts the head curve. The point A is called as the duty point. The values of head and discharge rate corresponding to the maximum efficiency are known as the normal or designed head and normal or designed discharge of the pump.

The head corresponding to zero or no discharge is known as the shut-off head of the pump. From H-Q curve, it is possible to determine whether the pump will handle the necessary quantity of liquid against a desired head or not and the effect of increase or decrease of head. The $P_B - Q$ curve gives us an idea regarding the size of motor (power) required to operate the pump at the required conditions and whether or not motor will be overloaded under any other operating conditions. The $\eta - Q$ curve shows the relationship between pump efficiency and capacity. It is clear from $\eta - Q$ curve that efficiency rises rapidly with discharge at low discharge rate, reaches a maximum in the region of the rated capacity and then falls.

The difference between power input and fluid power (power delivered to the fluid) represents the power lost in the pump (Fig. 9.17). It is due to fluid friction and shock losses (which are conversions of mechanical energy into heat), leakage, disk friction and bearing losses.

Characteristic curves for discharge with varying speed are plots of head versus discharge, power versus discharge and efficiency versus discharge, while characteristic curves for speed are plots of discharge versus speed, power versus speed and head versus speed.

Fig. 9.18 (a) shows the performance of a centrifugal pump under varying speeds. It shows that for a given speed, the head decreases with the discharge.

Fig. 9.18 (b) shows the performance of a centrifugal pump under different speeds. It shows that for a given speed, the power increases with the discharge.

Fig. 9.18 (c) shows the performance of a centrifugal pump under different speeds. It shows that for a given speed, the efficiency increases with the discharge and after a certain discharge, the efficiency decreases.

Fig. 9.18 (d) shows the performance of a centrifugal pump under constant head. It shows that the discharge increases with the speed.

Fig. 9.18 (e) shows the performance of a centrifugal pump under constant discharge. It shows that the head increases with the speed.

Fig. 9.18 (f) shows the performance of a centrifugal pump under constant head and discharge. It shows that the power increases with the speed.

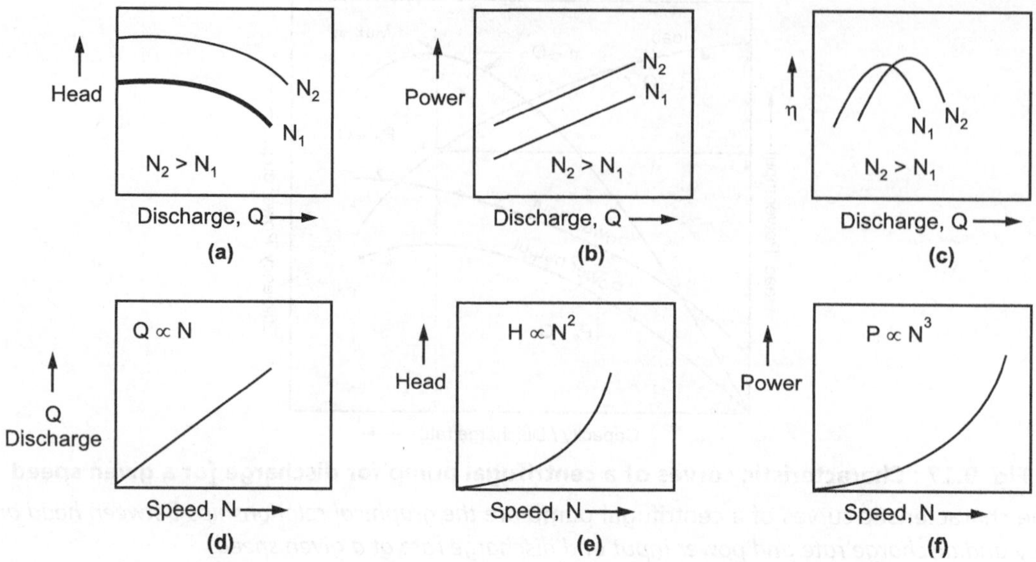

Fig. 9.18 : Characteristic curves for a centrifugal pump for discharge (a to c) and for speed (d to f)

9.6.3 Affinity Laws

In case of centrifugal pumps, the affinity laws are mathematical expressions relating the capacity (discharge), head (H) and break horse power (BHP or take P for convenience) to the impeller diameter (D) and to the pump or impeller speed (N).

Mathematically,

(i) $Q \propto D$

and $Q \propto N$

(Discharge is directly proportional to the impeller diameter/speed.)

$$\therefore \quad \frac{Q_2}{Q_1} = \frac{D_2}{D_1}$$

$$\therefore \quad Q_2 = Q_1 (D_2/D_1)$$

$$\frac{Q_2}{Q_1} = \left(\frac{N_2}{N_1}\right)$$

(ii) $H \propto D^2$

and $H \propto N^2$

i.e., Head is directly proportional to the square of impeller diameter/speed.

$$\therefore \quad \frac{H_2}{H_1} = \left(\frac{D_2}{D_1}\right)^2$$

and $$\frac{H_2}{H_1} = \left(\frac{N_2}{N_1}\right)^2$$

(iii) $P \propto D^3$

and $P \propto N^3$

i.e., Pump power is directly proportional to the cube of impeller diameter/speed.

$$\therefore \quad \frac{P_2}{P_1} = \left(\frac{D_2}{D_1}\right)^3$$

$$\frac{P_2}{P_1} = \left(\frac{N_2}{N_1}\right)^3$$

With these relations one can calculate Q, H or P at a given speed knowing Q, H or P at some another speed.

9.6.4 Centrifugal Pump Operation and Troubleshooting

For trouble-free operation and longer service life of centrifugal pumps, the following basic requirements should be met at all times :

(i) No cavitation of the pump occurs throughout the broad operating range.

(ii) A certain minimum continuous flow is always maintained during the operation.

Starting the pump ; Adopt the following procedure :

(i) Prime the pump by flooding the pump casing with liquid to be pumped before starting the pump in order to avoid damage to pump parts.

(ii) Start a pump motor and check rotation of the motor to make sure that it rotates in the same direction as that indicated by an arrow on the pump.

(iii) Check the pump to see that the liquid is flowing and all pipe joints and seals are leak-free.

(iv) Slowly open discharge valve until desired flow is obtained. Observe pressure gauges and if the pressure is not attained then stop the pump and prime again.

Stopping the pump :

To stop the pump, shut off power to the pump motor.

Effective troubleshooting requires

(i) An ability to observe changes in performance over time as well as (ii) the capacity to thoroughly investigate the causes of failure and take measures to prevent the problem recurrence, in the event of a failure. Design errors, poor operation and poor maintenance practices are the problems that are usually encountered with centrifugal pumps.

Commonly experienced troubles during the operation of the centrifugal pump and remedial measures to be taken for the same are given below :

(A) Pump fails to start pumping :

(i) The voltage may be low.

(ii) The pump may not be primed properly. Reprime the pump.

(iii) Total static head may be higher than that for which pump is designed. Check it by measuring distance between the suction liquid level and pump discharge. Add to this head loss due to friction (in pipe and pipe fittings).

(iv) Direction of rotation of the impeller may be wrong. Arrow on pump casing shows the proper direction of rotation.

(v) Suction lift may be too high. Check with a vacuum gauge or by actual measurement.

(vi) Strainer or suction line may be clogged. Check it.

(vii) Speed may be too low. Check speed with a techometer and compare with the speed cited on the name plate of pump.

(viii) Impeller may be clogged. Check carefully for solids or foreign matter lodged in the impeller.

Before starting the pump, check whether the pump shaft is free or not. Make it free.

(B) Pump is working but not upto the capacity and pressure :

(i) Air may be leaking into the pump (through suction line or stuffing box).

(ii) Speed may be too low.

(iii) Suction lift may be too high.

(iv) Discharge head may be higher than anticipated.

(v) Foot valve or end of suction pipe may not have sufficient submergence or it may be partially clogged.

(vi) Impeller may be partially clogged.

(vii) Wearing rings may be worn.

(viii) Impeller may be damaged, shaft may be loose, stuffing box packing may be defective.

(C) Pump starts and then stop pumping :

(i) Pump may be improperly primed or may be leaky suction line.

(ii) There may be air pockets in the suction line.

(iii) Suction lift may be too high.

(D) Pump takes too much power :

(i) Speed may be too high.

(ii) Head may be too low and pump delivers too much liquid.

(iii) Specific gravity of liquid may be too high.

(iv) Pump may be operating in wrong direction.

(v) Shaft may have bend, impeller may be rubbing on casing, stuffing box may be too tight, wearing rings may be worn out.

Centrifugal Pump Troubleshooting :

Problem	Possible cause	Suggested action
1. No enough liquid delivered	(i) Pump not primed.	(i) Prime the pump.
	(ii) Wrong direction of rotation.	(ii) Adjust the electrical motor connections (wiring) to correct the rotation.
	(iii) Insufficient NPSH.	(iii) Adjust system to provide correct NPSH.
	(iv) Suction lift too high.	(iv) Lower the pump until it is easily supplied with liquid, i.e., reduce lift by lowering the pump.
	(v) Suction or discharge plugged.	(v) Stop the pump and remove blockage.
	(vi) Air leak in supply or at seal area.	(vi) Check the system for air leaks and make necessary repairs. Replace the seal if required.
2. Motor overload	(i) Faulty electrical connections.	(i) Check the wiring and repair/replace as necessary.
	(ii) Too high flow rate.	(ii) Lower the flow rate by adding a discharge restriction.
	(iii) Seal binding.	(iii) Disassemble the pump and check for damage and crystallisation of any material on seals.
3. Vibration/Noise	(i) Pump not level.	(i) Level the pump.
	(ii) Piping not supported.	(ii) Support the piping as per installation guide.
	(iii) Starved suction/insufficient NPSH available/cavitation.	(iii) Adjust the system to provide correct NPSH.
	(iv) Impeller shaft loose or bent.	(iv) Disassemble the pump and inspect for damage.
	(v) Foreign material in the pump.	(v) Disassemble the pump, remove all foreign materials. Inspect for damage.
4. Not enough pressure	(i) Wrong direction of rotation.	(i) Adjust the motor electrical wiring to correct rotation.
	(ii) Excessive air in liquid.	(ii) Make necessary arrangements to remove air from the liquid before it reaches the pump.
	(iii) Air leak in supply or at seal area.	(iii) Check the system for air leaks and make necessary repairs. Replace seal if required.

SOLVED EXAMPLES

Example 9.1 : *Calculate the net positive suction head (NPSH) of a centrifugal pump using the following data :*

(i) Vapour pressure of the liquid = 26.66 kN/m²

(ii) Distance between the level of liquid in the reservoir and suction line = 1.2 m.

(iii) Density of the liquid = 865 kg/m³.

(iv) Friction in the suction line = 3.5 J/kg.

(v) Reservoir is open to atmosphere.

Solution : NPSH in terms of J/kg, i.e., in energy units is given by the following equation:

$$\text{NPSH} = \frac{P_a' - P_v}{\rho} - h_{fs} - g\,Z_a$$

where $g = 9.81\ \text{m/s}^2$

ρ = density of the liquid = 865 kg/m³

h_{fs} = friction in the suction line = 3.6 J/kg

Z_a = distance between the level of liquid in the reservoir and the suction line = 1.2 m

P_v = vapour pressure of the liquid = 26.66 kN/m² = 26660 N/m²

$P_{a'}$ = pressure over the liquid surface = 101325 N/m²

Substituting the values of various parameters in the above equation gives

$$\text{NPSH} = \left(\frac{101325 - 26660}{865}\right) - 3.5 - 9.81 \times 1.2$$

$$= 71.04\ \text{J/kg} = 71.04\,/g = 71.04/9.81 = \mathbf{7.24\ m}$$ ... **Ans.**

Example 9.2 : *Sulphuric acid is to be pumped at a rate of 3 kg/s through a 50 mm i.d. pipe over a straight run of 800 m and is then raised vertically 15 m. If the pump is electrically driven and has an efficiency of 50%, find the power required by the pump. Density of the acid = 1650 kg/m³. Viscosity of the acid = 8.6 mPa·s.*

Solution : Let us calculate u.

Mass flow rate of sulphuric acid = 3 kg/s

i.d. of pipe = D = 50 mm = 0.05 m

Cross-sectional area of pipe = $A = \pi/4\ D^2 = \frac{\pi}{4}(0.05)^2 = 1.963 \times 10^{-3}\ \text{m}^2$

Density of the sulphuric acid = 1650 kg/m³

Mass flow rate = $\dot{m} = \rho u A$

$3 = 1650 \times u \times 1.963 \times 10^{-3}$

u = velocity of fluid through distance pipe = 0.926 m/s

Let us calculate N_{Re} to get idea regarding the nature of flow.

$$N_{Re} = \text{Reynolds number} = \frac{Du\rho}{\mu}$$

where D = 0.05 m

u = 0.926 m/s

ρ = 1650 kg/m³

μ = 8.6 mPa.s = 8.6×10^{-3} Pa.s = 0.0086 kg/(m.s)

$$N_{Re} = \frac{0.05 \times 0.926 \times 1650}{0.0086} = 8883.14$$

Let us calculate f.

Since the Reynolds number is greater than 4000, the flow is turbulent. For turbulent flow, the relation between f and N_{Re} is

$$f = \frac{0.078}{(N_{Re})^{0.25}} = \frac{0.078}{(8883.14)^{0.25}}$$

$$f = 0.00803$$

Total length of piping = 800 + 15 = 815 m.

Let us calculate h_f.

$$h_f = \text{frictional loss in flow system} = \text{head lost due to friction} = \frac{4\,f.LV^2}{2\,g\,D}$$

$$h_f = \frac{4 \times 0.00803 \times (0.926)^2\,(815)}{2 \times 9.81 \times 50 \times 10^{-3}} = 22.9 \text{ m}$$

$$h_f \text{ in energy units} = h_f \text{ in m} \times g = 22.9 \times 9.81 = 224.65 \text{ J/kg}$$

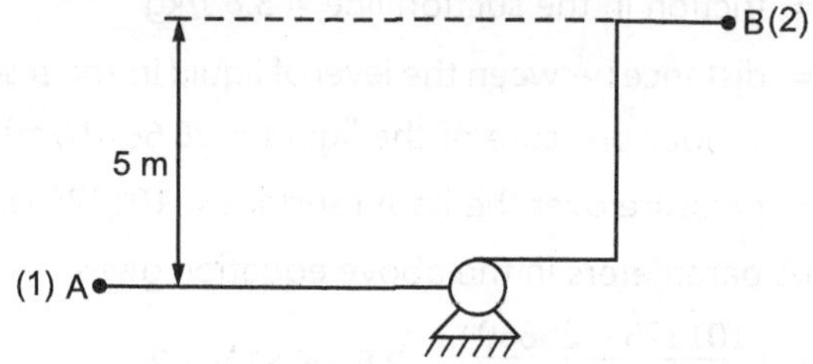

Fig. Ex. 9.2

Let us calculate W_p.

The Bernoulli's equation in terms of energy units (i.e., in J/kg) between stations 1 and 2 is

$$\frac{P_1}{\rho} + g\,Z_1 + \frac{u_1^2}{2} + \eta W_p = \frac{P_2}{\rho} + g\,Z_2 + \frac{u_2^2}{2} + h_f$$

$$(\alpha_1 = \alpha_2 = 1.0)$$

Stations 1 and 2 are open to atmosphere.

$\therefore$ $P_1 = P_2$

The velocity at station 1 (u_1) is negligible in comparison with the discharge velocity.

Assume station 1 as a datum level.

$\therefore$ $Z_1 = 0$ and $Z_2 = 15$ m

Thus, the Bernoulli equation reduces to

$$\eta W_p = g\,Z_2 + \frac{u_2^2}{2} + h_f$$

when $\eta = 0.5$ (efficiency = 50%)

$h_f = 224.65$ J/kg

$u = 0.926$ m/s

$Z_2 = 15$ m

$$0.5\,W_p = 9.81 \times 15 + \frac{(0.926)^2}{2} + 224.65$$

$$W_p = 744.46 \text{ J/kg}$$

$$\text{Mass flow rate} = \dot{m} = 3 \text{ kg/s}$$

Let us calculate the power required by the pump. The power required by the pump/the power input to the pump is given by

$$\text{Power required by the pump} = P_B = \dot{m} \times W_p$$
$$= 3 \times 744.46 = 2333 \text{ J/s} = 2333 \text{ W} = \mathbf{2.33 \text{ kW}}$$ ... **Ans.**

Example 9.3 : *Water is to be pumped at a rate of 8 m³/h from a large reservoir resting on the floor to the open top of an experimental absorption tower through a 50 mm i.d. pipe. The point of discharge is 5 metres above the floor and frictional losses in the entire flow system amount to 2.5 J/kg. At what height in the reservior the water be kept if the pump can develop a power of 94 W ?*

Solution : Volumetric flow rate of water = Q = 8 m³/h = 0.00222 m³/s

A = Cross-sectional area of pipe = $\pi/4\ D^2$

where D = 50 mm = 0.05 m

$$A = \frac{\pi}{4}(0.05)^2 = 1.963 \times 10^{-3}\ m^2$$

Let us calculate the velocity through the discharge pipe (u_2),

$$u_2 = Q/A = \text{velocity through the discharge pipe} = \frac{0.00222}{1.963 \times 10^{-3}}$$

$$u_2 = 1.131 \text{ m/s}$$

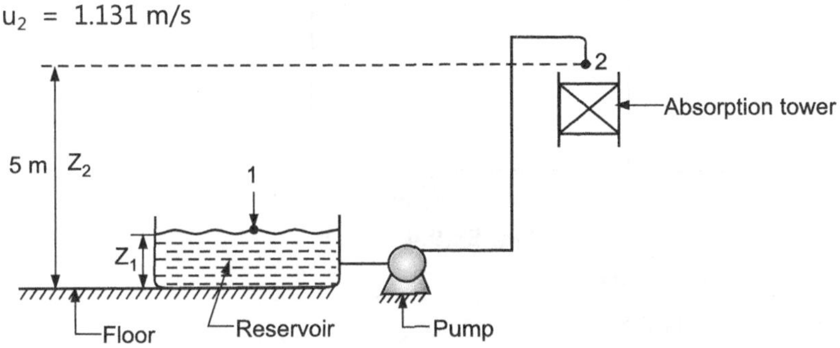

Fig. 9.20

The Bernoulli's equation between stations 1 and 2 is

$$\frac{P_1}{\rho} + g\,Z_1 + \frac{u_1^2}{2} + \eta W_p = \frac{P_2}{\rho} + g\,Z_2 + \frac{u_2^2}{2} + h_f$$

where P_1 and P_2 are pressures at stations 1 and 2.

Each term in the above equation has the units of J/kg, i.e., energy per unit mass.

P_1 = P_2 = 101.325 kPa (as stations 1 and 2 are open to atmosphere)

ρ = density of water = 1000 kg/m³

u_1 = velocity at station - 1 - very negligible in comparison with u_2, hence may be neglected.

u_2 = velocity at discharge = 1.131 m/s

h_f = frictional losses = 2.5 J/kg

Z_1 = height of discharge point from the floor = 5m

$\dot{m}\,W_p$ = power developed by the pump = 94 W = 94 J/s

$\dot{m}$ = mass flow rate of water in kg/s

$$\text{Mass flow rate} = \dot{m} = Q.\rho = 8 \times 1000 = 8000 \text{ kg/h} = \frac{8000}{3600} \text{ kg/s}$$

$$\dot{m} = 2.22 \text{ kg/s}$$

$$W_p = 94/\dot{m} = 94/2.22 = 42.34 \text{ J/kg}$$

$$\therefore \quad \eta W_p = 1 \times 42.34 = 42.34 \text{ J/kg}$$

The Bernoulli's equation after putting $P_1 = P_2$ and $u_1 = 0$ reduces to

$$g Z_1 + \eta W_p = g Z_2 + \frac{u_2^2}{2} + h_f$$

Substituting the values of the terms involved in the above equation

$$9.81 Z_1 + 42.34 = 9.81 \times 5 + \frac{(1.131)^2}{2} + 2.5$$

Solving for Z_1, we get

$$Z_1 = 1.0 \text{ m}$$

The height of water level in the reservoir from the floor is **1.0 m** ... **Ans.**

Example 9.4 : *A liquid of density 1150 kg/m³ is flowing from point A to point B which is 5 metres above the point A. The frictional losses in a pipeline of internal diameter 40 mm are 1 J/kg for a volumetric flow rate of 500 cm³/s. If points A and B are at the atmospheric pressure and the velocity at the point A is zero, calculate the pump work done.*

Solution :

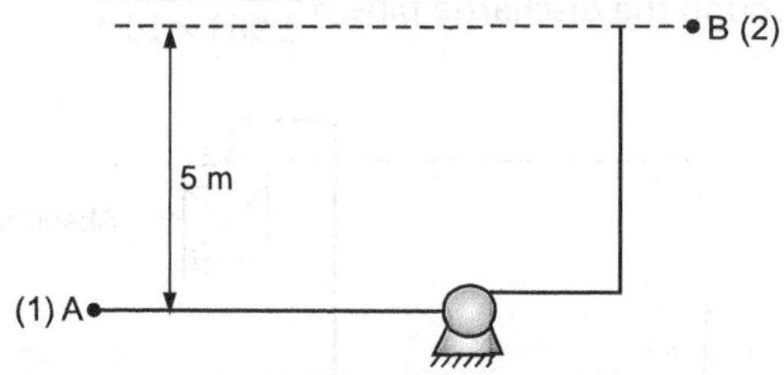

Fig. Ex.9.4

Pressure at A = Pressure at B = 101.325 kPa

Density of the liquid = ρ = 1150 kg/m³

h_f = frictional losses = 1 J/kg

Volumetric flow rate = Q = 500 cm³/s = 500×10^{-6} m³/s

Area of the pipe = $A = \frac{\pi}{4} D^2$

where D = 40 mm = 0.04 m

$$A = \frac{\pi}{4} (0.04)^2 = 1.26 \times 10^{-3} \text{ m}^2$$

Let us calculate u at B.

$$\text{Velocity at point B} = u_2 = Q/A = \frac{500 \times 10^{-6}}{1.26 \times 10^{-3}} = 0.40 \text{ m/s}$$

Given : Velocity at point A = $u_1 = 0$

Pressure at point A = P_1; Pressure at point B = P_2

The Bernoulli's equation between stations A and B is

$$\frac{P_1}{\rho} + g Z_1 + \frac{u_1^2}{2} + \eta W_p = \frac{P_2}{\rho} + g Z_2 + \frac{u_2^2}{2} + h_f$$

Let the point A (station 1) be at a datum line, then

$Z_1 = 0$, $Z_2 = 5$ m, $P_1 = P_2$

With this the Bernoulli equation reduces to

ηW_p = pump work = $u_2^2 / 2 + h_f + g Z_2 = (0.4)^2/2 + 1 + 9.81 \times 5 =$ **50.13 J/kg** ... **Ans.**

Example 9.5 : *Water is to be pumped from a tank, located on the ground, to a cooling tower. The tank is open to the atmosphere. The difference between the level of water in the tank and discharge point is 15 metres. The velocity of water through a 40 mm internal diameter discharge pipe is 3 m/s. In the pipe line there is a valve which is equivalent to 200 pipe diameters and a fitting equivalent to 150 pipe diameters. The length of the entire piping is 30 metres. Calculate the power requirement of the pump.*

Data : *Density of water* = *1000 kg/m³*

Viscosity of water = *0.0008 Pa.s*

Friction factor 'f ' = *0.004*

The pump efficiency is 60%.

Solution :

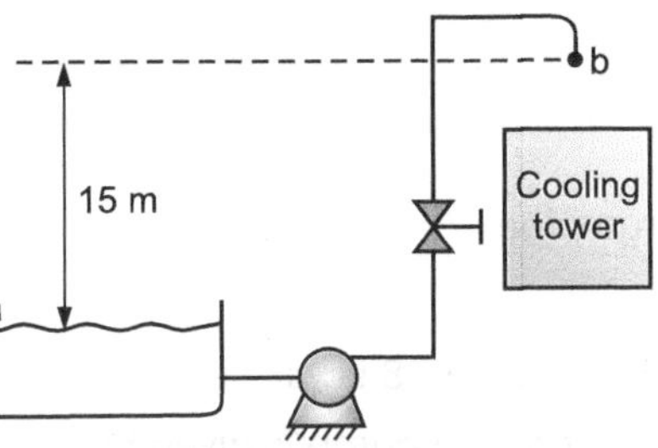

Fig. Ex. 9.5

Velocity through the discharge pipe = 3 m/s

I.D. of pipe = 40 mm

= 0.04 m

Equivalent length of the valve and pipe fitting = (1 × 200 × 0.04 + 1 × 150 × 04) = 14 m

Total length responsible for the friction loss = L = 30 + 14 = 44 m.

The friction loss is given by

$$h_f = \frac{4\,f\,L\,u^2}{2g\,D}$$

where f = 0.004

L = 44 m

D = 0.04 m

g = 9.81 m/s²

u = 3 m/s

$$h_f = \frac{4 \times 0.004 \times 44 \times (3)^2}{2 \times 9.81 \times 0.04} = 8.07 \text{ m}$$

h_f in terms of energy per unit mass is

$$h_f = 8.07 \times 9.81 = 79.17 \text{ J/kg}$$

The Bernoulli equation between stations a and b is

$$\frac{P_a}{\rho} + g\,Z_a + \frac{u_a^2}{2} + \eta W_p = \frac{P_b}{\rho} + g\,Z_b + \frac{u_b^2}{2} + h_f \qquad \text{... (1)}$$

P_a = P_b = 1 atmosphere as the stations a and b are open to the atmosphere

u_a ≈ 0, negligible as compared to u_b

u_b = velocity at discharge = 3 m/s

If point a is assumed as a datum, then Z_a = 0, Z_b = 15 metres, η = efficiency of the pump = 0.60

W_p = pump work

With these values of u_a, Z_a, P_a and P_b reduce to

$$\eta W_p = g Z_b + \frac{u_b^2}{2} + h_f$$

Substituting the values gives

$$0.6 W_p = 9.81 \times 15 + \frac{(3)^2}{2} + 79.17$$

$$W_p = \frac{147.15 + 4.5 + 79.17}{0.6} = 384.7 \text{ J/kg}$$

Mass flow rate of water = $\dot{m} = \rho$ u.A.

ρ = 1000 kg/m^3

u = u_b = 3 m/s

A = Area of the discharge pipe.

A = $\pi/4$ (D^2) = $\pi/4$ (0.04)^2=0.00126 m^2

$\dot{m}$ = mass flow rate = ρuA = 1000 × 3 × 0.00126 = 3.78 kg/s

The power required by the pump, i.e., the power input to the pump is given by

Power required by the pump = $P_B = \dot{m} \times W_p = 3.78 \times 384.7 \left(\frac{kg}{s}\right)$ (J/kg) = 1454 J/s = 1454 W = **1.45 kW** ... **Ans.**

Example 9.6 : *Calculate the power required for a pump used to recirculate cooling water from the following data :*

(i) Water flow rate = 5000 l/h

(ii) Straight length of the pipe = 12 m

(iii) Pipe line inside diameter = 25 mm

(iv) Equivalent length of the various pipe fittings = 1 m

(v) The height difference between discharge at top of cooling and level in tank = 10 m

(vi) Viscosity of water = 0.8 mPa.s

(vii) Efficiency of the pump = 60%.

Solution :

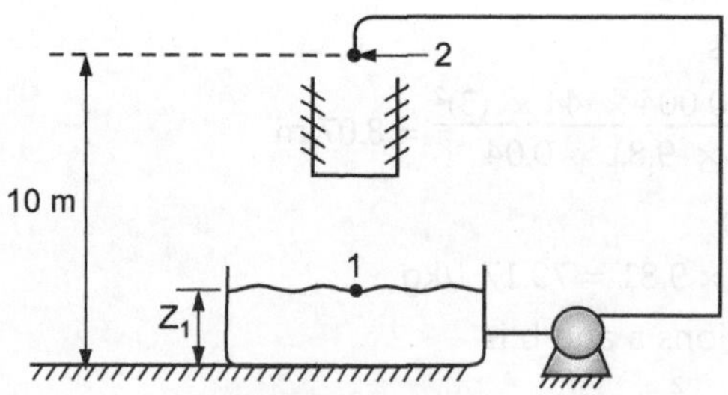

Fig. Ex. 9.6

The Bernoulli equation in terms of energy per unit mass between stations (1) and (2) with $\alpha_1 = \alpha_2 = 1.0$ is

$$\frac{P_1}{\rho} + g Z_1 + \frac{u_1^2}{2} + \eta W_p = \frac{P_2}{\rho} + g Z_2 + \frac{u_2^2}{2} + h_f \quad \text{... (1)}$$

$P_1 = P_2$ = 101325 N/m^2 as both stations are open to atmosphere.

u_1 = Negligible as compared to the velocity at station (2).

Z_1 = 0, Assuming a datum line as the surface of liquid in the tank (i.e., at station 1)

Z_2 = 10 m

η = Efficiency of the pump = 60% = 0.6

W_p = Pump work in J/kg

With this Equation (1) becomes :

$$\eta W_p = g Z_2 + \frac{u_2^2}{2} + h_f \quad \text{... (2)}$$

Volumetric flow rate = 5000 l/h

$$Q = 5000 \times 10^{-3}/3600 = 0.00139 \text{ m}^3/\text{s}$$

$$A = \frac{\pi}{4} D^2 = \text{Area of pipe in m}^2$$

$$D = 25 \text{ mm} = 0.025 \text{ m}$$

$$\therefore \quad A = \frac{\pi}{4} (0.025)^2 = 4.909 \times 10^{-4} \text{ m}^2$$

$$\therefore \quad u_2 = Q/A = 0.00139 / 4.909 \times 10^{-4} = 2.83 \text{ m/s}$$

Equivalent length of the piping system = Length of straight pipe + Equivalent length of fitting

$$L = 12 + 1 = 13 \text{ m}$$

h_f = Friction loss in terms of m of liquid between stations (1) and (2)

$$h_f = \frac{4 f L V^2}{2g D}$$

f = Friction factor

L = 13 m

u = 2.83 m/s

D = 0.025 m

$$N_{Re} = \frac{D u_2 \rho}{\mu} = \frac{0.025 \times 2.83 \times 1000}{0.0008} = 88437.5$$

Since N_{Re} is greater than 4000, the flow is turbulent and thus f is given by

$$f = \frac{0.078}{(N_{Re})^{0.25}}$$

$$f = \frac{0.078}{(88437.5)^{0.25}} = 0.0045$$

$$\therefore \quad h_f = \frac{4 \times 0.0045 \times 13 \times (2.83)^2}{2 \times 9.81 \times 0.025} = 3.82 \text{ m}$$

h_f in terms of energy per unit mass is

$$h_f = h_f \text{ in m} \times g$$

$$h_f = 3.82 \times 9.81 = 37.47 \text{ J/kg}$$

Substituting the values of u_2, h_f, η in Equation (2) gives

$$0.6 W_p = 9.81 \times 10 + \frac{(2.83)^2}{2} + 37.47$$

$$W_p = 233 \text{ J/kg}$$

Mass flow rate, $\dot{m} = \rho u_2 A = 1000 \times 2.83 \times 4.909 \times 10^{-4} = 1.39$ kg/s

The power required by the pump/supplied to the pump is given by

Power required by the pump = P_B = $\dot{m} W_p = 1.39 \times 233$ (kg/s) (J/kg) = 324 J/s = **324 W** **... Ans.**

Example 9.7 : *A pump pumps a solution having a specific gravity 1.84 from a storage tank of large cross section through a 8 cm ID pipe. The velocity in the suction pipe is 1 m/s. The pump discharges through a 5 cm ID pipe to an overhead tank. The end of the discharge line is 15 metres above the level of the solution in the tank. Friction loss in the entire system are 3 metres of the solution. Find the pressure developed by the pump and the theoretical power required to do this pumping.*

Solution : Let us calculate the velocity through the discharge pipe.

Mass flow rate through the suction pipe

= Mass flow rate through the discharge pipe

$$\therefore \quad \rho\, u_1\, A_1 = \rho\, u_2\, A_2$$

where $D_1 = 8$ cm $= 0.08$ m, $D_2 = 5$ cm $= 0.05$ m and $u_1 = 1$ m/s

$$\text{Velocity through the discharge pipe} = u_2 = u_1\,(A_1/A_2) = 1\left[\frac{\pi/4 \times (0.08)^2}{\pi/4 \times (0.05)^2}\right]$$

$$\therefore \quad u_2 = 2.56 \text{ m/s}$$

Given : $Z = 15$ m and $h_f = 3$ m.

Head developed by the pump in terms of m of solution $= \Delta H = \eta\, W_p$

$$= \frac{u_2^2}{2g} + Z + h_f = \frac{(2.56)^2}{2 \times 9.81} + 15 + 3$$

$$\Delta H = \mathbf{18.334\ m} \qquad \text{... Ans. (1)}$$

$\dot{m}$ = mass flow of rate of the solution= $\rho\, u_2\, A_2 = 1840 \times (2.56) \times \pi/4 \times (0.05)^2 = 9.25$ kg/s

The theoretical power required/power delivered to the fluid/fluid power is given by

$$\text{Theoretical power required} = P_f = \dot{m}\,\Delta H\, g$$

$$= 9.25 \times 18.334 \times 9.81 = 1664 \text{ J/s} = 1664 \text{ W} = \mathbf{1.66\ kW} \qquad \text{... Ans. (2)}$$

Example 9.8 : *Water at 203 K (20°C) is pumped at a rate of 1.0 kg/s from a storage tank through a pipe line of internal diameter 30 mm and 100 m long. The pipeline consists of two fully open globe valves and three 90° elbows. Water is discharged into an overhead tank through spray nozzles. The discharge is at a height of 20 m above the level of water in the storage tank. The required pressure at the nozzle entrance is 300 kPa gauge. Calculate the theoretical power requirement for the pump.*

Equivalent lengths in terms of pipe diameters :

Fully open globe valve = 300 D,

90° elbow = 30 D

Friction factor, $f = 0.078 / (N_{Re})^{0.25}$

ρ of water = 996 *kg/m³,*

μ of water = 0.975 mPa.s

Solution :

Mass flow rate of water $= \dot{m} = 1.0$ kg/s

$D = 30$ mm $= 0.03$ m

Cross-sectional area of the pipe $= A = \pi/4 \times (0.03)^2 = 7.07 \times 10^{-4}\ m^2$

$\rho = 996\ kg/m^3$, $\mu = 0.975$ mPa.s $= 0.975 \times 10^{-3}$ Pa.s

Let us calculate u, N_{Re}, f, total length, head and then P_f.

We have : $\dot{m} = \rho u A$

Velocity of water = u = $\dot{m} / \rho A$ = 1.0 / 996 × 7.07 × 10^{-4} = 1.42 m/s

$$N_{Re} = \frac{Du\rho}{\mu} = \frac{0.03 \times 1.42 \times 996}{0.975 \times 10^{-3}} = 45317$$

Since N_{Re} > 4000, the flow is turbulent and thus f is given by

$$f = 0.078 / (N_{Re})^{0.25} = 0.078 / (45317)^{0.25} = 0.0054$$

L_e of 2 globe valves = 2 × 300 D = 2 × 300 × 0.03 = 18 m

L_e of 3, 90° elbows = 3 × 30 D = 3 × 30 × 0.03 = 2.7 m

Equivalent length of the entire piping = L = L of straight pipe + globe valves + elbow

= 100 + 18 + 2.7 = 120.7 m

The frictional head loss is given by

$$\text{Frictional head loss} = \frac{4 f L u^2}{2g D} = \frac{4 \times 0.0054 \times 120.7 \times (1.42)^2}{2 \times 9.81 \times 0.03} = 8.9 \text{ m}$$

$$\text{Pressure head} = \frac{300 \times 10^3}{996 \times 9.81} = 30.7 \text{ m}$$

Total head = ΔH = Z + h_f + Pressure head = 20 + 8.9 + 30.7 = 59.6 m

Theoretical power required by the pump = P_f = $\dot{m}$ × head × g = 1.0 × 59.6 × 9.81 = 585 J/s = **585 W** ... **Ans.**

Example 9.9 : *A pump delivers water from a holding tank at atmospheric pressure (101.325 kPa) to a process equipment at 450 kPa at a flow rate of 6.2 l/s. The process equipment is located 10 m higher than the holding tank. Calculate the power requirement of the pump if the fluid friction and changes in kinetic energy are negligible.*

Data : *Density of water = 995 kg/m³*

Efficiency of the pump = 70%

Solution : P_2 = 450 kPa = 450 × 10³ Pa or N/m²

P_1 = 101.325 kPa = 101325 Pa or N/m²

Volumetric flow rate = Q = 6.2 l/s = 6.2 × 10^{-3} m³/s

Density of water = ρ = 995 kg/m³

Mass flow rate of water = $\dot{m}$ = 6.2 × 10^{-3} × 995 = 6.17 kg/s

$$\text{Pressure head} = \frac{\Delta P}{\rho g} = \frac{450 \times 10^{-3} - 101325}{995 \times 9.81} = 35.72 \text{ m}$$

Potential head = ΔZ = 10 m

Neglecting changes in the kinetic energy and fluid friction, we get

Total head = Pressure head + Potential head = 35.72 + 10 = 45.72 m

The efficiency of the pump is 70%.

The power required by the pump is given by

Power required by the pump = P_B = (mass flow rate × head × g)/η

= (6.17 × 45.72 × 9.81) / 0.70 = 3953 J/s = 3953 W = 3.953 kW ≃ **4 kW** ... **Ans.**

Example 9.10 : *Water is to be pumped at a rate of 55.5 l/s from a river to an overhead tank of a chemical factory situated at a height of 25 m from the river bed. The total length of pipeline is 1500 m and a mild steel pipe of 300 mm inside diameter to be used for the said purpose. Calculate the head loss due to friction and power required by the pump neglecting fitting losses.*

Data : *ρ of water = 992 kg/m³ and μ of water = 0.8 mPa.s*

Efficiency of the pump = 60%.

Take a vertical distance of 25 m for the calculation purpose.

Use the relation $f = 0.0014 + 0.125 / (N_{Re})^{0.32}$

Solution : D = 300 mm = 0.30 m, L = 1500 m

ρ = 992 kg/m³, μ = 0.8 m Pa.s = 0.8×10^{-3} Pa.s

Cross-sectional area of the pipe = A = $\pi/4\ (0.30)^2 = 0.0707\ m^2$

Volumetric flow rate of water = Q = 55.5 *l*/s = $55.5 \times 10^{-3}\ m^3/s$

Let us calculate u.

Velocity of water through the pipe = u = Q/A = $55.5 \times 10^{-3}/0.0707 = 0.785$ m/s

Let us calculate N_{Re}.

$$N_{Re} = \frac{Du\rho}{\mu} = \frac{0.3 \times 0.785 \times 992}{0.8 \times 10^{-3}} = 292020$$

Let us calculate f and h_f.

Given :
$$f = 0.0014 + \frac{0.125}{(N_{Re})^{0.32}} = 0.0014 + \frac{0.125}{(292020)^{0.32}} = 0.00363$$

The head loss due to friction is given by

$$h_f = \frac{4\,f\,L\,u^2}{2g\,D} = \frac{4 \times 0.00363 \times 1500 \times (0.785)^2}{2 \times 9.81 \times 0.30} = \mathbf{2.28\ m\ of\ H_2O} \quad \textbf{... Ans.}$$

Let us calculate the total head.

$$\text{Total head} = Z + h_f = 25 + 2.28 = 27.28\ m$$

$$\text{Mass flow rate} = \dot{m} = 55.5 \times 10^{-3} \times 992 = 55.056\ kg/s$$

Let us calculate P_B.

For the efficiency of 60%, the power required by the pump is given by

$$P_B = (\dot{m} \times \text{head} \times g) / \eta = 55.056 \times 27.28 \times 9.81 / 0.6$$
$$= 24566\ J/s = 24566\ W = \mathbf{24.56\ kW} \quad \textbf{... Ans.}$$

Example 9.11 : *A mild steel pipeline of 200 mm inside diameter is to be installed for transporting 16.7 kg/s of molasses. The pipeline is 1000 m long and the delivery end is to be 5 m higher than the intake. Calculate the power requirement for the pump if the efficiency of the pump is 60%.*

Data : *ρ of molasses = 1600 kg/m³ and μ of molasses = 0.5 Pa.s*

Solution : D = 200 mm = 0.20 m, L = 1000 m, ρ = 1600 kg/m³, μ = 0.5 Pa.s

Mass flow rate of molasses = $\dot{m}$ = 16.7 kg/s

Cross-sectional area of the pipeline = A = $\pi/4\ (0.2)^2 = 0.0314\ m^2$

We have, $\dot{m} = \rho u A$

Let us calculate u, N_{Re}, f, h_f, ΔH and then P_B.

Velocity of molasses through the pipeline = u = $\dot{m}/\rho A$ = 16.7 / 1600 × 0.0314 = 0.332 m/s

$$N_{Re} = Du\rho/\mu = 0.20 \times 0.332 \times 1600 / 0.5 = 212.48$$

Since N_{Re} is less than 2100, the flow is laminar.

For laminar flow, f and N_{Re} are related by

$$f = 16 / N_{Re} = 16 / 212.48 = 0.0753$$

The head loss due to friction is given by

$$h_f = \frac{\Delta P}{\rho g} = \frac{32\,\mu\,u\,L}{D^2.\,\rho\,g}$$

$$h_f = \frac{32 \times 0.5 \times 0.332 \times 1000}{(0.2)^2 \times 1600 \times 9.81} = 8.46 \text{ m}$$

Total head loss = $Z + h_f = 5 + 8.46 = 13.46$ m

Pump efficiency = η = 60% or 0.60

Power required by the pump = $P_B = (\dot{m} \times \text{head} \times g)/\eta = 16.7 \times 13.46 \times 9.81 / 0.60$

= 3675 J/s = 3675 W = 3.675 kW = **3.7 kW** ... **Ans.**

Example 9.12 : *A centrifugal pump with an efficiency of 65% is driven by an electric motor having an efficiency of 90%. The pump delivers water at a rate of 4 kg/s against the total head of 25 m. Calculate the power required by the motor and the power delivered by the motor to the pump.*

Solution : Mass flow rate of water = $\dot{m}$ = 4 kg/s

Total head = ΔH = 25 m

Efficiency of the pump = $\eta = 0.65$

Efficiency of the motor = $\eta_m = 0.95$

Power delivered to the fluid = Theoretical power = Fluid power

= P_f = mass flow rate × head × g = 4 × 25 × 9.81 = 981 J/s = 981 W

Power input to the pump by the motor = Power supplied to the pump by the motor

Power delivered by the motor to the pump = $P_B = P_f/\eta$ = 981/0.65 = **1509 W** ... **Ans.**

P_m = Power taken by the motor from the power line

= Power requirement of the pump-motor set = Power required by the motor

The motor efficiency is given by

$$\eta_m = P_B/P_m$$

$P_m = P_B/\eta_m$ = 1509/0.90 = **1677 W** ... **Ans.**

Example 9.13 : *Crude oil is to be transported at a rate of 7500 m³/day from an oil field to a refinery which is located at 750 km away from the oil field through a 400 mm i.d. mild steel pipe.*

(a) Calculate the theoretical power requirement for the pump.

(b) Since the maximum allowable pressure at any section of the pipeline is 2.94 MPa, it will be required to install pumping stations at suitable intervals/distances along the pipeline. Each station increases the pressure which drops to 165 kPa at the inlet of the next pumping station. Find the minimum number of pumping stations required.

Data : *Density of oil = 870 kg/m³*

Viscosity of oil = 48 mPa.s

Solution :

$$\text{Volumetric flow rate} = Q = 7500 \text{ m}^3\text{/day} = \frac{7500}{24 \times 3600} = 0.087 \text{ m}^3\text{/s}$$

ρ = 870 kg/m³, μ = 48 mPa.s = 48 × 10^{-3} Pa.s, D = 400 mm = 0.40 m

Mass flow rate = $\dot{m}$ = 0.087 × 870 = 75.69 kg/s

L = 750 km = 750000 m

Cross-sectional area of the pipe = A = $\pi/4\ (0.40)^2$ = 0.126 m²

Velocity of crude oil = u = Q/A = 0.087 / 0.126 = 0.69 m/s

$$N_{Re} = \frac{Du\rho}{\mu} = \frac{0.40 \times 0.69 \times 870}{48 \times 10^{-3}} = 5002.5$$

Since $N_{Re} > 4000$, the flow is turbulent. For this flow, f may be calculated by

$$f = 0.078 / (N_{Re})^{0.25} = 0.078 / (5002.5)^{0.25} = 0.0093$$

The frictional head loss is given by

$$h_f = \frac{4 f L u^2}{2g D} = 4 \times 0.0093 \times 750000 \times (0.69)^2 / 2 \times 9.81 \times 0.40 = 1692.6 \text{ m}$$

Theoretical power required = fluid power = $P_f = \dot{m} \times \text{head} \times g$

Theoretical power required $= 75.69 \times 1692.6 \times 9.81 = 1256787$ J/s $= 1256787$ W $=$ **1256.8 kW** **... Ans.**

Maximum allowable pressure at any section = 2.94 MPa

$= 2.94 \times 10^6$ Pa

Let the pressure increased by one station be the maximum allowable pressure which is 2.94×10^6 Pa

Pressure at the inlet of the next station $= 165$ kPa $= 165 \times 10^3$ Pa

Pressure drop between two stations $= 2.94 \times 10^6 - 165 \times 10^3 = 2775000$ Pa

Frictional head loss between two stations $= \frac{\Delta P}{\rho g} = \frac{2775000}{870 \times 9.81} = 325.14$ m

Total frictional head loss from the oil field to the refinery = 1692.6 m

Number of pumping stations required $= 1692.6 / 325.14 = 5.2 \approx$ **5** **... Ans.**

Example 9.14 : *It is required to install a pipeline of 200 mm inside diameter to transport 3800 m³ of oil per day. The pipeline is 32 km long and the delivery end is 30 m higher than the intake. The pressure drop due to friction (i.e., frictional pressure drop) in the pipeline is estimated to be 5.3 MPa. Determine the power requirement of the pump-motor set.*

Data : *ρ of oil = 897 kg/m³ and μ of oil = 50 mPa.s*

Overall efficiency of the pump-motor set = 60%.

Solution :

Volumetric flow rate of oil $= Q = 3800 \text{ m}^3/\text{day} = \frac{3800}{24 \times 3600} = 0.044 \text{ m}^3/\text{s}$

D = 200 mm = 0.2 m, L = 2 km = 2000 m

$\rho = 897 \text{ kg/m}^3$, $\mu = 50 \text{ mPa.s} = 50 \times 10^{-3}$ Pa.s

Cross-sectional area of the pipeline $= A = \pi/4\,(0.2)^2 = 0.0314 \text{ m}^2$

Velocity of oil through the pipeline $= u = Q/A = 0.044 / 0.0314 = 1.40$ m/s

Frictional head loss $= h_f = \frac{\Delta P}{\rho g}$

where $\Delta P = 5.3 \text{ MPa} = 5.3 \times 10^6$ Pa (i.e., N/m²)

$h_f = 5.3 \times 10^6 / 897 \times 9.81 = 602.3$ m

Therefore, total head is

Total head $= \Delta H = Z + h_f = 602.3 + 30 = 632.3$ m

Mass flow rate of oil $= \dot{m} = 0.044 \times 897 = 39.47$ kg/s

η_o = pump-motor efficiency = 60% or 0.60

The power requirement of the pump-motor set of 60% overall efficiency is

Power requirement $= P_m = P_f/\eta_o = (\text{mass flow rate} \times \text{head} \times g) / \eta_o$

$= 39.47 \times 632.3 \times 9.81 / 0.6 = 408045$ J/s $= 408045$ W $=$ **408 kW** **... Ans.**

Example 9.15 : *Benzene is pumped at a rate of 9.09 m³/h from a reservoir at atmospheric pressure. The gauge pressure at the end of discharge line is 345 kN/m².g. The discharge is 3.5 m and the pump suction is 1.3 m above the level in the reservoir. The diameter of the discharge line is 40 mm. The friction in the suction line is 3.45 kN/m² and that in the discharge line is 37.9 kN/m². The efficiency of the pump is 60%. The density of benzene is 865 kg/m³ and its vapour pressure at the temperature of pumping [310 K (37°C)] is 26.2 kN/m². Calculate :*

(i) The developed head of the pump,

(ii) The power input to the pump and

(iii) The net positive suction head.

Solution :

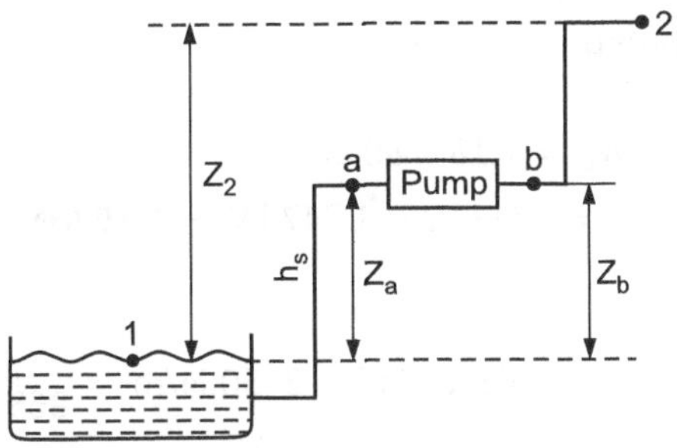

Fig. Ex. 9.15

Volumetric flow rate = Q = 9.09 m³/h = 0.002525 m³/s

D = diameter of discharge pipe = 40 mm = 0.04 m

Cross-sectional area of the pipe = A = $\pi/4 \times (0.04)^2 = 0.00126$ m²

Velocity of benzene through the discharge pipe

$$u_2 = \frac{Q}{A} = \frac{0.002525}{0.00126} = 2 \text{ m/s}$$

Assume station 1 to be at the level of benzene in the reservoir and station 2 to be at the end of the discharge line.

Assume the benzene level in the reservoir as a datum line. Then,

$Z_1 = 0$ and $u_1 = 0$.

The Bernoulli's equation between stations 1 and 2 is

$$\frac{P_1}{\rho} + g Z_1 + \frac{u_1^2}{2} + \eta W_p = \frac{P_2}{\rho} + g Z_2 + \frac{u_2^2}{2} + h_f$$

Each term in the above equation has the units of J/kg (i.e., of energy)

Therefore, let us find h_f in J/kg.

$$h_f = \text{total friction loss} = h_{f_s} + h_{f_d} = 3.45 + 37.9$$

$$= 41.35 \text{ kN/m}^2 = 41350 \text{ N/m}^2 = \frac{41350}{865} = 47.8 \text{ J/kg}$$

[ρ of benzene = 865 kg/m³]

Given : P_1 = 345 kN/m².g. Convert it into absolute pressure.

∴ $P_2 = 345 + 101.325 = 446.325$ kN/m² = 446325 N/m²

$P_1 = 101.325$ kN/m² = 101325 N/m²

$Z_2 = 3.5$ m, $\eta = 0.60$

Let us find the developed head.

Thus, the developed head in J/kg by the pump with $Z_1 = 0$ and $u_1 = 0$ becomes

$$\eta W_p = \frac{P_2}{\rho} + g Z_2 + \frac{u_2^2}{2} + h_f - \frac{P_1}{\rho}$$

$$= \frac{446325}{865} + 9.81 \times 3.3 + \frac{(2)^2}{2} + 47.8 - \frac{101325}{865} = \mathbf{481\ J/kg}$$ **... Ans. (i)**

$$W_p = \frac{481}{\eta} = \frac{481}{0.60} = 801.67 \text{ J/kg}$$

$\dot{m}$ = mass flow rate of benzene = 0.002525 × 865 = 2.18 kg/s

Let us find the power input to the pump.

Power input to the pump = $P_B = \dot{m} W_p$ = 2.18 × 801.67

= 1747 J/s = 1747 W = 1.747 kW ≈ **1.75 kW** **... Ans. (ii)**

Let us find NPSH.

Friction in the suction line = h_{fs} = 3.45 kN/m² = 3450 N/m² = $\frac{3450}{865 \times 9.81}$ = 0.41 m

P_v = vapour pressure of benzene = 26.2 kN/m² = 26200 N/m² = $\frac{26200}{865 \times 9.81}$ = 3.1 m

P_1 = pressure at the surface of reservoir= 101325 N/m² = $\frac{101325}{865 \times 9.81}$ = 11.94 m

Z_a = 1.3 m

Available NPSH in terms of m of the flowing fluid is given by

$$\text{NPSH} = P_1 - P_v - Z_a - h_{f_s} = 11.34 - 3.1 - 1.3 - 0.41 = \mathbf{6.53\ m}$$ **... Ans. (iii)**

Example 9.16 : *A centrifugal pump pumps brine from the bottom of the supply tank and delivers it into the bottom of another tank. The level of the brine in the receiving tank is 50 m above that in the supply tank. The tanks are connected by a 180 mm pipe of length 200 m. The flow rate of brine is 0.05 m³/s. The pipeline between the tank has two gate valves, and 8 other pipe fittings. What is the energy cost for running this pump for a 24-h day ?*

Data : *Density of brine = 1180 kg/m³, viscosity of brine = 1.2 mPa.s, one gate valve is equivalent to 7 pipe diameters and each of the fittings is equivalent to 60 pipe diameters.*

Energy costs Rs. 0.80 per kW.h and the overall efficiency of the pump-motor set is 60%.

Solution : Consider stations 1 and 2 at the brine surfaces in the supply and delivery tanks respectively.

∴ $P_1 = P_2$ and $u_1 = u_2$, $Z_1 = 0$ and $Z_2 = 50$ m

D = 180 mm = 0.18 m, L = 200 m, ρ = 1180 kg/m³, μ = 1.2 mPa.s = 1.2 × 10⁻³ Pa.s

L_e = equivalent length of the piping system

= 200 + 2 × 7 × 0.18 + 8 × 60 × 0.18 = 288.92 m

Volumetric flow rate = Q = 0.05 m³/s

Cross-sectional area of the pipe = A = π/4 (0.18)² = 0.02545 m²

Velocity of brine through the pipe = u = Q/A = 0.05 / 0.02545 = 1.965 m/s

Let us calculate N_{Re}.

$$N_{Re} = D u \rho/\mu = 0.18 \times 1.965 \times 1180/1.2 \times 10^{-3} = 347805$$

Since N_{Re} > 4000, the flow is turbulent.

For turbulent flow, f may be related to N_{Re} by

$$f = 0.078 / (N_{Re})^{0.25} = 0.078 / (347805)^{0.25} = 0.0032$$

The head loss due to friction is

$$h_f = 4\,f\,L\,u^2 / 2g\,D = \frac{4 \times 0.0032 \times 288.92 \times (1.965)^2}{2 \times 9.81 \times 0.18} = 4 \text{ m}$$

$$\text{Total head} = Z + h_f = 50 + 4 = 54 \text{ m}$$

$$\text{Mass flow rate } \dot{m} = 0.05 \times 1180 = 59 \text{ kg/s}$$

Overall efficiency of the pump-motor set = η_o = 0.60.

Power requirement for the pump-motor set is

$$P_m = P_f/\eta_o = (\text{mass flow rate} \times \text{head} \times g) / \eta_o$$

$$= 59 \times 54 \times 9.81 / 0.6 = 52091 \text{ W} = 52.1 \text{ kW}$$

$$\text{Energy required per day} = 52.1 \times 24 = 1250.4 \text{ kWh}$$

$$\text{Energy cost per day} = 1250.4 \times \frac{0.8}{1} = \textbf{Rs. 1000.32}$$ **... Ans.**

Example 9.17 : *600 cm³/s of water at 320 K (47°C) is pumped through a 40 mm internal diameter pipe over a length of 150 m in a horizontal direction and up through a vertical height of 10 m. In the piping system there is a one control valve which may be taken as equivalent to 200 pipe diameters and other pipe fittings equivalent to 60 pipe diameters. Also in the system there is a heat exchanger across which there is a loss in head of 1.5 m of water. What power must be delivered to the pump, if the pump efficiency is 60% ?*

Data : *ρ of water = 1000 kg/m³*

μ of water = 1.0 (mN.s)/m²

Solution : D = 40 mm = 0.04 m, Volumetric flow rate = Q = 600 cm³/s = 6.0×10^{-4} m³/s

Cross-sectional area of the pipe = A = $\pi/4 \times (0.04)^2 = 1.26 \times 10^{-3}$ m²

Velocity of water, u = Q/A = $6.0 \times 10^{-4}/1.26 \times 10^{-3}$ = 0.476 m/s

$\mu = 1$ (mN.s)/m² = 1×10^{-3} (N.s)/m² ≡ kg/(m.s)

Let us calculate N_{Re}.

Reynolds number, $N_{Re} = Du\rho/\mu = 0.04 \times 0.476 \times 1000 / 1 \times 10^{-3} = 19046$

Let us calculate f. Since $N_{Re} > 4000$, the flow is turbulent and for this type of flow, we have

$f = 0.078 / (N_{Re})^{0.25} = 0.078 / (19046)^{0.25} = 0.0066$

Equivalent length of valve = 200 D = 200 × 0.04 = 8 m

Equivalent length of fittings = 60 D = 60 × 0.04 = 2.4 m

Equivalent length of entire piping = L = 150 + 10 + 8 + 2.4 = 170.4 m

The head loss due to friction, h_f is given by

$$h_f = 4\,f\,L\,u^2/2\,g\,D = 4 \times 0.0066 \times 170.4 \times (0.476)^2 / (2 \times 9.81 \times 0.04) = 1.3 \text{ m}$$

$$\text{Head load across exchanger} = h_{ex} = 1.5 \text{ m}$$

$$\text{Total head} = Z + h_f + h_{ex} = 10 + 1.3 + 1.5 = 12.8 \text{ m}$$

$$\text{Mass flow rate} = \dot{m} = Q \times \rho$$

$$\text{Mass flow rate} = \dot{m} = 6.0 \times 10^{-4} \times 1000 = 0.60 \text{ kg/s}$$

Efficiency of the pump = 60% or 0.60

∴ Actual power delivered to the pump = $P_B = P_f/\eta$

$$= (\text{mass flow rate} \times \text{head} \times g)/\eta = (0.6 \times 12.8 \times 9.81) / 0.60 = \textbf{126 W}$$ **... Ans.**

Example 9.18 : *Calculate the power required to pump oil at a rate of 4000 cm³/s through a 50 mm pipeline 100 m long. The outlet of the pipeline is 15 m higher than the inlet.*

Data : *Specific gravity of oil = 0.85*

Viscosity of oil = 3 (mN.s)/m²

The efficiency of the pump is 50%.

Solution : D = 50 mm = 0.05 m, L = 100 m (total), $Q = 4000\ cm^3/s = 4000 \times 10^{-6}\ m^3/s$

$$\rho \text{ of oil} = \text{specific gravity of oil} \times \rho \text{ of water} = 0.85 \times 1000 = 850\ kg/m^3$$

$$\text{Cross-sectional area of the pipe} = A = \pi/4 \times (0.05)^2 = 1.96 \times 10^{-3}\ m^2$$

$$\text{Velocity of oil, } u = Q/A = \frac{4000 \times 10^{-6}}{1.96 \times 10^{-3}} = 2.04\ m/s$$

$$\mu \text{ of oil} = 3\ (mN.s)/m^2 = 3 \times 10^{-3}\ (N.s)/m^2 = kg/(m.s)$$

$$\text{Reynolds number, } N_{Re} = Du\rho/\mu = \frac{0.05 \times 2.04 \times 850}{3 \times 10^{-3}} = 28900$$

Since $N_{Re} > 4000$, the flow is turbulent.

For turbulent flow, we have $f = 0.078/(N_{Re})^{0.25} = 0.078/(28900)^{0.25} = 0.00598 \simeq 0.006$

The head loss due to friction h_f is

$$h_f = 4\,f\,L\,u^2/2gD = 4 \times 0.006 \times 100 \times (2.04)^2 / (2 \times 9.81 \times 0.05) = 10.2\ m$$

$$\text{Total head} = Z + h_f = 15 + 10.2 = 25.2\ m$$

$$\text{Mass flow rate} = \dot{m} = 4000 \times 10^{-6} \times 850 = 3.4\ kg/s$$

$$\text{Power required by the pump} = P_B = P_f/\eta$$

$$= (\dot{m} \times \text{head} \times g)/\eta = (3.4 \times 25.2 \times 9.81) / 0.5 = 1681\ W = \mathbf{1.68\ kW}$$ ... **Ans.**

Example 9.19 : *Dilute sulphuric acid is to be pumped through a 25 mm lead pipe at a rate of 4000 cm³/s. It is desired to raise it to a height of 25 m. The pipe is 30 m long and includes two right angled bends. Calculate the theoretical power required.*

Data : *Specific gravity of acid = 1.531, Kinematic viscosity of acid = 0.425 cm²/s, Density of water = 1000 kg/m³.*

Assume that 0.8 velocity heads are lost through each 90° bend.

Solution : D = 25 mm = 0.025 m, L = 30 m (total)

Volumetric flow rate = $Q = 4000\ cm^3/s = 4 \times 10^{-3}\ m^3/s$

Cross-sectional area = $A = \pi/4 \times (0.025)^2 = 4.909 \times 10^{-4}\ m^2$

$$\text{Velocity, } u = Q/A = \frac{4.0 \times 10^{-3}}{4.909 \times 10^{-4}} = 8.15\ m/s$$

$\rho = 1531\ kg/m^3$, [as ρ = specific gravity × density of water]

$\upsilon = 0.425 \times 10^{-4}\ m^2/s$

$\mu = 0.425 \times 10^{-4} \times 1531 = 0.065\ (N.s)/m^2 = kg/(m.s)$

[As $\upsilon = \mu/\rho$]

$$N_{Re} = \frac{Du\rho}{\mu} = \frac{0.025 \times 8.15 \times 1531}{0.065} = 4799$$

Since $N_{Re} > 4000$, the flow is turbulent. For turbulent flow, the Fanning friction factor f, may be calculated by

$$f = 0.078 / (N_{Re})^{0.25} = 0.078 / (4799)^{0.25} = 0.0094$$

Head loss due to friction is given by

$$h_f = \frac{4\,f\,L\,u^2}{2\,g\,D} = \frac{4 \times 0.0094 \times 30 \times (8.15)^2}{2 \times 9.81 \times 0.025} = 152.75 \text{ m}$$

Given : Head loss in each bend = $0.80\ u^2/2g$

Head loss in the bend = $h_{ff} = 2\,[0.8\,u^2 / 2g] = 2 \times 0.8 \times (8.15)^2 / (2 \times 9.81) = 5.42$ m

Total head = $Z + h_f + h_{ff} = 25 + 152.75 + 5.42 = 183.17$ m

Theoretical power required = Power delivered to fluid = P_f = mass flow rate × head × g

Mass flow rate = $\dot{m} = Q \times \rho = 4.0 \times 10^{-3} \times 1531 = 6.124$ kg/s

Theoretical power required = $\dot{m}$ × head × g = 6.124 × 183.17 × 9.81 = 11004 W = **11 kW** ... **Ans.**

Example 9.20 : *4.5 m^3/h of water is to be pumped through a 25 mm i.d. pipe, 30 m long, to a tank 12 m higher than its reservoir. Calculate the power required. Also find the pump rating if the efficiency of the pump is 60%.*

Data : Viscosity of water = 1.30 (mN.s)/m^2

Density of water = 1000 kg/m^3

Solution : Volumetric flow rate = Q = 4.5 m^3/h = 1.25×10^{-3} m^3/s,

D = 25 mm = 0.025m , L = 30 m (total) , ρ = 1000 kg/m^2

μ = 1.30 (mN.s)/m^2 = 1.30×10^{-3} (N.s)/m^2 = 1.30×10^{-3} Pa.s ≡ kg/(m.s)

Cross-sectional area = A = $\pi/4\,(0.025)^2 = 4.909 \times 10^{-4}$ m^2

$$\text{Velocity, } u = Q/A = \frac{1.25 \times 10^{-3}}{4.909 \times 10^{-4}} = 2.55 \text{ m/s}$$

$N_{Re} = Du\rho/\mu = 0.025 \times 2.55 \times 1000 / 1.30 \times 10^{-3} = 49038$

Since N_{Re} > 4000, the flow is turbulent. For turbulent flow, the Fanning friction factor, 'f ' may be calculated by

$$f = \frac{0.078}{(N_{Re})^{0.25}} = \frac{0.078}{(49038)^{0.25}} = 0.0052$$

The head loss due to friction is given by

$$h_f = \frac{4\,f\,L\,u^2}{2\,g\,D}$$

$$h_f = \frac{4 \times 0.0052 \times 30 \times (2.55)^2}{2 \times 9.81 \times 0.025} = 8.3 \text{ m of } H_2O$$

Total head = $Z + h_f = 12 + 8.3 = 20.3$ m

Mass flow rate = Volumetric flow rate × density

Mass flow rate = $(1.25 \times 10^{-3}) \times 1000 = 1.25$ kg/s

Theoretical power required = P_f = Mass flow rate × head × g = 1.25 × 20.3 × 9.81 = **249 W** ... **Ans.**

Pump rating i.e. power supplied to the pump

$P_B = P_f/\eta = 249 / 0.6 =$ **415 W** ... **Ans.**

Example 9.21 : *Sulphuric acid of strength 75% is to be pumped for a distance of 0.8 km through a 50 mm i.d. pipe at a rate of 3.0 kg/s and then raised vertically 15 m by the pump. The efficiency of the pump is 50%. Find the power will be required for this duty.*

Data : *Density of acid = 1650 kg/m^3*

Viscosity of acid = 8.6 (mN.s)/m^2

Solution :

D = 50 mm = 0.05 m, L = 0.80 km = 800 m (total)

ρ = 1650 kg/m^3, μ = 8.6 (mN.s)/m^2 = 8.6 × 10^{-3} (N.s)/m^2

Cross-sectional area of the pipe = $\pi/4 \times (0.05)^2 = 1.963 \times 10^{-3}$ m^2

Velocity, u = $\dot{m}/(\rho A)$ = 3 / (1650 × 1.963 × 10^{-3}) = 0.93 m/s

$$\text{Reynolds number, } N_{Re} = \frac{Du\rho}{\mu} = \frac{0.05 \times 0.93 \times 1650}{8.6 \times 10^{-3}} = 8921.5$$

As N_{Re} is greater than 4000, the flow is turbulent.

For turbulent flow, Fanning friction factor 'f' can be calculated as :

$$f = 0.078 / (N_{Re})^{0.25} = 0.078 / (8291.5)^{0.25} = 0.008$$

Head loss due to friction is given by :

$$h_f = \Delta P/\rho.g = 4\, f\, L\, u^2 / (2\, g\, D)$$

$$h_f = 4 \times 0.008 \times 800 \times (0.93)^2 / (2 \times 9.81 \times 0.05) = 22.6 \text{ m}$$

Total head = 22.6 + 15 = 37.6 m

Power delivered to fluid = P_f = mass flow rate × head × g = 3.0 × 37.6 × 9.81 = 1106 (N.m)/s = J/s = W

Theoretical power required = 1106 W

Pump has an efficiency of 50%, i.e., η = 0.6

∴ Power required by the pump or power input to the pump = $P_B = P_f/\eta$

= 1106/0.50 = 2212 W = **2.212 kW** **... Ans.**

Example 9.22 : *1.3 kg/s of commercial grade sulphuric acid is to be pumped through a 25 mm diameter pipe, 30 m long, to a tank which is 12 m higher than the acid reservoir. Calculate the actual power required by the pump if the efficiency of the pump is 55%.*

Data : *Density of acid = 1840 kg/m^3*

Viscosity of acid = 0.025 (N.s)/m^2

Solution : Acid flow rate = $\dot{m}$ = 1.3 kg/s

D = 25 mm = 0.025 m, L = 30 m (total)

ρ = 1840 kg/m^3, μ = 0.025 (N.s)/m^2

$$\text{Cross-sectional area} = A = \frac{\pi}{4}(0.025)^2 = 4.909 \times 10^{-4} \text{ m}^2$$

Let us calculate u.

$$\text{Velocity } u = \dot{m}/\rho A = \frac{1.3}{1840 \times 4.909 \times 10^{-4}} = 1.44 \text{ m/s}$$

Let us calculate N_{Re} to get idea of the nature of flow.

$$N_{Re} = \frac{D\,u\,\rho}{\mu} = \frac{0.025 \times 1.44 \times 1840}{0.025} = 2650$$

Since N_{Re} > 2100 and < 4000, the flow is in a transition region and for practical purpose it is treated as turbulent.

The Fanning friction factor, 'f', may be calculated by

$$f = 0.078/(N_{Re})^{0.25} = 0.078/(2650)^{0.25} = 0.0109$$

The head loss due to friction is given by

$$h_f = \Delta P / \rho.g = 4\, f\, L\, u^2 / 2\, g\, D$$

$$= 4 \times 0.0109 \times 30 \times (3.32)^2 / (2 \times 9.81 \times 0.025) = 29.4 \text{ m}$$

Total head = Potential head + Frictional head = $Z + h_f = 12 + 29.4 = 41.4$ m

Power delivered to fluid = mass flow rate × head × g = $1.3 \times 41.4 \times 9.81 = 528$ W

Pump has an efficiency of 55%. Therefore, $\eta = 0.55$.

Power required by the pump = $P_B = P_f/\eta = 528/0.55 = 960$ W = **0.96 kW** ... **Ans.**

Example 9.23 : *A single acting reciprocating pump has a stroke of 300 mm length and a piston diameter of 150 mm. The delivery and suction heads are 20 m and 4 m respectively. The pump works at 60 r.p.m. Find the horse power required to drive the pump if the efficiency of the pump is 90%.*

Solution : Stroke length = L = 300 mm = 0.30 m

Diameter of piston = d = 150 mm = 0.15 m

$$\text{Area} = A = \frac{\pi}{4} d^2 = \frac{\pi}{4} (0.15)^2 = 0.0177 \text{ m}^2$$

Delivery head = H_d = 20 m

Suction head = H_s = 4 m

Speed = N = 60 rpm

Theoretical discharge of the pump is

$$Q_{th} = \frac{LAN}{60} \text{ m}^3/\text{s} = \frac{0.3 \times 0.0177 \times 60}{60} = 5.31 \times 10^{-3} \text{ m}^3/\text{s}$$

The theoretical power required by the pump is

$$P = \frac{\rho\ Q_{th}\ [H_s + H_d]}{75} = \frac{\rho\ Q_{th}\ H}{75}, \text{ h.p.} = \frac{1000 \times 5.31 \times 10^{-3} \times [20 + 4]}{75}$$

$$P_{theoretical} = 1.6992 \approx 1.7 \text{ hp}$$

Efficiency of the pump = 90% or 0.90

The actual power required by the pump is

$$P_{actual} = \frac{1.7}{0.9} = 1.888 \approx \mathbf{1.89\ hp} \qquad \text{... Ans.} \left(\because\ \rho_p = \frac{P_{act}}{P_{th}}\right)$$

Example 9.24 : *A single acting reciprocating pump having a cylinder diameter of 150 mm and a stroke of 300 mm delivers water through a height of 25 m at a speed of 60 rpm. Find the theoretical power required and the theoretical discharge of the pump, if the actual discharge is 4.8 l/s. Calculate the percentage slip.*

Solution : Diameter of cylinder or bore = 150 mm = 0.15 m

$$\text{Area of cylinder} = A = \frac{\pi}{4} d^2 = \frac{\pi}{4} (0.15)^2$$

$$A = 0.0177 \text{ m}^2$$

Length of stroke = L = 300 mm = 0.3 m

Actual discharge = Q_{act} = 4 l/s

Speed = N = 60 rpm

Theoretical discharge of the pump,

$$Q_{th} = \frac{LAN}{60}, \text{ m}^3/\text{s} = \frac{0.3 \times 0.0177 \times 60}{60} = \mathbf{5.31 \times 10^{-3}\ m^3/s} \qquad \text{... Ans.}$$

$$= \mathbf{5.31\ l/s} \qquad \text{... Ans.}$$

$$\text{Coefficient of discharge} = C_d = \frac{Q_{act}}{Q_{th}} = \frac{4}{5.31} = 0.753$$

$$\text{Slip of the pump} = Q_{th} - Q_{act} = 5.31 - 4 = 1.31\ l/s$$

$$\text{The percentage slip of the pump} = \frac{Q_{th} - Q_{act}}{Q_{th}} \times 100 = \frac{5.31 - 4}{5.31} \times 100 = \mathbf{24.67\%} \qquad \textbf{... Ans.}$$

$$\text{Total head of the pump} = H = H_s + H_d = 25\ m$$

$$\text{Theoretical discharge} = 5.31\ l/s = 5.31 \times 10^{-3}\ m^3/s$$

$$\text{Theoretical power required to drive the pump} = \rho g\, Q_{th}\, [H]$$

$$= 1000 \times 9.81 \times 5.31 \times 10^{-3} \times 25 = 1302\ J/s = \mathbf{1302\ W} \qquad \textbf{... Ans.}$$

$$= \mathbf{1.302\ kW} \qquad \textbf{... Ans.}$$

OR :

$$P = \frac{\rho Q_{th} H}{75}\ hp = \frac{1000 \times 5.31 \times 10^{-3} \times 25}{75} = \mathbf{1.77\ hp} \qquad \textbf{... Ans.}$$

$$\text{Power} = \rho \cdot g \cdot Q_{th} \cdot H$$

$$\text{Unit of power} \Rightarrow \frac{kg}{m^3} \times \frac{m}{s^2} \times \frac{m^3}{s} \times m$$

We know that : $N.m = J \quad \therefore \quad m = J/N$

and $N \Rightarrow (kg.m)/s^2$

$$\therefore \quad m = \frac{J.s^2}{kg.m}$$

$$\text{Unit of power} \Rightarrow \frac{kg}{m^3} \cdot \frac{m}{s^2} \cdot \frac{m^3}{s} \cdot \frac{J.s^2}{kg.m} \Rightarrow \frac{J}{s} = W$$

Example 9.25 : *A single acting reciprocating pump having a cylinder diameter of 150 mm and a stroke of 300 mm length discharges 200 l/min of water at 40 r.p.m. Find the theoretical discharge in l/min and the percentage slip of the pump.*

Solution : Diameter of cylinder or bore = d = 150 mm = 0.15 m

$$\text{Length of stroke} = L = 300\ mm = 0.30\ m$$

$$\text{Area of cylinder/bore} = \frac{\pi}{4} d^2 = \frac{\pi}{4} (0.15)^2 = 0.0177\ m^2$$

$$\text{Speed} = N = 40\ r.p.m.$$

$$\text{Theoretical discharge of the pump} = Q_{th} = \frac{LAN}{60}\ , m^3/s = \frac{0.3 \times 0.0177 \times 40}{60} = 3.54 \times 10^{-3}\ m^3/s = 3.54\ l/s$$

$$Q_{th} = 3.54 \times 60 = \mathbf{212.4\ l/min} \qquad \textbf{... Ans.}$$

$$\text{Actual discharge} = Q_{act} = 200\ l/s$$

$$\text{Coefficient of discharge} = C_d = \frac{\text{Actual discharge}}{\text{Theoretical discharge}} = \frac{200}{212.4} = 0.9416$$

The slip of the pump is, $\text{slip} = Q_{th} - Q_{act} = 212.4 - 200 = 12.4\ l/min$

$$\%\ \text{slip of the pump} = \left(\frac{Q_{th} - Q_{act}}{Q_{th}}\right) \times 100 = \left(\frac{212.4 - 200}{212.4}\right) \times 100 = 5.838 \approx \mathbf{5.84} \qquad \textbf{... Ans.}$$

Example 9.26 : *A double acting reciprocating pump has a stroke length of 400 mm and a piston of diameter 150 mm. The delivery and suction heads are 26 m and 4 m respectively. Find the theoretical power required if the pump is working with 50 r.p.m. and pumping water.*

Solution : Stroke length = L = 400 mm = 0.40 m

Diameter of piston = d = 150 mm = 0.15 m

$$\text{Area} = A = \frac{\pi}{4} \cdot d^2 = \frac{\pi}{4}(0.15)^2 = 0.0177\ m^2$$

Delivery head = H_d = 26 m

Suction head = H_s = 4 m

Total head = H = $H_d + H_s$ = 26 + 4 = 30 m

Speed = N = 50 r.p.m.

Theoretical discharge from the double acting pump = Q_{th}

$$Q_{th} = \frac{2LAN}{60},\ m^3/s = \frac{2 \times 0.40 \times 0.0177 \times 50}{60} = 0.0118\ m^3/s = 11.8\ l/s$$

Theoretical power required to drive the pump = P

$$P = \frac{\rho\, Q_{th}\, H}{75} = \frac{\rho\, Q_{th}\,[H_s + H_d]}{75} = \frac{1000 \times 0.0118 \times 30}{75} = \mathbf{4.72\ hp} \quad \text{... Ans.}$$

OR :

$$P = \rho g Q_{th} H = 1000 \times 9.81 \times 0.0118 \times 30$$

$$= 3472.7\ J/s = 3472.7\ W = 3.4727 \approx \mathbf{3.47\ kW} \quad \text{... Ans.}$$

Example 9.27 : *A single acting reciprocating pump has a plunger diameter of 15 cm and a stroke length of 22.5 cm. The pump delivers 115 l of water per minute at 30 r.p.m. Find the theoretical discharge of the pump and slip of the pump.*

Solution : Diameter of plunger = d = 15 cm = 0.15 m

$$\text{Area} = A = \frac{\pi}{4} \cdot d^2 = \frac{\pi}{4}(0.15)^2 = 0.0177\ m^2$$

c

Stroke length = L = 22.5 cm = 0.225 m

Actual discharge = Q_{actual} = 115 l/min

Speed = N = 30 r.p.m.

$$\text{Theoretical discharge} = \frac{LAN}{60},\ m^3/s = \frac{0.225 \times 0.0177 \times 30}{60} = 1.991 \times 10^{-3}\ m^3/s = \mathbf{1.991\ l/s} \quad \text{... Ans.}$$

$$Q_{th} = \mathbf{119.5\ l/min} \quad \text{... Ans.}$$

Slip of the pump = $Q_{th} - Q_{actual}$ = 119.5 − 115 = 4.5 l/min

$$\%\ \text{slip of the pump} = \frac{Q_{th} - Q_{actual}}{Q_{th}} \times 100 = \left(\frac{119.5 - 115}{119.5}\right) \times 100 = \mathbf{3.76} \quad \text{... Ans.}$$

Example 9.28 : *A double acting reciprocating pump has a cylinder diameter of 15 cm and stroke of 30 cm. It is used to raise water through a total height of 30 m. Calculate the theoretical power required to drive the pump if the rank rotates at 60 r.p.m.*

Solution : Cylinder diameter = d = 15 cm = 0.15 m

$$A = \frac{\pi}{4} \cdot d^2 = \frac{\pi}{4} \times (0.15)^2 = 0.0177\ m^2$$

Stroke length = L = 30 cm = 0.30 m

Speed = N = 60 r.p.m.

Theoretical discharge from the double acting pump = Q_{th}

$$Q_{th} = \frac{2LAN}{60}, m^3/s = \frac{2 \times 0.30 \times 0.0177 \times 60}{60} = 0.01062 \ m^3/s$$

$$\text{Total head} = H = 30 \ m$$

$$\rho \text{ of water} = 1000 \ kg/m^3$$

Theoretical power required to drive the pump = P

$$P = \frac{\rho \cdot Q_{th} N}{75} = \frac{1000 \times 0.01062 \times 30}{75} = 4.248 \approx \mathbf{4.25 \ hp}$$ **... Ans.**

OR :

$$P = \rho \cdot g \ Q_{th} H = 1000 \times 9.81 \times 0.01062 \times 30$$

$$= 3125.47 \approx 3125 \ J/s = 3125 \ W = \mathbf{3.125 \ kW}$$ **... Ans.**

Practice Questions

1. Give the classification of pumps.
2. Give the principle of a centrifugal pump.
3. Give the principle of a reciprocating pump.
4. Draw a neat sketch of centrifugal pump and list its parts.
5. Draw a neat sketch of reciprocating pump and list its parts.
6. Draw characteristic curves of centrifugal pump at a given speed.
7. Explain in brief construction of a gear pump.
8. Compare centrifugal pump with reciprocating pump.
9. Give applications of plunger pump and gear pump.
10. Give applications, merits and demerits of centrifugal pump.
11. Give relation between power (P_f) and total head (ΔH) for centrifugal pump.
12. Write on operation of centrifugal pump.
13. Draw a neat sketch of diaphragm pump and state its applications.
14. Define cavitation and state its causes.
15. What are the effects of cavitation ?
16. Define NPSH.
17. If a centrifugal pump is not pumping enough fluid (problem of troubleshooting) then mention various causes and respective suggested actions.

❑❑❑

CHAPTER TEN

GAS PUMPING DEVICES [Transportation of Gases]

10.1 DEVICES OF TRANSPORTATION OF GAS/AIR

Like liquids, gases must also be moved through pipelines and process equipments using proper fluid moving devices. **Fans, blowers, and compressors** are used for transportation of gases based upon the service conditions.

The pumping devices for gases can be operated at higher speeds as the density of gas is considerably less than the density of liquid. As the viscosity of a gas is low, there is a greater tendency for leakage to occur, thus, gas compressors are designed with small clearances between the moving parts.

In gas compression, a large proportion of energy of compression appears as heat in the gas, thereby the temperature of the gas increases significantly. The increased temperature of the gas may limit the operation of the compressor due to this we have to make a provision for cooling of the gas in order to maintain reasonable temperatures. Thus, gas compression is often carried out in a number of stages with interstage cooling in the compressors.

Fans : A fan is a machine that moves large amounts of a gas with low increase in pressure.

Blowers : A blower is a machine for moving volume of a gas with moderate increase of pressure.

Compressors : A compressor is a machine for raising a gas to a higher level of pressure.

The main objective in case of compressors is to increase the static pressure. Therefore, the mechanical energy held by the gas is in the form of pressure energy whereas the main objective of fans and blowers is a high flow of gas and hence the mechanical energy of rotor/impeller is utilised to increase the kinetic energy of the gas and consequently a rise in the static pressure is less.

10.2.1 Induced Draft Fan and Forced Draft Fan

- In an induced draft system, the fan is at the exit/discharge end of flow path of the system and consequently the system is under negative pressure, i.e., at a pressure less than the atmospheric pressure as the air/gas is drawn out of the system by the fan.

  Examples : Cooling towers and boilers.

- In a forced draft system, the fan is located at the inlet of the flow path of the system and thus the system is under positive pressure, i.e., at a pressure more than the atmospheric pressure as the gas/air is forced into the system by the fan.

  Examples : Cooling towers and boilers.

- Draft fans are of two types :

  (i) Forced draft fan (FD fan) and (ii) Induced draft fan (ID fan).

- **Induced Draft Fan :** It is a fan that maintains a system under negative pressure (i.e. below atmospheric pressure) by pulling out air/gas through the system.

Examples :

(i) A fan used to remove/draw out flue gases from boiler into the atmosphere. It is located prior to the boiler chimney.

(ii) A fan is located at the top of a cooling tower to draw or pull air through the cooling tower. Axial fans are commonly used.

- **Forced Draft Fan (FD fan) :** It is a fan which maintains a positive pressure in a system by supplying pressurised air to the system.

  It produces a positive pressure inside a system.

Examples :

(i) A fan used to force outside air into a boiler to provide hot air for the combustion of a fuel thruogh a preheater. It maintains flue gases at a pressure more than the atmospheric pressure.

(ii) A fan placed at the base of a cooling tower which forces air into the cooling tower. Centrifugal fans are commonly used.

Fans :

Fans are used for moving gases when the pressure heads of less than 30 kPa are involved. They are either centrifugal type or axial flow type. The centrifugal type fans depend on the centrifugal force for moving the gas, while axial flow type fans impart energy to the gas as it flows parallel to the central axis of the fan. The inlet and outlet volumes of gases are essentially equal due to very low operating pressures and hence are simply movers of gas. Fans are employed industrially for ventilation works, supplying air to dryers, supplying draft to boilers, removal of fumes, etc.

10.2.2 Types of Fans

These fans are classified as :

(i) Axial flow fans/Axial fans.

(ii) Centrifugal fans.

An axial fan moves air or gas parallel to the axis of rotation.

A centrifugal fan moves air or gas perpendicular to the axis of rotation.

Axial fans are suited for low pressure, high flow applications.

Centrifugal fans are suited for high pressure, low flow applications.

1. Axial fans :

These fans use axial forces for the movement of air and increase its pressure.

An axial fan consists of a fan casing and impeller consisting of blades mounted on a hub. The hub is mounted on a shaft.

Blades rotating around an axis draw air in parallel to the axis and force air out in the same direction.

Axial fans are used in aircraft, helicopters, wind tunnels, cooling towers, general ventilation, paint spray booth exhaust, etc.

2. Centrifugal fans :

Principle : Centrifugal fans use centrifugal force to impart movement to the air/gas and increase its pressure.

Construction and Working :

A centrifugal fan has a fan wheel/impeller which is composed of a number of blades, mounted on a hub. The hub is fixed on a drive shaft that passes through the casing/housing.

In the centrifugal fan, the air/gas enters the fan in an axial direction and leaves the fan in a radial direction.

When a impeller rotates, the air/gas near the impeller is thrown-off from the impeller at high velocity due to centrifugal force and then moves into the casing. In the casing the gas/air decelerated and static pressure increases. The gas/air is finally guided to the exit via outlet duct.

Applications : Centrifugal fans are used for high pressure applications. They are used in ventilation systems for buildings, heating, drying, air conditioning, air pollution, steam boilers, power plants, cooling towers, exhausting corrosive and hazardous gases from chemical plants, for cooling ovens in glass industry, chemical plants, etc.

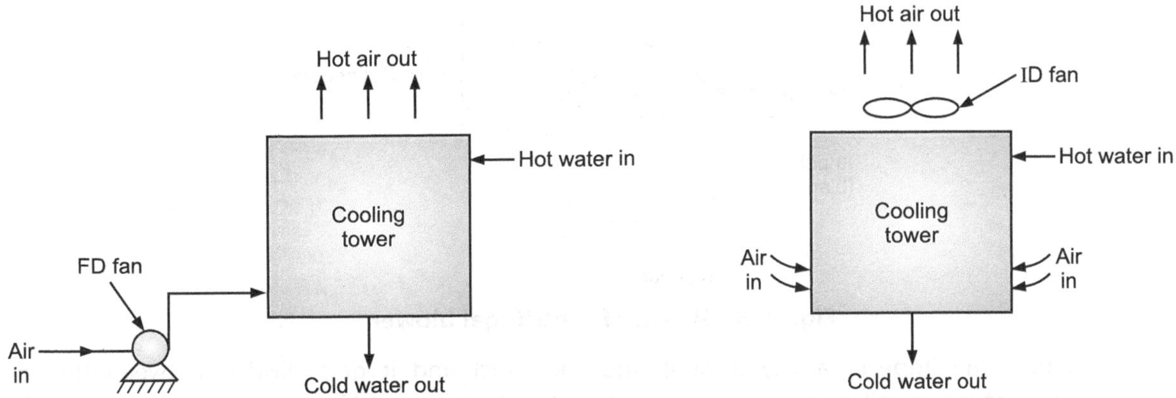

(A) Use of FD fan **(B) Use of ID fan**

Fig. 10.1 (A) and (B) : ID and FD fans in cooling tower operation (Block diagram)

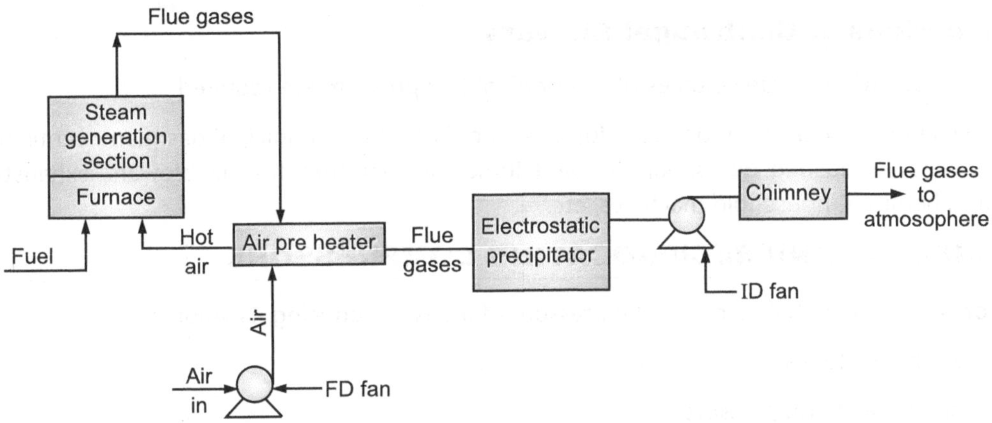

Fig. 10.2 : Use of ID and FD fans in boiler operation (Block diagram)

10.3.1 Centrifugal Blowers

A single-stage centrifugal blower bears a superfacial resemblance to a centrifugal pump. The centrifugal blower differs from a centrifugal pump in that the gas handled in it is compressible, whereas the liquid handled in the pump is incompressible. Also, centrifugal blowers require large diameter impellers and high speed of operations as compared to centrifugal pumps since very high heads of low density fluid (gas/air) is needed for generating pressure ratios.

Principle : The principle of a centrifugal blower is the same as that of the centrifugal pump.

A spinning impeller causes increase in kinetic energy of a gas/air by centrifugal force which then used to increase pressure of gas/air.

Construction :

A centrifugal blower consists of :

(i) Impeller (vaned/bladed).

(ii) Casing.

(ii) Inlet and outlet ducts.

An impeller is keyed to a shaft and enclosed in a volute casing. The shaft is connected to a drive (e.g. an electric motor). Inlet and outlet ducts (connections) are provided on the casing perpendicular to each other. The outlet duct is tangential to the rotation of the impeller. (Refer Fig. 10.3).

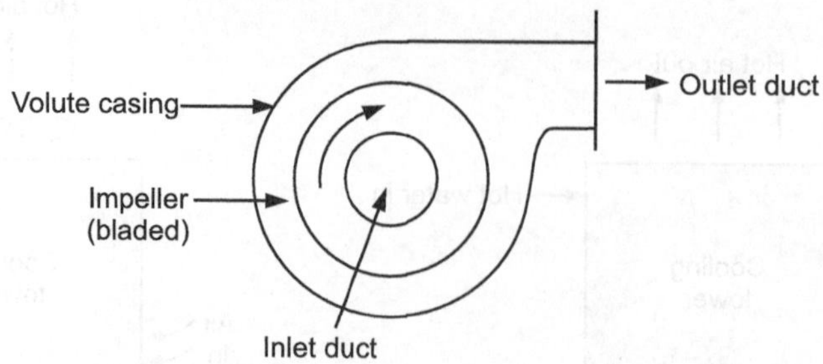

Fig. 10.3 : Sketch of centrifugal blower

Gas or air enters the impeller axially through the inlet duct and it gets divided between the impeller blades/vanes. The spinning impeller throws the gas outward with increased kinetic energy. The gas leaving the impeller at high velocity then flows into the surrounding volute casing which converts kinetic energy into pressure energy and thereby increase static pressure of the gas and is finally discharged through the outlet duct.

10.3.2 Applications of Centrifugal Blowers

- Blowers are used where large volumes of gas or air at low pressure are required.
- These are used in sewage aeration, blast furnace, cupola furnace, municipal gas plant, ventilation, pollution control, dryers, exhausting gases, vapours and fumes, incinerators, hot air blowers, exhausting corrosive hazardous chemicals, household machines, etc.

10.4 CENTRIFUGAL AND RECIPROCATING COMPRESSORS

A compressor is a machine that increases the pressure of a gas by reducing its volume.

Compressors are of two types :

(a) Positive displacement compressors :

(i) Reciprocating compressors and (ii) Rotary compressors.

(b) Dynamic compressors :

(i) Centrifugal compressors (radial flow compressors), (ii) Axial flow compressors.

- Positive displacement compressors work by mechanically changing the volume of the gas/air whereas dynamic compressors (e.g., centrifugal compressors) work by mechanically changing the velocity of the gas/air (the working fluid).
- Compressors can be single-stage or multistage machines depending on the maximum delivery pressure. For a compression ratio upto 5, single-stage compressors are used and for the compression ratio (the ratio of delivery pressure to suction pressure) greater than 5, multistage compressors are used.

10.4.1.1 Centrifugal Compressors

Principle : These compressors work on the principle of conversion of kinetic energy into pressure energy.

Construction : A centrifugal compressor is a dynamic compressor/turbo compressor which consists of

(i) Stationary casing/volute.

(ii) Rotating impeller.

(iii) Diffuser.

The casing houses a rotating impeller. The impeller is mounted on a shaft which is directly connected to a prime mover. The impeller consists of a set of blades. The diffuser may be vaned or vaneless.

Fig. 10.4 shows a schematic diagram of a centrifugal compressor.

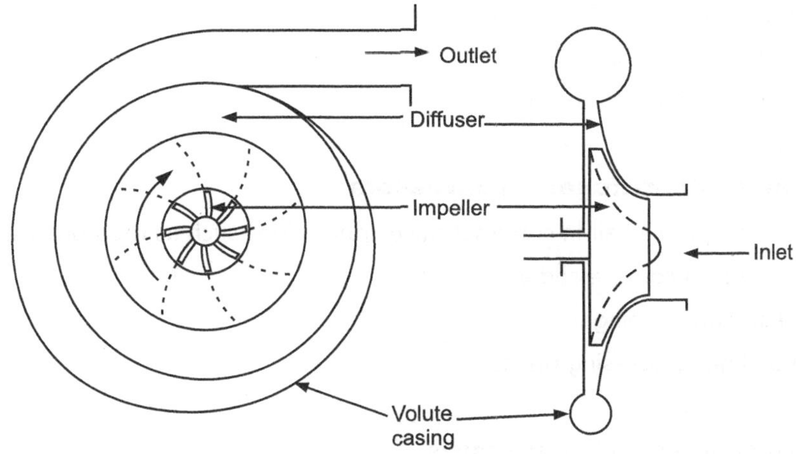

Fig. 10.4 : Centrifugal compressor (Schematic diagram)

(Note : You can draw any view in the examination)

Working :

As the impeller starts rotating, the vanes draw low pressure gas/air from the centre of the impeller and push outward to its periphery i.e., to open end of the impeller by centrifugal force. The radial movement of gas/air results in a pressure rise and the generation of kinetic energy. The gas exits the impeller vanes at very high velocity and enters into the diffuser where kinetic energy of the gas is converted into pressure energy. The volute casing collects the gas from the diffuser and directs to the outlet.

Nearly half of the total pressure is achieved in the impeller and the remaining in the diffuser.

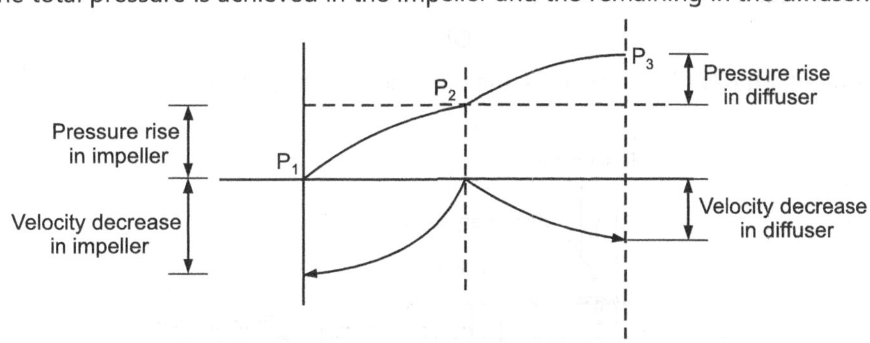

Fig. 10.5 : Velocity - pressure variation in a compressor

10.4.1.2 Advantages of Centrifugal Compressor

Wide operating range :

(i) Low weight and easy to manufacture.

(ii) Suitable for continuous compressed air supply.

(iii) Oil free in nature.

(iv) Relatively energy efficient.

(v) Does not require special foundation.

(vi) High flow rate than the positive displacement compressor.

(vii) Reliable and low maintenance costs.

(viii) Low initial cost.

10.4.1.3 Disadvantages of Centrifugal Compressor

(i) Unsuitable for very high compression, limited pressure.

(ii) Problem of surging and choking.

(iii) Sophisticated vibration mounting needed.

(iv) Limited compression ratios.

10.4.1.4 Applications of Centrifugal Compressors

- To provide oil free compressed air in some sensitive applications such as food processing.
- To meet high demands of compressed air.
- Refrigeration and air conditioning.
- Oil refineries, natural gas processing plants.
- Gas turbines.
- In automobile, turbochargers and superchargers.
- Off-shore platform applications.
- Petrochemicals (e.g., manufacture of ammonia, urea, ethylene oxide).
- Chemical industries.

10.4.2.1 Reciprocating Compressors

- A reciprocating compressor is a positive displacement compressor that uses a piston (driven by a crankshaft) to compress a gas and deliver it at high pressure.

Principle :

- Displacement of piston in cylinder causes rise in pressure.

OR

- The compression of a gas/air is achieved by the displacement of piston.

Construction :

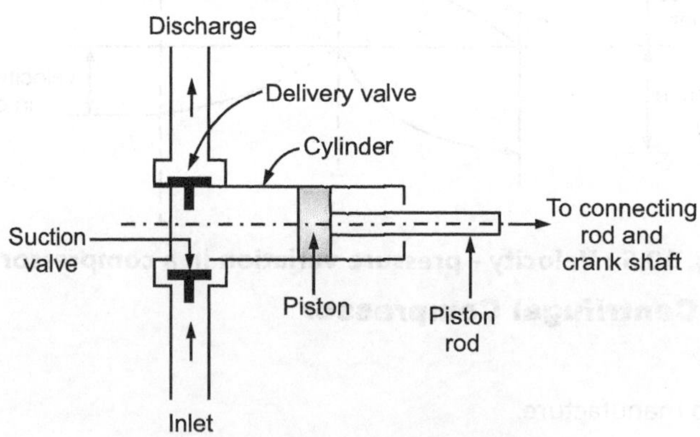

Fig. 10.6 : Reciprocating compressor

A reciprocating compressor mainly consist of :

(i) Cylinder.

(ii) Piston and connecting rod.

(iii) Crankshaft with a drive.

(iv) Suction/inlet valve (check valve).

(v) Discharge/outlet valve (check valve).

Reciprocating compressors may be **single acting** (where compression is only on one side of the piston, i.e., the gas/air is compressed only on one side of the piston) or **double acting** (where compression is on both sides of the piston i.e., the gas/air is compressed alternately on both sides of the piston).

In these compressors, a piston is the compressing and displacing element. The piston has a reciprocating motion within a cylinder.

The cylinder is equipped with suction and delivery valves.

A connecting rod transforms the rotary motion of the crankshaft into the reciprocating motion of the piston in the cylinder.

The compressor cylinder is usually water jacketed in order for removing heat of compression. These compressors are usually belt driven from an electric motor.

Reciprocating compressors are available as a single-stage or multistage units. Multistage units are used to achieve the required compression ratio. In multistage compressors, the gas is cooled between the stages as this interstage cooling reduces the temperature of the gas approximately to the suction temperature. The interstage cooling is necessary for safe operation as it keeps the temperature within the safe operating limits and it also results in reduction in the power required for compression.

Working :

During the suction stroke, the motion of the piston reduces the pressure inside the cylinder than that in the suction line. Due to this, the suction valve opens and the gas flows into the cylinder. During this stroke, the discharge valve remains closed. During the discharge stroke, the piston moves forward in the cylinder, compressing the gas. In this stroke the suction valve remains closed. As the piston is moving forward, the gas is getting compressed, pressure is increasing and when the pressure in the cylinder exceeds the pressure in the discharge, the valve opens. The compressed gas is then delivered through the discharge line during the remaining discharge stroke.

10.4.2.2 Applications of Reciprocating Compressors

Refrigeration plants.

Gas pipelines.

Natural gas processing plants.

Chemical industries.

Compressed air for automobiles.

Petrochemicals (e.g., manufacture of LDPE).

Petroleum refining (e.g., desulphurization and reforming).

Hydrogen stations.

Gas filling stations, compressed air to power pneumatic tools.

10.4.3.1 Comparison of Reciprocating and Centrifugal Compressors

Reciprocating Compressor	Centrifugal Compressor
1. Reciprocating compressors are slow speed machines.	1. Centrifugal compressors are high speed machines.
2. Reciprocating compressors cannot be directly connected to a prime mover.	2. Centrifugal compressors can be directly coupled to a prime mover.
3. Reciprocating compressors can develop pressures upto 1000 kg f/cm^2.	3. Centrifugal compressors can develop pressures upto 10 kg f/cm^2.
4. Reciprocating compressors are used when low rate of flow and high pressure are required.	4. Centrifugal compressors are used when high rate of flow and low pressure are required.

... Contd.

5. In these compressors, the compression process may be achieved as isothermal (theoretically) by providing sufficient cooling and due to of low speed of operation.	5. In these compressors, the compression process may be achieved as adiabatic due to the high speed of operation.
6. Handle dry gas.	6. Can also handle wet gas.
7. They work at a constant flow.	7. They work at a constant pressure.
8. These work by mechanically changing the volume of the working fluid-gas/air.	8. These work by mechanically changing the velocity of the working fluid-gas/air.

10.4.3.2 Comparison of Reciprocating and Centrifugal Compressors based on some Parameters

Parameters	Reciprocating compressor	Centrifugal compressor
Pressure	Very high	Medium to high
Flow rate	Low	High
Efficiency	Low	High
Maintenance cost	Very high	Low
Pulsation	Yes	No
Surging	No	Yes

10.5 FANS, BLOWERS AND COMPRESSORS, COMPARISON ON THE BASIS OF RANGE

Fans, blowers and compressors are differentiated by a measure called the specific ratio/pressure ratio.

The American Society for Mechanical Engineers (ASME) use the specific ratio to define fans, blowers and compressors. The specific ratio is the ratio of the discharge/outlet pressure to the suction/inlet pressure.

Difference between fans, blowers and compressors :

Machine	Specific ratio	Pressure rise (mm WC)
Fans	Upto 1.1	1136
Blowers	1.11 to 1.2	1136 to 2066
Compressors	More than 1.20	–

WC = Water (dumn)

Thus, fans have the lowest pressure ratio/specific ratio, blowers have the ratio in between that of fans and compressors and compressors have the highest pressure ratio.

10.6.1 Vacuum Service / Vacuum Generating Devices

A vacuum is *any system pressure below the atmospheric pressure* [760 mm Hg (torr)]. A vacuum pump is *a device which takes the suction at a pressure below the atmospheric and discharges at the atmospheric pressure.* (Vacuum refers to a subatmospheric pressure). For carrying out an operation under vacuum, pumping device is required to create a vacuum, and to maintain the low pressure. The compression ratios in vacuum producing devices are very high as compared to those in compressors. The operations under vacuum are very common in chemical industry, especially in performing operations such as distillation, evaporation, etc.

Every vacuum producing device takes in gas at low pressure where the volume is very large and discharges at atmospheric pressure. Commonly used vacuum producing devices are :

1. Rotary vacuum pumps (liquid ring type) and
2. Steam jet ejectors.

10.6.2.1 Steam Jet Ejectors

- An ejector is a device used to pump gases and vapours from a system in order to create a vacuum in the system.
- The purpose of an ejector is same as that of vacuum pump.
- An ejector is a device in which the kinetic energy of primary/motive fluid (it is a high pressure fluid, e.g., steam) is used to pump or suck secondary fluid (e.g., gases and vapours) from a process equipment with a subsequent creation of vacuum in the process equipment.

Principle : They work on the nozzle-venturi principle.

An ejector is a device that produces vacuum in a system using a motive fluid. If steam is used as a motive fluid then the ejector is called as a steam jet ejector. A dry saturated high pressure stream is the normal requirement.

The kinetic energy of steam is used to pump process gases and vapours.

Construction :

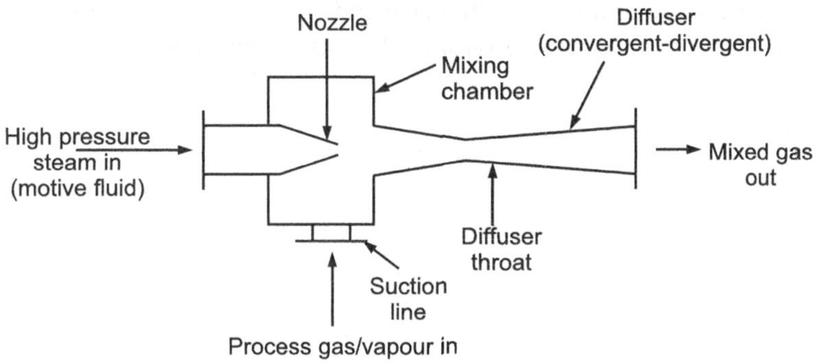

Fig. 10.7 : Steam jet ejector (single-stage)

A steam jet ejector consists of :

(i) Nozzle.

(ii) Mixing chamber.

(iii) Diffuser.

(iv) Suction fluid (gas or vapour from process) inlet.

(v) Motive fluid (steam) inlet.

(vi) Mixed fluid outlet.

Working :

- The suction line of the ejector is connected to the process equipment which is to be kept under low pressure, i.e., which is to be operated under vacuum.
- A high pressure steam (motive fluid) is accelerated from its initial velocity to a high velocity as it passes through the nozzle (i.e., in the nozzle, steam velocity increases) with a subsequent decrease in its pressure. Thus, a low pressure, high velocity flow prevails in the mixing chamber (i.e., a vacuum is created in the mixing chamber) and because of this, the process gas/vapour from a process equipment gets sucked into the mixing chamber and gets mixed with the steam. During the mixing fluid, the steam gets retarded and the vapours/gas gets accelerated.
- The mixed fluid/mixture from the mixing chamber enters the diffuser (convergent-divergent diffuser) where its pressure is increased by rapid deceleration of the mixed fluid (i.e., where its velocity is converted into pressure energy). It means that the mixed fluid is compressed to a high pressure in the diffuser. The high pressure mixed fluid is then released to the atmosphere or some closed system.

10.6.2.2 Advantages of Steam Jet Ejectors (Compared to Vacuum Pumps)

Relatively low initial cost.

High efficiency.

Less maintenance requirements.

Reliability.

Rugged and simple construction.

Simple operation.

No moving parts.

Easy to install and operate.

No source of power is required.

10.6.2.3 Applications of Steam Jet Ejector

- Steam jet ejectors are used to create a wide range of vacuum conditions.
- Steam jet ejectors are used in oil refineries, chemical industry, fertilizer plants, petrochemical industry, power plants, food industry, edible/vegetable oil industry, steel and allied industries, etc. for carrying out process operations such as :

  High vacuum distillation.

  Deodorisation.

  Drying.

  Crystallisation.

  Degassing.

  Filtration.

  Vacuum packaging.

  Deaeration and cooling.

10.6.3.1 Liquid Ring Vacuum Pumps

- Liquid ring vacuum pumps are mechanical devices that are used to create a vacuum in the system to which they are connected by making use of a suitable sealing liquid.
- In most of the applications, the sealing liquid is water and hence they are known as water ring vacuum pumps (or water ring water piston rotary vacuum pumps).

10.6.3.2 Water Ring Vacuum Pumps

Principle : Water ring vacuum pumps (or any liquid ring vacuum pump) operate according to the positive displacement principle.

Construction :

A water ring vacuum pump consists of impeller (vaned/bladed).

Casing :

Inlet/discharge ports.

Inlet connection – gas/vapour inlet.

Discharge connection – gas/vapour + water outlet.

Operating liquid inlet/water inlet.

It consists of a multiblade impeller fixed on a shaft and located eccentrically within a cylindrical casing. The casing is partially filled with water (a sealing liquid). Inlet/outlet connections and water inlet connections are located on the casing.

Working :

- Before starting the pump, it should be partially filled with water. During the pump operation, water should be continuously supplied through the water inlet. It circulates within the casing and leaves with the exit gases/vapours through the discharge connection.
- During operation the gases/vapours from a process equipment are sucked through the inlet connection into the pump and pushed out through the discharge connection along with water into the atmosphere.
- As the shaft turns, the rotating impeller throws off the water by centrifugal force towards the walls of the casing forming a ring of water. The water ring thus formed is concentric with the casing and eccentric with the impeller and also rotates within the casing.
- In the uppermost portion of the casing, the water ring is nearest to the impeller and completely fills the impeller cell (space between the two adjacent blades) with water (Refer Fig. 10.8) and in the lowermost portion, the impeller cells are free of water, i.e., the impeller blades are not dipped into the water.
- During the first half rotation, the water ring moves away from the impeller hub acting as a piston. The free/void space in the cells increases and thus the gases/vapours are sucked through the inlet port at the side of the impeller.

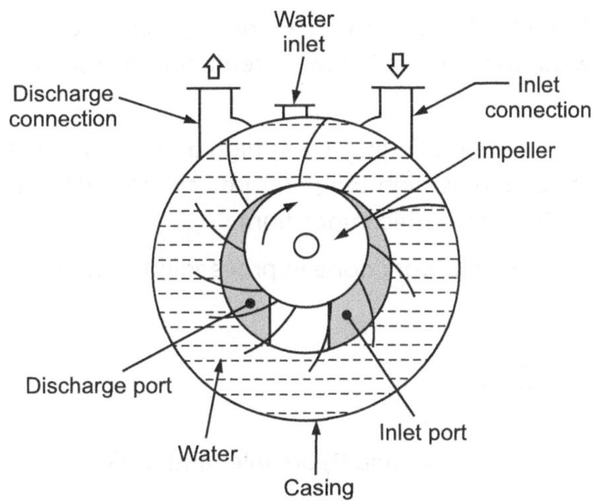

Fig. 10.8 : Water ring vacuum pump (OR Liquid ring vacuum pump)

- During the second half rotation, the water ring moves towards the impeller hub (water ring acting as a piston). Because of this the free space in the cells decreases and consequently the gases/vapours are compressed and pushed out through the discharge port along with some water. The gases/vapours along with water are discharged into the atmosphere through the discharge connection. The inlet port is joined by an internal passage to the inlet connection and the outlet port is joined by an internal passage to the outlet connection.

10.6.3.3 Applications of Water Ring Vacuum Pumps

Water ring vacuum pumps are used to carry plant operations such as distillation, filtration, evaporation, crystallisation, degassing and drying, under negative pressures in chemical, polymer, food, pharmaceutical, plastics and rubber industries [In plastic and rubber industries for degassing].

(Operations under vacuum are used to reduce the operating temperatures to reduce heat loads or to avoid thermal decomposition of heat sensitive materials.)

10.7 THEORY OF COMPRESSION

The relationship between absolute pressure and volume for any continuous compression process is given by

$$PV^n = C$$

For an isothermal compression, n = 1 and for an adiabatic compression (i.e., wherein heat is neither added nor removed), n = γ = the ratio of the specific heat at constant pressure to that at constant volume. [$\gamma = C_p/C_v$].

For an adiabatic and frictionless processes, the change in the pressure of an ideal gas is given by

$$\left(\frac{P_2}{P_1}\right) = \left(\frac{V_1}{V_2}\right)^{\gamma/\gamma - 1} \quad \text{... (10.1)}$$

$$\left(\frac{P_2}{P_1}\right) = \left(\frac{T_2}{T_1}\right)^{\gamma/\gamma - 1} \quad \text{... (10.2)}$$

$$\left(\frac{T_2}{T_1}\right) = \left(\frac{P_2}{P_1}\right)^{\gamma - 1/\gamma} \quad \text{... (10.3)}$$

where P_2/P_1 is known as the compression ratio. T_1, P_1, and V_1 are the temperature, pressure and volume of the gas at the inlet, respectively and T_2, P_2 and V_2 are the corresponding quantities at the outlet of the compressor.

From Equation (10.3), it is clear that for a given gas, the temperature ratio increases with an increase in compression ratio. In case of blowers, this ratio is below 3, temperature rise is not large and thus do not need cooling provisions.

In case of compressors, the compression ratio may be as high as 10 or more, the temperature rise is very large and therefore, need a provision for cooling as high temperatures create problems with lubricants, stuffing boxes, material of construction and with the fluid itself (decomposition).

For ideal and frictionless compression, the work done in proceeding from P_1 to P_2 is given by

$$W = \int_{P_1}^{P_2} \frac{dP}{\rho} \quad \text{... (10.4)}$$

where ρ is density of the gas to be compressed, P_1 and P_2 are inlet and outlet absolute pressures respectively.

For isothermal compression :

$$\frac{P}{\rho} = \frac{P_1}{\rho_1} \quad \text{... (10.5)}$$

Rearranging Equation (10.5), we get

$$\rho = \frac{\rho_1}{P_1} P \quad \text{... (10.6)}$$

$$W = \int_{P_1}^{P_2} \frac{dP}{\frac{\rho_1}{P_1}(P)} \quad \text{... (10.7)}$$

Integrating Equation (10.5), we get

$$W = \frac{P_1}{\rho_1} \ln \frac{P_2}{P_1} \quad \text{... (10.8)}$$

For adiabatic compression :

$$\frac{P_1}{\rho^{\gamma}} = \frac{P_1}{\rho_1^{\gamma}} \qquad \text{... (10.9)}$$

$$\rho^{\gamma} = \rho_1^{\gamma}\left(\frac{P}{P_1}\right) \qquad \text{... (10.10)}$$

$$\rho = \rho_1 (P/P_1)^{\frac{1}{\gamma}} \qquad \text{... (10.11)}$$

$$W = \int_{P_1}^{P_2} \frac{dP}{\rho} \qquad \text{... (10.12)}$$

Substituting the value of ρ from Equation (10.10) into Equation (10.12), we get

$$W = \frac{P_1^{1/\gamma}}{\rho_1} \int_{P_1}^{P_2} \frac{dP}{P^{1/\gamma}} \qquad \text{... (10.13)}$$

$$W = \frac{P_1^{1/\gamma}}{\rho_1} \left[\int_{P_1}^{P_2} P^{-1/\gamma} . dp\right] \qquad \text{... (10.14)}$$

Integrating Equation (10.12), we get

$$W = \frac{P_1^{1/\gamma}}{\rho_1} \left[\frac{P^{-1/\gamma + 1}}{\left(\frac{-1}{\gamma} + 1\right)}\right]_{P_1}^{P_2} \qquad \text{... (10.15)}$$

$$W = \frac{P_1^{1/\gamma}}{\rho_1} \frac{-\gamma}{(\gamma - 1)} [(P_2^{-1/\gamma + 1} - P_1^{-1/\gamma + 1})] \qquad \text{... (10.16)}$$

Taking out $P_1^{1-1/\gamma}$ common from the bracket of Equation (10.16), we get

$$W = \frac{\gamma . P_1^{1/\gamma} . P_1^{1-1/\gamma}}{\rho_1 (\gamma - 1)} \left[\left(\frac{P_2}{P_1}\right)^{1-1/\gamma} - 1\right] \qquad \text{... (10.17)}$$

$$W = \frac{P_1 . \gamma}{P_1 (\gamma - 1)} \left[\left(\frac{P_2}{P_1}\right)^{1-\frac{1}{\gamma}} - 1\right] \qquad \text{... (10.18)}$$

Thus, for a given compression ratio and suction condition, the work required in an isothermal compression is less as compared to that in an adiabatic compression. This is one reason for using cooling provisions in multistage compressors.

Practice Questions

1. Define fan, blower and compressor.
2. Give the specific ratio as per ASME of fans, blowers and compressors.
3. Give the principle, construction and applications of centrifugal blower.
4. Define ID fan and FD fan.
5. Define FD system and ID system.
6. Give classification of fans and applications of each type.
7. Give the principle of steam jet ejector.

8. Give applications of FD fan.
9. Give applications of ID fan.
10. Give applications of steam jet ejector.
11. Give applications of water ring vacuum pumps.
12. What do you mean by fluidisation ?
13. Define minimum fluidisation velocity.
14. Give the principle of water ring vacuum pump.
15. Draw a neat sketch of steam jet ejector.
16. Define liquid ring vacuum pump.
17. Distinguish between centrifugal and reciprocating compressor.
18. Draw a labelled sketch of centrifugal compressor.
19. Give applications of centrifugal compressor.
20. List applications of reciprocating compressors.
21. Give three differences between axial and centrifugal fans.

❑❑❑

APPENDIX

TRY YOURSELF

1. Why the faces of the jaws are corrugated ?
2. The crushers operate by

   (a) impact (b) compression (c) attrition
3. If the movable jaw is pivoted at top, the maximum movement is at the bottom. This is a case with

   (a) Blake crusher (b) Dodge crusher (c) Gyratory crusher
4. How the Blake jaw crusher is usually protected against damage due to tramps, etc ?
5. How smooth roll crusher is protected against damage due to tramps, etc. ?
6. Why the Dodge crusher is less widely used ?
7. If the particle is to be drawn between the rolls and crushed, the coefficient of friction, μ should be

   (a) Less than tan α (b) Greater than tan α (c) Smaller than tan α
8. Grinding means subdividing the solids to a finer product than crushing. State whether the statement is true or false.
9. The centrifuging in a ball mill occurs at a speed called as

   (a) normal speed (b) critical speed (c) operating speed.
10. When the mill is centrifuging –

    (a) maximum grinding takes place

    (b) little or no grinding takes place

    (c) normal grinding takes place.
11. The operating speed of a ball mill must be less than .... speed.
12. To avoid centrifuging in the ball mill, the operating speed of the mill should be

    (a) equal to critical speed.

    (b) more than critical speed.

    (c) less than critical speed.
13. In case of ball mill, the solid particles are reduced in size by

    (a) compression (b) attrition (c) impact
14. The work required for crushing the material is proportional to the logarithm of the ratio between the initial and final diameters. This is the statement of

    (a) Rittinger's law (b) Kick's law (c) Bond's law
15. In which one of the following machines, the particles are broken by sets of swing hammers ?

    (a) Ball mill (b) Hammer mill (c) Rod mill
16. Which one of the following sentences is true in case of grinding machines ?

    (a) Residence time is larger and throughput is smaller.

    (b) Residence time is less and throughput is larger.
17. What do you mean by open circuit grinding ?
18. Partially ground material is sent to a size separation unit from where the oversize is returned to the machine for reground. State, is this the case with open or closed circuit grinding ?

19. The screening method depends primarily on
    (a) the size of particles.
    (b) specific gravity of the particles.
    (c) magnetic properties of the particles.
20. Which one of the following machines make use of differences in the electrical properties of materials to effect a separation ?
    (a) Magnetic separator (b) Electrostatic separator (c) Froth floatation machine
21. The Tyler standard screen series is based on
    (a) 240 mesh screen (b) 200 mesh screen (c) 150 mesh screen
22. State only the two methods of reporting the screen analysis of a sample.
23. What do you understand by the term mesh number ?
24. In case of screen analysis, the notation 10/14 means through ...... mesh and on .... mesh.
25. The method of separating the particles according to .... alone is called screening.
26. Which of the following is a revolving screen ?
    (a) trommel (b) grizzly (c) shaking screen
27. Rapid vibrations with a small amplitude keep the material moving and prevent binding. State true or false.
28. Which of the following are two fractions obtained from a screen ?
    (a) cross flow (b) over flow (c) parallel flow (d) underflow
29. Which of the following are opposing factors in a screening operation ?
    (a) capacity (b) density (c) sharpness (d) effectiveness
30. The moisture content of a feed material adversely affects the screening operation. State true or false and justify.
31. What the pulsation of liquid causes in case of a jig ?
32. Generally with a hydraulic jig, the original feed material is divided into
    (a) two separate fractions
    (b) three separate fractions
    (c) four separate fractions.
33. .... solids are unfloatable.
    (a) Hydrophobic (b) Hydrophilic.
34. In the early stage of filtration, the rate of filtration is –
    (a) high (b) low (c) medium.
35. In constant rate filtration :
    (a) ΔP is minimum at start and maximum at the end of filtration run.
    (b) ΔP is constant throughout the run.
    (c) ΔP is maximum at start and minimum at the end.
36. A slurry no. 1 composed of solid A and liquid B. A slurry no. 2 composed of solid A and liquid 'C'. The viscosity of B is greater than that of 'C'. All other conditions being same in both the cases. State in which case, time of filtration will be more ?
37. Pressure filters operate with –
    (a) super atmospheric pressure on the upstream side of filter medium.
    (b) atmospheric pressure on the upstream side of filter medium.
    (c) subatmospheric pressure on the upstream side of filter medium.
38. The vacuum filters are limited to a maximum filtering pressure of .... atmosphere. (one, five, more than one).

39. Addition of filter aids to the slurry before filtration is done to –
    (a) increase the porosity of cake.
    (b) decrease the porosity of cake.
    (c) increase the mass of cake.
40. Which one of the following is the most effective washing technique in filter presses ?
    (a) simple washing (b) thorough washing (c) partial washing
41. What is the purpose of casting buttons on the sides of plates and frames in washing press ?
42. Which one of the following filters is entirely automatic in action ?
    (a) leaf filter (b) rotary drum filter (c) plate and frame filter press
43. What is the purpose of gently agitating the slurry in a trough in a rotary drum filter ?
44. Which one of the following filters is used for filtration of large quantities of a free filtering material ?
    (a) leaf filter (b) rotary drum filter (c) filter press
45. In centrifuge machines, the filtration operation is carried by using –
    (a) a centrifugal force (b) a gravity force (c) a pressure force
46. .... is any rotating machine in which a centrifugal force is utilized for the phase separation.
47. In centrifugal filtration, which one of the following operations is done at high speed ?
    (a) loading (b) unloading (c) spinning to dryness
48. The separation of solids from a suspension in a liquid by gravity alone is called ....
49. The sedimentation operations may be carried out batchwise or continuously on industrial scale in an equipment called ....
50. State the two functions of a thickener.
51. State the two main types of impellers.
52. A propeller is an ....
    (a) axial flow, low speed impeller.
    (b) radial flow, high speed impeller.
    (c) axial flow, high speed impeller.
53. .... are very effective impellers over a wide range of viscosities (upto 10^4 cP).
    (a) Propellers (b) Turbines (c) Paddles
54. To improve the rate of mixing and minimise vortex formation, ... are usually incorporated in vertical vessels.
55. Addition of .... in vertical vessels considerably increase the power requirement.
56. State the reasons for incorporation of baffles in vertical vessels.
57. If solids are to be kept in suspension, a gap of about 25 mm is kept between the baffle and tank ? Why ?
58. Which one of the following is the power number ?
    (a) $\frac{N D_a^2 P}{\mu}$ (b) $\frac{N^2 D_a}{\rho}$ (c) $\frac{P g_c}{N^3 D_a^5 \rho}$

    State whether it is a dimensionless number or not.
59. Generally, in a mechanically agitated vessel for gas-liquid system, the relation between the turbine diameter and tank diameter is –
    (a) diameter of turbine is one-third the tank diameter.
    (b) diameter of turbine is two-third the tank diameter.
    (c) diameter of turbine is three-fourth the tank diameter.

60. Name the three blades used for double arm kneaders.
61. A ribbon blender mix solids by mechanical shuffling. State true or false.
62. A tumbling mixer mix solids by repeatedly lifting and dropping the material and rolling it over. State true or false.
63. .... are very effective for thin pastes and for powders that do not flow readily.
64. State the function of plows in Muller mixer.
65. Which one of the following blades is most widely used ?

    (a) dispersion (b) sigma (c) double-naben blade
66. A fluid includes liquids, gases and vapours. State true or false.
67. When the density of a fluid is not appreciably affected by moderate changes in pressure and temperature, the fluid is said to be ....
68. In case of compressible fluids, ... is sensitive to changes in temperature and pressure.
69. The most important physical property of a fluid which affects the stress distribution is the .... of the fluid.

    (a) density (b) viscosity (c) thermal conductivity.
70. For Newtonian fluids, the ratio of the shear stress to the shear rate is constant and is equal to the .... of the fluid.
71. Of the following, state which are Newtonian and non-Newtonian fluids.

    (a) Gases (b) Sewage sludge (c) Sand-filled emulsions

    (d) True solutions.
72. Manometer is a simple device used for measuring .... differences.

    (a) force (b) density (c) pressure (d) temperature
73. The density of a manometric fluid should be .... than that of flowing fluid.

    (greater/lower/smaller)
74. Which one of the following is the equation of continuity ? The terms involved are having their usual meaning.

    (a) $\dot{m} = \rho u A$ = constant.

    (b) $\frac{P}{\rho} + gZ + \frac{u^2}{2g}$ = constant.
75. Methanol is flowing through a certain pipeline for which the Reynolds number is 15,000. State whether the flow is laminar or turbulent.
76. The flow with the unchanging velocity distribution is called ....

    (a) potential flow.

    (b) unsteady flow.

    (c) fully developed flow.
77. The Bernoulli's equation relates –

    (a) pressure at a point in the fluid to its position and velocity.

    (b) pressure at a point in the fluid to its position and viscosity.

    (c) pressure at a point in the fluid to its state and velocity.

78. Match correctly from rows A and B.

    A → Pressure energy, kinetic energy, potential energy.

    B → gZ, P/ρ, $u^2/2$

79. Which are the four variables grouped by Reynolds into a dimensionless group / number ?
80. The ratio of inertia force to viscous force is called ....
    (a) Nusselt number
    (b) Reynolds number
    (c) Prandtl number.
81. The relationship between the maximum local velocity and the average velocity is given by
    (a) $u = 0.75\ u_{max}$
    (b) $u = 0.50\ u_{max}$
    (c) $u = 0.25\ u_{max}$
82. The Hagen–Poiseuille equation is useful for the determination of
    (a) viscosity (b) density (c) heat capacity of the fluid.
83. State the relationship between 'f' (friction factor) and N_{Re} (Reynolds number) for laminar flow.
84. A log-log plot of f v/s $16/N_{Re}$ results in a straight line with a slope equal to ....
    (a) minus 16 (b) plus 16 (c) minus one (d) one
85. The equivalent length of pipe fittings is usually expressed as a certain number of pipe ....
    (a) threads (b) diameters (c) joints (d) lengths
86. What do you understand by the term vena-contracta ?
87. In the orifice meter, the .... of fluid increases and the .... therefore decreases.
    (a) potential energy (b) pressure energy
    (c) kinetic energy (d) heat energy
88. Which one of the following is a variable area meter ?
    (a) venturi meter (b) rotameter (c) orifice meter
89. In case of .... meter, pressure drop over the meter is constant.
    (a) venturi (b) rota (c) orifice
90. The pitot tube is used to measure a local or point ....
91. In venturi meter, the converging cone angle is of the order of .... degrees.
    (a) 15 - 20 (b) 5 - 7 (c) 7 - 10
92. In the diverging section of a venturi meter, a large proportion of kinetic energy is converted back to .... energy.
93. Which one of the following devices (flow measuring) can be used for exploring the velocity distribution across the pipe section ?
    (a) rotameter (b) pitot-tube (c) current meter.
94. State various means of fluid transport.
95. BWG number for tubing ranges from 24 to 7. State which of these numbers represent a very heavy tubing ?
96. The pressure drop through a globe valve is much .... than through a gate valve of the same size.
    (a) greater (b) lesser (c) lower

97. In .... pumps, kinetic energy is imparted to the liquid which is then converted efficiently into the pressure energy.

98. The conversion of kinetic energy into pressure energy is more efficient with ....

(a) volute type casing (b) diffuser type casing

99. What phenomenon occurs when the pressure in the suction line of a centrifugal pump is less than the vapour pressure of the liquid ?

100. The .... is the amount by which the pressure at a suction point of the pump is in excess of the vapour pressure of the liquid.

101. The centrifugal pump depends upon the .... of a fluid to develop pressure.

102. If NPSH is equal to .... cavitation will occur. Thus, NPSH must be greater than ....

103. The head of centrifugal pump .... continuously as the capacity is decreased.

(a) decreases (b) increases (c) becomes less

104. In rotary pumps, the chamber moves from .... to .... and back to the inlet.

105. The need to priming is eliminated by providing –

(a) negative suction head.

(b) positive suction head.

(c) positive discharge head.

106. If air enters an impeller, the pump is not capable to drive the liquid through the delivery pipe – why ?

107. What is the purpose of providing an air vessel at the discharge of a reciprocating pump ?

108. Which of the following pumps is used for boiler feed water applications ?

(a) gear pump (b) plunger pump (c) metering pump

109. Which of the following pumps may be used for feeding reactants, catalysts, inhibitors to reactors at controlled rates ?

(a) plunger pump (b) metering pump (c) centrifugal pump

110. .... pumps are commonly employed in the chemical industry for handling high viscosity liquids.

111. Gear pump is not suited for liquids that carry solids in suspension. State true or false.

112. Gear pumps are not suited for liquids that carry solids in suspension due to .... (Complete the sentence.)

113. Name the pump which is commonly used for corrosive liquids, toxic liquids or those containing suspensions of abrasive solids or those containing suspensions of abrasive solids.

114. Pneumatically actuated diaphragm pumps utilise a compressed plant .... for pumping.

115. Gas compressors are designed with small clearances between the moving parts. Why ?

116. Gas compression is often carried out in a number of stages with interstage cooling. Why ?

117. In the Nash Hystor pump, the sealing .... is supplied at a pressure equal to the discharge pressure of the gas.

(a) gas (b) vapour (c) liquid.

118. Reciprocating compressors are .... speed machines.

(a) slow (b) high (c) critical.

119. In case of reciprocating compressors, the compression process may be achieved as isothermal (theoretically). How it is achieved ?

120. .... is any system pressure below that of the atmospheric pressure (101.325 kPa).

121. The vacuum pump is any compressor which takes suction at pressure .... atmospheric and discharges at atmospheric pressure.

    (a) equal to (b) above (c) below

122. State two commonly employed vacuum producing devices.

123. In case of blower, the .... is below 3 and in case of compressors, it may be as high as 10 or more.

124. The blowers do not need cooling provisions. State true or false and explain.

125. Centrifugal pumps are equally suitable for handling suspensions with high solids content. State true or false. State also the type of impeller used for the same case.

ANSWERS

1. The faces of the crushing jaw are usually corrugated for concentrating the pressure on relatively small areas.
2. Compression.
3. Blake jaw crusher.
4. The crusher is protected against damage by making one of the toggles in the driving mechanism relatively weak (i.e., toggle is made in two pieces).
5. The smooth roll crusher is protected by spring loading, i.e., by mounting the bearings of one roll shaft against coiled springs.
6. The Dodge crusher is less widely used because of its tendency to choke due to minimum movement of the jaw at the bottom.
7. Greater than tan (α).
8. True.
9. Critical speed.
10. Little or no grinding takes place.
11. Critical speed.
12. Less than the critical speed.
13. Impact.
14. Kick's law.
15. Hammer mill.
16. Residence time is larger and throughput is smaller.
17. Open-circuit grinding is the one in which the material is passed only once through the machine and no attempt is made to return the oversize material to it for further reduction.
18. Closed-circuit grinding.
19. Size of particles.
20. Electrostatic separator.
21. 240 mesh screen.
22. (i) Differential screen analysis.

    (ii) Cumulative screen analysis.
23. It is the number of openings per linear inch counting from the centre of any wire to a point exactly one inch distant.
24. 10, 14.

25. Size.
26. Trommel.
27. True.
28. Overflow and underflow.
29. Capacity and effectiveness.
30. True. The moisture content of a feed adversely affects the screening operation as the damp particles are prone to stick to the screen surface and to each other.
31. The pulsation or jigging up and down of liquid causes the heavy material to work down to the bottom of the bed and the lighter material to rise to the top.
32. Four separate fractions.
33. Hydrophilic.
34. High.
35. ΔP is minimum at start and maximum at the end of filtration run.
36. The slurry no. 1.
37. Super atmospheric pressure on the upstream side of filter medium.
38. One.
39. Increases the porosity of the cake.
40. Thorough washing.
41. For ease in identification and quick proper assembling the press.
42. Rotary drum filter.
43. The slurry is gently agitated to prevent the setting of solids.
44. Rotary drum.
45. Centrifugal force.
46. A centrifuge or centrifugal.
47. Spinning to dryness.
48. Sedimentation.
49. Thickeners.
50. (a) To produce a clear liquid.
    (b) To produce the given degree of thickening of the suspension.
51. Axial flow impellers and radial flow impellers.
52. Axial flow, high speed impeller.
53. Turbines.
54. Baffles.
55. Baffles.
56. To improve the rates of mixing and to minimise vortex formation, baffles are usually incorporated in vertical vessels.
57. To minimise accumulation of solids on or behind baffles, a gap of about 1 inch should be kept between the baffle and the tank.
58. $\dfrac{P}{N^3 D_a^5 \rho}$

59. Diameter of turbine is one-third the tank diameter.
60. (i) Sigma.
    (ii) Dispersion.
    (iii) Double.
61. True.
62. True.
63. Ribbon blenders.
64. Flow guide the solids under the muller wheels during mixing or to an opening in the pan floor for discharge of the mixer at the end of cycle.
65. Sigma blade.
66. True.
67. Incompressible.
68. Density.
69. Viscosity.
70. Viscosity.
71. Newtonian fluids – gases and true solutions.
    Non-Newtonian fluids – sewage sludge and sand filled emulsions.
72. Pressure.
73. Greater.
74. $\dot{m} = \rho u A$ = constant.
75. Turbulent.
76. Fully developed flow.
77. Pressure at a point in the fluid to its position and velocity.
78. Pressure energy – P/ρ.
    Kinetic energy – $u^2/2$
    Potential energy – $g Z$
79. Diameter of the pipe, density of the fluid, viscosity of the fluid, and average velocity of the fluid.
80. Reynolds number.
81. $u = 0.50 u_{max}$
82. Viscosity.
83. $f = 16/N_{Re}$
84. Minus one
85. Diameters
86. Vena-contracta : The cross-section of minimum area at which jet changes from a contraction to an expansion located at a short distance from the sudden contraction.
87. (i) Kinetic energy.
    (ii) Pressure energy.
88. Rotameter.
89. Rota.
90. Velocity.
91. 15 – 20.
92. Pressure.
93. Pitot tube.

(i) Centrifugal force.
(ii) Displacement.
(iii) Mechanical impulse.
(iv) Transfer of momentum from another fluid and
(v) Gravity.

95. 7 – very heavy.
96. Greater.
97. Centrifugal.
98. Diffuser type casing.
99. Cavitation.
100. N.P.S.H. (Net positive suction head).
101. Inertia.
102. Zero, Zero.
103. Increases.
104. Inlet, Outlet.
105. Positive suction head.
106. If air enters the impeller, the pressure developed is reduced by a factor equal to the ratio of density of air to the density of liquid, i.e., this head in meters of air represents a very small pressure in terms of liquid.
107. The purpose of providing an air vessel at the discharge of a reciprocating pump is to even out the flow in the delivery line and to reduce the energy requirement at the beginning of each stroke.
108. Plunger pump.
109. Metering pump.
110. Gear pumps.
111. True.
112. Close clearance between gear-wheels and teeth.
113. Diaphragm pump.
114. Air.
115. As the velocity of a gas is lower, there is greater tendency for leakage to occur, thus, the gas compressors are designed with small clearances between the moving parts.
116. In case of gas compression, a large proportion of energy of compression appears as heat in the gas, thereby the temperature of gas increases considerably. This increased temperature of a gas may limit the operation of compressor unless provision is made for cooling of the gas to maintain a reasonable temperature.
117. Liquid.
118. Slow.
119. By providing sufficient cooling and because of low speed.
120. Vacuum.
121. Below.
122. (a) Rotary vacuum pumps (liquid ring type)
    (b) Steam jet ejectors.
123. Compression ratio.
124. True. In case of blower, the compression ratio is below 3, the temperature rise is not large and thus, do not need cooling provisions.
125. True. Open or semi-open impeller.

❑❑❑